SOLUTIONS TO CHEMISTRY TEST BANKS

SIVARAJAN PANDIAN

Published by

Sivarajan Pandian

CONTENTS

CHAPTER 1

THE FUNDAMENTALS OF MEASUREMENT

1-1. The prefix 'nano' is equivalent to multiplying by which of the following?

(a) 10^{-9} (b) 10^{-6} (c) 10^{-3} (d) 10^{3} (e) 10^{9}

Multiple	Prefix	Symbol	
10^{-24}	yocto	y	
10^{-21}	zepto	z	
10^{-18}	atto	a	
10^{-15}	femto	f	
10^{-12}	pico	p	
10^{-9}	nano	n	
10^{-6}	micro	μ	
10^{-3}	milli	m	
10^{-2}	centi	c	
10^{-1}	deci	d	
10	deca	da	
10^{2}	hecto	h	
10^{3}	kilo	k	
10^{6}	mega	M	
10^{9}	giga	G	
10^{12}	tera	T	
10^{15}	peta	P	
10^{18}	exa	E	
10^{21}	zeta	Z	
10^{24}	yotta	Y	Answer(a)

1-2. How many 'Claus'-es are there in a centi-Claus?

(a) 0.01   (b) 0.1   (c) 10   (d) 100

centi-Claus $= 10^{-2}$ Claus $= 0.01$ Claus   Answer(a)

1-3. If a nanometer is equal to 10^{-9} meter and a centimeter is equal to 10^{-2} meter, how many nanometers are there in one centimeter?

(a) 10^{-11} (b) 10^{-7} (c) 10^7 (d) 10^{11} (e) None of the above

1 nanometer $= 10^{-9}$ meter   1 centimeter $= 10^{-2}$ meter

1 centimeter $= (10^{-2}/10^{-9}) = 10^7$ nanometers   Answer(c)

1-4. How many picometers are there in one Angstrom?

1 Angstrom (Å) $= 10^{-10}$ meter   1 picometer $= 10^{-12}$ meter

1 Angstrom (Å) $= (10^{-10}/10^{-12}) = 10^2$ picometers

1-5. How many Angstroms are there in one nanometer?

1 nanometer $= 10^{-9}$ meter   1 Angstrom (Å) $= 10^{-10}$ meter

1 nanometer $= (10^{-9}/10^{-10}) = 10$ Angstroms (Å)

1-6. What are the derived SI units for the following quantities:

(a) Speed   (b) Area   (c) Volume   (d) Density

Quantity	Name of unit	Symbol
Speed	meters per second	m/s
Area	square meter	m^2
Volume	cubic meter	m^3
Density	kilograms per cubic meter	kg/m^3

1-7. Calcium carbonate decomposes on heating to form calcium oxide and carbon-di-oxide:

$CaCO_3$ (s) $\rightarrow$ CaO (s) $+ CO_2$ (g)

Which of the following would cause the weight loss during the decomposition of a sample of $CaCO_3$ to appear too large?

(a) The $CaCO_3$ contained some CaO before it was heated
(b) The crucible was wet when it was weighed before the $CaCO_3$ was heated
(c) The crucible was heated for too short a time
(d) The student read the last weight as 40.3047 g instead of the actual value of 30.3047 g
(e) All of these would cause the weight loss to appear too large

a : The weight loss would appear too small (CaO would not lead to weight loss).

b : The weight loss would appear too large because the water loss due to the water in the crucible would increase the weight loss.

c : The weight loss would appear too small because the decomposition was not complete.

d : The weight loss would appear too small because the student recorded a larger weight for CaO instead of the actual value.   Answer(b)

1-8. When 0.10274 gram of sugar is added to 3.05 grams of water, the answer should be reported to how many significant figures?

    (a) 2   (b) 3   (c) 4   (d) 5   (e) 6

Rule : When measurements are added or subtracted, the answer can contain no more decimal places than the least accurate measurement.

    3.05 g H_2O Least accurate measurement contains two decimal places.

+0.10274 g sugar

3.15 g solution (can not contain more than two decimal places). 3 significant figures. Answer(b)

1-9. How many significant figures can be used to describe the speed of a car that travels 20.24 kilometers in 15.1 minutes?

    (a) 1   (b) 2   (c) 3   (d) 4   (e) 5

Rule : When measurements are muliplied or divided, the answer can contain no more significant figures than the least accurate measurement.

Speed of the car = {20.24 (SF 4)×1000 (not a measurement)}/{15.1(Least accurate measurement, SF 3)×60 (not a measurement)} m/s

           = 22.3 m/s (SF 3)   Answer(c)

1-10. Calculate the charge on a single electron if the charge on a mole of electrons is 96485.34 C and there are 6.022142×10^{23} particles in a mole.

    (a) 1.602176×10^{-19} C   (b) 1 C   (c) 96484.56 C   (d) 5.810343×10^{28} C

    (e) None of the above

Charge on a single electron = $(96485.34)/(6.022142 \times 10^{23}) = 1.602176 \times 10^{-19}$ C   Answer(a)

1-11. Calculate how long it would take for 5 billion people to count out the number of molecules in a mole (6×10^{23}) if each person counted one molecule per second.

(a) 4 years   (b) 4000 years   (c) 4 million years   (d) 4 billion years

(e) 400 billion years

Number of molecules counted by 1 person/year = $60 \times 60 \times 24 \times 365$

Number of molecules counted by 5 billion persons/year = $5 \times 10^9 \times 60 \times 60 \times 24 \times 365$

Total number of molecules = 6×10^{23}

Time it would take = $(6 \times 10^{23})/(5 \times 10^9 \times 60 \times 60 \times 24 \times 365)$ = 4 million years   Answer(c)

1-12. It takes 13.60 electron volts of energy to remove an electron from a neutral hydrogen

atom in the gas phase. If 1 eV is equal to 1.602×10^{-19} J, how much energy, in KJ, would it take

to remove an electron from a mole of hydrogen atoms in the gas phase?

   (a) 1.18×10^{-20} kJ   (b) 2.18×10^{-18} kJ   (c) 7100 J   (d) 1312 kJ/mol

   (e) 7091 kJ

$1 eV = 1.602 \times 10^{-19}$ J

Energy required to remove an electron from one hydrogen atom = $13.60 \times 1.602 \times 10^{-19}$ J

Energy required to remove an electron from a mole of hydrogen atoms

= $6.022 \times 10^{23} \times 13.60 \times 1.602 \times 10^{-19}$ J/mol   = $6.022 \times 10^{23} \times 13.60 \times 1.602 \times 10^{-19} \times 10^{-3}$ kJ/mol

= 1312 kJ/mol   Answer(d)

1-13. LD_{50} for a drug is the dose that will be lethal for 50% of the population given that amount

of drug. LD_{50} for aspirin in rats is 1.75 grams per kilogram of body weight. How many aspirin

tablets weighing 325 mg each would the average 70-kg human have to consume to achieve this

dose?

   (a) About 4 tablets   (b) About 40 tablets   (c) About 400 tablets

   (d) About 4000 tablets   (e) About 400000 tablets

LD_{50} for aspirin in the average human = 70×1.75 g

Weight of one tablet = 325×10^{-3} g

Number of tablets required = $(70 \times 1.75)/(325 \times 10^{-3})$ = 377 tablets   Answer(c)

1-14. What is 60 GPa in atm?

   (a) 1.0×10^{-4} atm   (b) 5.9×10^3 atm   (c) 5.9×10^6 atm   (d) 1.0×10^9 atm

   (e) None of the above

$1 \text{ GPa} = 10^9 \text{ Pa}$

$60 \text{ GPa} = 60 \times 10^9 \text{ Pa} = 60 \times 10^9 \times 10^{-3} \text{ kPa} = 60 \times 10^6 \text{ kPa}$

$1 \text{ atm} = 101.325 \text{ kPa}$

$60 \text{ GPa} = (60 \times 10^6)/(101.325) = 5.9 \times 10^5 \text{ atm}$ Answer(e)

1-15. The reagent grade 'ethyl alcohol' we use in the lab has a density of 0.7892 g/mL at 20 °C. The formula for ethyl alcohol is CH_3CH_2OH and the molecular weight is 46.08 g/mol. Calculate the volume of reagent grade ethyl alcohol that contains 1.5 moles of this alcohol.

   (a) 24.2 mL   (b) 54.5 mL   (c) 87.6 mL   (d) 389 mL

   (e) None of the above

Weight of 1.5 moles of ethyl alcohol = 1.5×46.08 g   Weight of 1 mL of alcohol = 0.7892 g

Volume of the reagent that contains 1.5 moles of ethyl alcohol

$= (1.5 \times 46.08)/(0.7892) = 87.6 \text{ mL}$ Answer(c)

1-16. The following data show how the pressure of a gas depends on its volume when the amount of gas and the temperature of the gas are held constant. Which of the following would give a straight line that would fit the equation y = mx + c?

Pressure (atm) :   0.0586   0.0856   0.1420   0.2890

Volume (L) :   1.28   0.876   0.528   0.259

   (a) Only P vs V   (b) Only P vs 1/V   (c) Both P vs V and P vs 1/V

   (d) Both P vs 1/V and V vs 1/P   (e) P vs V, P vs 1/V, and V vs 1/P

P vs V

	P	V
	$x_1 = 0.0586$	$y_1 = 1.280$
	$x_2 = 0.0856$	$y_2 = 0.876$
	$x_3 = 0.1420$	$y_3 = 0.528$
	$x_4 = 0.2890$	$y_4 = 0.259$

If P vs V would give a straight line that would fit the equation y = mx + c,

$(y_2 - y_1)/(x_2 - x_1)$ should be equal to $(y_4 - y_1)/(x_4 - x_1)$.

$(y_2 - y_1)/(x_2 - x_1) = (-0.404/0.027) = -14.963$

$(y_4 - y_1)/(x_4 - x_1) = (-1.021/0.2304) = -4.431$

P vs V will not fit the equation $y = mx + c$.

P vs 1/V

P	1/V
$x_1 = 0.0586$	$y_1 = 0.781$
$x_2 = 0.0856$	$y_2 = 1.142$
$x_3 = 0.1420$	$y_3 = 1.894$
$x_4 = 0.2890$	$y_4 = 3.861$

If P vs 1/V would give a straight line that would fit the equation $y = mx + c$,

$(y_2 - y_1)/(x_2 - x_1)$ should be equal to $(y_4 - y_1)/(x_4 - x_1)$.

$(y_2 - y_1)/(x_2 - x_1) = (0.361/0.027) = 13.370$

$(y_4 - y_1)/(x_4 - x_1) = (3.080/0.2304) = 13.368$

P vs 1/V will fit the equation $y = mx + c$.

1/P vs V

1/P	V
$x_1 = 17.06$	$y_1 = 1.280$
$x_2 = 11.68$	$y_2 = 0.876$
$x_3 = 7.04$	$y_3 = 0.528$
$x_4 = 3.46$	$y_4 = 0.259$

If 1/P vs V would give a straight line that would fit the equation $y = mx + c$,

$(y_2 - y_1)/(x_2 - x_1)$ should be equal to $(y_4 - y_1)/(x_4 - x_1)$.

$(y_2 - y_1)/(x_2 - x_1) = (-0.404/-5.38) = 0.075$

$(y_4 - y_1)/(x_4 - x_1) = (-1.021/-13.6) = 0.075$

1/P vs V will fit the equation $y = mx + c$.   Answer(d)

1-17. The following data show how the volume of a gas depends on its temperature when the amount of gas and the pressure of the gas are held constant. Which of the following would give a straight line that would fit the equation $y = mx + c$.

Volume (mL) :	246	295	332	390	443
Temperature (°C) :	27	87	132	202	267

(a) Only V vs T   (b) Only V vs 1/T   (c) Both V vs T and V vs 1/T

(d) Both V vs 1/T and T vs 1/V   (e) V vs T, V vs 1/T, and T vs 1/V

V vs T

V	T
$x_1 = 246$	$y_1 = 27$
$x_2 = 295$	$y_2 = 87$
$x_3 = 332$	$y_3 = 132$
$x_4 = 390$	$y_4 = 202$
$x_5 = 443$	$y_5 = 267$

If V vs T would give a straight line that would fit the equation $y = mx + c$,

$(y_2 - y_1)/(x_2 - x_1)$ should be equal to $(y_5 - y_1)/(x_5 - x_1)$.

$(y_2 - y_1)/(x_2 - x_1) = (60/49) = 1.22$

$(y_5 - y_1)/(x_5 - x_1) = (240/197) = 1.22$

V vs T will fit the equation $y = mx + c$.

V vs 1/T

V	1/T
$x_1 = 246$	$y_1 = 0.037$
$x_2 = 295$	$y_2 = 0.011$
$x_3 = 332$	$y_3 = 0.00758$
$x_4 = 390$	$y_4 = 0.00495$
$x_5 = 443$	$y_5 = 0.00375$

If V vs 1/T would give a straight line that would fit the equation $y = mx + c$,

$(y_2 - y_1)/(x_2 - x_1)$ should be equal to $(y_5 - y_1)/(x_5 - x_1)$.

$(y_2 - y_1)/(x_2 - x_1) = (-0.026/49) = -5.3 \times 10^{-4}$

$(y_5 - y_1)/(x_5 - x_1) = (-0.033/197) = -1.7 \times 10^{-4}$

V vs 1/T will not fit the equation $y = mx + c$.

1/V vs T

	1/V		T
	$x_1 = 4.06 \times 10^{-3}$		$y_1 = 27$
	$x_2 = 3.39 \times 10^{-3}$		$y_2 = 87$
	$x_3 = 3.01 \times 10^{-3}$		$y_3 = 132$
	$x_4 = 2.56 \times 10^{-3}$		$y_4 = 202$
	$x_5 = 2.26 \times 10^{-3}$		$y_5 = 267$

If 1/V vs T would give a straight line that would fit the equation $y = mx + c$,

$(y_2 - y_1)/(x_2 - x_1)$ should be equal to $(y_5 - y_1)/(x_5 - x_1)$.

$(y_2 - y_1)/(x_2 - x_1) = (60/-6.7 \times 10^{-4}) = -8.96 \times 10^4$

$(y_5 - y_1)/(x_5 - x_1) = (240/-1.8 \times 10^{-3}) = -1.33 \times 10^5$

1/V vs T will not fit the equation $y = mx + c$.   Answer(a)

1-18. The density of sugar is 1.59 g/mL. Calculate the volume of a teaspoon of sugar if this

quantity of sugar weighs 4.50 grams.

   (a) 0.140 mL   (b) 0.353 mL   (c) 2.83 mL   (d) 6.090 mL   (e) 7.16 mL

Volume of 1.59 g of sugar = 1 mL

Volume of 4.50 g of sugar = 4.50/1.59 = 2.83 mL   Answer(c)

1-19. The iron in steel has a density of 7.86 g/cm^3. When iron rusts, it forms hydrated Fe_2O_3

that has a density of 5.12 g/cm^3. What happens to the volume of iron as iron rusts?

   (a) Expands   (b) It contracts   (c) The volume must remain the same

Volume of 7.86 g of iron = 1 mL

Volume of 7.86 g of hydrated Fe_2O_3 = 7.86/5.12 = 1.54 mL

So the volume of the iron increases as it rusts.   Answer(a)

1-20. The first silver dollar minted in the Americas was 3.95 cm in diameter and contained

27.07 g of silver. If the density of silver is 10.5 g/cm^3, what was the height of each coin?

   (a) 0.00791 cm   (b) 0.0316 cm   (c) 0.0526 cm   (d) 0.210 cm   (e) None of these

Volume of silver dollar = (27.07/10.5) cm^3

Volume of cylinder = $\pi r^2 h = \pi(d/2)^2 h = \pi d^2 h/4$

h = (27.07×4)/(10.5×π×3.95×3.95) = 0.210 cm $\hspace{4cm}$ Answer(d)

1-21. A sample of sucrose ($C_{12}H_{22}O_{11}$) that weighs 9.35 g is placed in a 15.00 mL flask. The remaining volume of the flask is filled with benzene – a liquid in which sucrose does not dissolve. The density of benzene is 0.879 g/cm^3. The benzene and sucrose together weigh 17.33 g. What is the density of sucrose?

$\quad$ (a) 0.532 g/cm^3 $\quad$ (b) 0.623 g/cm^3 $\quad$ (c) 1.03 g/cm^3 $\quad$ (d) 1.16 g/cm^3 $\quad$ (e) 1.58 g/cm^3

Weight of benzene = 17.33 – 9.35 = 7.98 g $\hspace{2cm}$ Volume of benzene = 7.98/0.879 = 9.08 mL

Volume of sucrose = 15.00 – 9.08 = 5.92 mL

Density of sucrose = 9.35/5.92 = 1.58 g/cm^3 $\hspace{4cm}$ Answer(e)

1-22. The following liquids don't mix with water. Which of these liquids would float on top of water?

$\quad$ (i) $\quad$ Benzene, C_6H_6, d = 0.877 g/cm^3
$\quad$ (ii) $\quad$ Carbon-di-sulfide, CS_2, d = 1.263 g/cm^3
$\quad$ (iii) $\quad$ Chloroform, $CHCl_3$, d = 1.483 g/cm^3
$\quad$ (iv) $\quad$ Diethyl ether, $(C_2H_5)_2O$, d = 0.714 g/cm^3
$\quad$ (v) $\quad$ Gasoline, d = 0.675 g/cm^3

$\quad$ (a) i,iv, and v $\quad$ (b) ii and iii $\quad$ (c) i,ii,and iii

$\quad$ (d) None of these liquids would float on top of water

$\quad$ (e) All of these liquids would float on top of water

The density of water is 1.0 g/cm^3. Only liquids with a density <1.0 g/cm^3 and which don't mix with water will float on top of water. So benzene,diethylether, and gasoline will float on top of water. $\hspace{2cm}$ Answer(a)

1-23. If a beaker that contains 250 mL of chloroform ($CHCl_3$) weighs 524 grams, what is the weight of the beaker? ($CHCl_3$: d = 1.483 g/cm^3)

$\quad$ (a) About 50 g $\quad$ (b) About 100 g $\quad$ (c) About 150 g $\quad$ (d) About 250 g $\quad$ (e) About 500 g

Weight of chloroform = 1.483×250 = 370.75 g

Weight of beaker = 524 – 370.75 = 153.25 g $\hspace{4cm}$ Answer(c)

1-24. Until recently, materials had to be cooled to temperatures on the order of 4 K to become superconducting. A ceramic has been discovered that becomes superconducting at 92 K. Which of the following temperatures corresponds to 92 K?

(a) 32 °F   (b) – 181.15 °C   (c) 212 °C   (d) 181 °C   (e) 365 °C

Conversion from K to °C : $T_{(K)} - 273.15 = T_{(°C)}$

Conversion from K to °F : $\{(T_{(K)} - 273.15) \times 9/5\} + 32 = T_{(°F)}$

92 K corresponds to $(92 - 273.15)$ °C $= -181.15$ °C

92 K corresponds to $\{(92 - 273.15) \times 9/5\} + 32 = -294.07$ °F   Answer(b)

1-25. Normal body temperature is 98.6 °F. What is the equivalent temperature in Kelvin?

(a) 236 K   (b) 293 K   (c) 310.15 K   (d) 393.45 K   (e) 418 K

Conversion from °F to K : $\{(T_{(°F)} - 32) \times 5/9\} + 273.15 = T_{(K)}$

98.6 °F corresponds to : $\{(98.6 - 32) \times 5/9\} + 273.15 = 310.15$ K   Answer(c)

1-26. At what temperature are the readings on the Celsius and Fahrenheit scales the same? At what temperature are the magnitudes the same but the signs different?

Conversion from °C to °F : $\{T_{(°C)} \times (9/5)\} + 32 = T_{(°F)}$

(i) $T_{(°F)} = T_{(°C)}$

$\{T_{(°F)} \times (9/5)\} + 32 = T_{(°F)}$

$- 0.8\ T_{(°F)} = 32$

$T_{(°F)} = T_{(°C)} = -40$ -40 °F corresponds to -40 °C

(ii) $-T_{(°F)} = T_{(°C)}$

$\{- T_{(°F)} \times (9/5)\} + 32 = T_{(°F)}$

$2.8\ T_{(°F)} = 32$

$T_{(°F)} = 32/2.8 = 11.42857$ 11.42857 °F corresponds to -11.42857 °C

1-27. What is the melting point of NaCl in °F, if NaCl melts at 801 °C?

(a) 463 °F   (b) 477 °F   (c) 1441.8 °F   (d) 1473.8 °F   (e) 1499 °F

Conversion from °C to °F : $\{T_{(°C)} \times (9/5)\} + 32 = T_{(°F)}$

801 °C corresponds to $\{801 \times (9/5)\} + 32 = 1473.8$ °F   Answer(d)

1-28. Fahrenheit defined 0° as the lowest temperature that would be achieved by adding salt to ice and found that ice itself freezes at 32° on this scale. Unfortunately, Fahrenheit's measurements were wrong. The lowest temperature that can be achieved by adding salt to ice is

– 21.2 ^{o}C. Calculate the equivalent temperature on the Fahrenheit scale.

   (a) −38.23 ^{o}F   (b) – 11.82 ^{o}F   (c) – 6.16 ^{o}F   (d) 20 ^{o}F   (e) 32 ^{o}F

Conversion from ^{o}C to ^{o}F : $\{(T_{(^{o}C)} \times 9/5) + 32\} = T_{(^{o}F)}$

−21.2 ^{o}C corresponds to $\{-21.2 \times (9/5)\} + 32 = -6.16$ ^{o}F   Answer(c)

1-29. Calculate the temperature on the Fahrenheit scale that corresponds to zero on the Rankin

scale.

Conversion from ^{o}R to ^{o}F : $T_{(^{o}R)} - 459.67 = T_{(^{o}F)}$

O ^{o}R corresponds to $0 - 459.67 = -459.67$ ^{o}F

1-30. Which of the following is not correct?

   (a) Extensive properties, such as mass and volume depend on the size of the sample
   (b) Intensive properties are characteristic properties of a substance, which do not depend on the size of the sample
   (c) The ratio of a pair of extensive properties, which both depend on the size of the sample, is often an intensive property that is a characteristic property of the substance
   (d) Density is an extensive property
   (e) Temperature is an intensive property

Extensive property depends on the amount of matter.

Examples : Mass, Volume

Intensive property does not depend on the amount of matter.

Examples: Temperature, Melting point, Boiling point

The ratio of two extensive properties of the same object or system is an intensive property

that is a characteristic property of the object or system.

Example: Density (mass/volume)

Density is an intensive property (ratio of a pair of extensive properties, mass/volume).

                                                                                          Answer(d)

CHAPTER 2

ELEMENTS AND COMPOUNDS

2-1. Which of the following correctly describes one of the differences between compounds and mixtures?

    (a) Compounds contain more than one element but mixtures do not
    (b) Compounds have a constant composition but mixtures do not
    (c) It is easier to separate the elements in a compound than it is to separate the elements in a mixture
    (d) All of these statements are true
    (e) None of these statements are true

Compounds have a constant composition. Mixtures do not have a constant composition.

Answer(b)

2-2. Which of the following is neither a metal nor a non-metal?

    (a) Na   (b) Mg   (c) Al   (d) Si   (e) P

Na,Mg,and Al are metals. P is a non-metal. Si (semimetal) is neither a metal nor a non-metal.

Answer(d)

2-3. Classify the following elements as either metals, non-metals, or semimetals:

S,Sb,Sc,Se,Si,Sm,Sn, and Sr.

Semimetals : Sb,and Si

Metals : Sc,Sn,Sr,and Sm

Non-metals : S,Se

2-4. Which of the following compounds is composed of molecular ions?

    (a) NH_4NO_3 (b) CaO   (c) H_2SO_4 (d) NO_2 (e) CO_2

NH_4NO_3 is composed of NH_4^+ and NO_3^- ions.   Answer(a)

2-5. Which of the following should conduct an electric current when dissolved in water?

    (i) $MgCl_2$ (ii) CO_2 (iii) CH_3OH (iv) KNO_3 (v) Ca_3P_2

    (a) i and ii   (b) ii and iii   (c) i and iv   (d) i,iv and v   (e) None of the above

$MgCl_2$, KNO_3, and Ca_3P_2 are ionic compounds which will conduct electricity when dissolved in water. Answer(d)

2-6. Which of the following compounds would you expect to be ionic, based on the elements' positions in the periodic table?

(i) ZnS   (ii) AlCl$_3$ (iii) BH$_3$ (iv) H$_2$S   (v) NH$_3$

(a) i and ii   (b) ii and iii   (c) i and iv   (d) iii, iv, and v   (e) None of the above

Zn and Al are metals. S and Cl are non-metals. As a rule of thumb, metals often react with non-metals to form ionic compounds.   Answer(a)

2-7. Which of the following compounds should be covalent?

(i) CH$_4$ (ii) CO$_2$ (iii) SrCl$_2$ (iv) NaH   (v) SF$_4$

(a) i and ii   (b) ii and iii   (c) i,ii, and v   (d) iii,iv, and v

C,H,O,S, and F are non-metals. As a rule of thumb non-metals combine with other non-metals to form covalent compounds. So CH$_4$, CO$_2$, and SF$_4$ are covalent compounds.   Answer(c)

2-8. Calculate the number of silicon atoms in the empirical formula of the mineral beryl, Be$_3$Al$_2$(SiO$_3$)$_6$.

(a) 1   (b) 3   (c) 6   (d) 18   (e) 29

Molecular formula : Be$_3$Al$_2$(SiO$_3$)$_6$

Empirical formula : Be$_3$Al$_2$(SiO$_3$)$_6$ Number of silicon atoms = 6×1 = 6   Answer(c)

2-9. Fluoride toothpastes convert the mineral apatite in tooth enamel into fluoroapatite, Ca$_5$(PO$_4$)$_3$F. Calculate the number of O atoms in the empirical formula of this compound.

(a) 1   (b) 3   (c) 4   (d) 12   (e) 21

Molecular formula : Ca$_5$(PO$_4$)$_3$F

Empirical formula : Ca$_5$(PO$_4$)$_3$F   Number of O atoms = 3×4 = 12   Answer(d)

2-10. Which of the following statements correctly describes the difference between the ^{12}C, ^{13}C, and ^{14}C isotopes of carbon?

(a) They have the same number of protons but different number of electrons
(b) They have the same number of neutrons but different number of electrons
(c) They have the same number of electrons but different number of protons
(d) They have the same number of protons but different number of neutrons
(e) They have the same number of protons,electrons, and neutrons but different atomic masses

Isotope   Number of protons   Number of electrons   Number of neutrons

^{12}C	6	6	6
^{13}C	6	6	7
^{14}C	6	6	8

Answer(d)

2-11. Calculate the number of protons in the average potassium atom if the atomic weight of potassium is 39 amu and the atomic number of this element is 19.

(a) 19 (b) 20 (c) 39 (d) 58

Number of protons in K = atomic number of K = 19   Answer(a)

2-12. Calculate the number of electrons on a P^{3-} ion if the atomic weight of P is 31 amu and the atomic number of this element is 15.

(a) 12 (b) 15 (c) 18 (d) 31 (e) 35

Number of electrons on neutral P = atomic number of P = 15

Number of electrons on P^{3-} ion = 15+3 = 18   Answer(c)

2-13. Which of the following correctly describes a Zn^{2+} ion that has a mass of 65 amu?

(a) It contains 28 electrons, 30 protons, and 35 neutrons
(b) It contains 30 electrons, 30 protons, and 35 neutrons
(c) It contains 32 electrons, 30 protons, and 35 neutrons
(d) It contains 35 electrons, 35 protons, and 30 neutrons
(e) It contains 37 electrons, 35 protons, and 30 neutrons

The atomic number of Zn = 30   Number of electrons in neutral Zn = 30

Number of electrons in Zn^{2+} = 30 – 2 = 28   Answer(a)

2-14. Which of the following symbols indicates an ion with 12 protons and 10 electrons?

(a) Mg^{2+} (b) C^{4+} (c) O^{2-} (d) None of these

Number of electrons = 10 = Number of protons – 2

So the oxidation number is +2.   Answer(a)

2-15. All of the following have the same number of electrons except:

(a) K^+ (b) Ca^{2+} (c) Al^{3+} (d) S^{2-} (e) Cl^-

Species	Number of electrons
K^+	19 – 1 = 18
Ca^{2+}	20 – 2 = 18

Al^{3+} $\qquad\qquad\qquad\qquad\qquad\qquad\qquad$ $13 - 3 = 10$

S^{2-} $\qquad\qquad\qquad\qquad\qquad\qquad\qquad$ $16 + 2 = 18$

Cl^- $\qquad\qquad\qquad\qquad\qquad\qquad\qquad$ $17 + 1 = 18$ $\qquad\qquad\qquad\qquad$ Answer(c)

2-16. Which of the following atoms have the same number of neutrons?

$\qquad$ (i) ^{233}Th $\quad$ (ii) ^{235}U $\quad$ (iii) ^{238}U $\quad$ (iv) ^{238}Np

$\qquad$ (a) i and ii $\quad$ (b) i,ii, and iv $\quad$ (c) ii and iii $\quad$ (d) iii and iv

Atom	Number of protons	Number of neutrons
^{233}Th	90	143
^{235}U	92	143
^{238}U	92	146
^{238}Np	93	145 $\quad$ Answer(a)

2-17. Fluoride toothpastes convert the mineral apatite in tooth enamel into fluoroapatite

$Ca_5(PO_4)_3F$. If fluoroapatite contains Ca^{2+} and PO_4^{3-} ions, what is the charge on the fluoride ion

in this compound?

$\qquad$ (a) +1 $\quad$ (b) 0 $\quad$ (c) –1 $\quad$ (d) –2 $\quad$ (e) –3

Let Y be the charge on the fluoride ion.

$(5\times2)+(-3\times3)+Y = 0$ $\qquad\qquad$ $Y = -1$ $\qquad\qquad\qquad\qquad$ Answer(c)

2-18. Aluminum chlorohydrate is added to antiperspirants to stop people from sweating. If this

compound contains neutral H_2O molecules, and Al^{3+}, OH^-, Cl^- ions, and the formula of this

compound is $Al_Y(OH)_5Cl.2H_2O$, what is the value of Y in this formula?

$\qquad$ (a) 0.5 $\quad$ (b) 1 $\quad$ (c) 1.5 $\quad$ (d) 2 $\quad$ (e) 3

$(3Y)+(-1\times5)+(-1\times1) = 0$ $\qquad\qquad$ $Y = +2$ $\qquad\qquad\qquad\qquad$ Answer(d)

2-19. Verdigris is a green pigment used in the manufacture of paint. The simplest formula for

this compound is $Cu_3(OH)_2(OAc)_4$. What is the charge on the copper ions in this compound if

the other ions are the OH^- (hydroxide) and OAc^- (acetate) ions?

$\qquad$ (a) –2 $\quad$ (b) 0 $\quad$ (c) +1 $\quad$ (d) +2 $\quad$ (e) +3

Let Y be the charge on the copper ion.

$3Y+\{(-1\times2)+(-1\times4)\} = 0$ $\qquad\qquad$ $Y = +2$ $\qquad\qquad\qquad\qquad$ Answer(d)

2-20. Predict the formulas for neutral compounds containing the following pairs of ions:

(a) Mg^{2+} and NO_3^- (b) Fe^{3+} and SO_4^{2-} (c) Na^+ and CO_3^{2-}

a : $Mg(NO_3)_2$ b : $Fe_2(SO_4)_3$ c : Na_2CO_3

2-21. Predict the formulas for neutral compounds containg the following pairs of ions:

(a) H^+ and O_2^{2-} (b) Zn^{2+} and PO_4^{3-} (c) K^+ and $PtCl_6^{2-}$

a : H_2O_2 b : $Zn_3(PO_4)_2$ c : K_2PtCl_6

2-22. What is the formula of the neutral compound formed when phosphorus reacts with sulfur if phosphorus is in the +5 oxidation state and sulfur is in the −2 oxidation state?

(a) PS   (b) PS_2 (c) PS_3 (d) P_2S_5 (e) P_5S_2

	a: P S	b : P S_2	c : P S_3	d : P_2 S_5	e : P_5 S_2
oxidation state :	+2 −2	+4 −2	+6 −2	+5 −2	+4/5 −2

Answer(d)

2-23. If the formula for magnesium nitride is Mg_3N_2 what would be the formulas for potassium nitride and aluminum nitride?

Magnesium nitride : Mg^{2+} and N^{3-} : Mg_3N_2

Potassium nitride: K^+ and N^{3-} : K_3N

Aluminum nitride : Al^{3+} and N^{3-} :   AlN

2-24. Compounds that contain the O^{2-} ion are called oxides; those that contain the O_2^{2-} ion are called peroxides. If the formula for potassium oxide is K_2O, what would be the formula for potassium peroxide?

(a) KO   (b) K_2O (c) KO_2 (d) K_2O_2

Potassium oxide : K_2O : K^+ and O^{2-}

Potassium peroxide : K^+ and O_2^{2-} : K_2O_2 Answer(d)

2-25. What is the charge on the $[Co(NO_2)_6]^Y$ ion if this complex is formed by combining Co^{3+} and NO_2^- ions?

(a) +3   (b) 0   (c) −3   (d) −6   (e) −12

Charge on the $[Co(NO_2)_6]^Y$ ion $= 3 + (-1 \times 6) = -3$ Answer(c)

2-26. Use the positions of Ba and O in the periodic table and predict the oxidation state of Fe in the so-called 'super iron' compound $BaFeO_4$.

   (a) +2   (b) +3   (c) +4   (d) +6   (e) +8

Let X be the oxidation state of Fe in $BaFeO_4$.

$$Ba \quad Fe \quad O_4$$

Oxidation state :   +2   X   −2

$2+X−8 = 0$ $X = +6$ Answer(d)

2-27. What is the oxidation sate of manganese in $LiMnO_4$?

   (a) +2   (b) +3   (c) +4   (d) +5   (e) +7

Let X be the oxidation state of manganese in Li Mn O_4

$$Li \quad Mn \quad O_4$$

Oxidation state: +1   X   −2

$1+X−8 = 0$ $X = 7$ Answer(e)

2-28. The active ingredient in Rolaids has the formula $NaAl(OH)_2CO_3$. Calculate the oxidation state of of the aluminum atom in this compound.

   (a) −3   (b) 0   (c) +1   (d) +2   (e) +3

Let Y be the oxidation state of aluminum in $NaAl(OH)_2CO_3$.

$$Na \quad Al \quad (OH)_2 \quad CO_3$$

Oxidation state :   +1   Y   −2 +1   +4 −2

$1+Y+(−2×2)+(1×2)+4+(−2×3) = 0$ $Y = +3$ Answer(e)

2-29. Use the positions of lithium and fluorine in the periodic table to predict the oxidation state of the phosphorus atom in $LiPF_6$?

   (a) +1   (b) +2   (c) +3   (d) +4   (e) +5

Let X be the oxidation state of phosphorus in $LiPF_6$.

$$Li \quad P \quad F_6$$

Oxidation state: +1   X   −1

$1+X−6 = 0$ $X = +5$ Answer(e)

CHAPTER 3

THE MOLE

INVALUABLE INFORMATION

Avogadro's number (N) $\qquad$: 6.022142×10^{23} /mol

Atomic mass unit (amu) $\qquad$: 1 amu = $1.6605389 \times 10^{-24}$ g = $1.6605389 \times 10^{-27}$ kg

$\qquad$: 1 g = 6.022142×10^{23} amu

3-1. What is the atomic weight of an element for which a single atom weighs 3.95×10^{-22} g?

(a) 6.56×10^{-46} (b) 1.31×10^{-45} (c) 3.95×10^{-22} (d) 238 (e) 476

The weight of one mole of atoms: $(3.95 \times 10^{-22}) \times (6.022 \times 10^{23})$ = 238 g/mol $\qquad$ Answer(d)

3-2. Suppose that a superheavy element with atomic number 120 is found on an expedition to one of the moons of Jupiter. Due to its scarcity only 2×10^{14} atoms could be isolated. If this sample weighed 100 ng, what was the element's atomic mass in amu?

(a) Less than 1 amu (b) Between 1 amu and 100 amu (c) Between 100 amu and 250 amu

(d) Between 250 amu and 400 amu (e) More than 400 amu

Atomic Mass = $(100 \times 10^{-9}) \times (6.022 \times 10^{23})/(2 \times 10^{14})$ = 301 amu $\qquad$ Answer(d)

3-3. What is the molecular weight of aspirin, $C_9H_8O_4$? (AW : H = 1.01, C = 12.01,

O = 16.00 amu)

(a) Less than 50 g/mol (b) Between 50 and 80 g/mol (c) Between 80 and 120 g/mol

(d) Between 120 and 160 g/mol (e) More than 160 g/mol

Molecular weight of aspirin = $(9 \times 12.01)+(8 \times 1.01)+(4 \times 16.00)$ = 180.17 g/mol $\qquad$ Answer(e)

3-4. Calculate the molecular weight of cocaine, assuming a molecular formula of $C_{17}H_{21}NO_4$

(AW : H = 1.00, C = 12.01, N = 14.01, O = 16.00 amu).

(a) Between 0 and 100 g/mol (b) Between 100 and 250 g/mol

(c) Between 250 and 300 g/mol (d) Between 300 and 350 g/mol (e) More than 350 g/mol

Molecular weight of cacaine = $(17×12.01)+(21×1.01)+14.01+(4×16.00) = 303.39$ g/mol

Answer(d)

3-5. Calculate the molecular weight of sucrose (cane sugar) that has the formula $C_{12}H_{22}O_{11}$ (AW : H = 1.01, C = 12.01, O = 16.00 amu).

   (a) Less than 100 g/mol   (b) Between 100 and 150 g/mol

   (c) Between 150 and and 200 g/mol   (d) Between 250 and 350 g/mol

   (e) More than 350 g/mol

Molecular weight of sucrose = $(12×12.01)+(22×1.01)+(11×16.00) = 342.34$ g/mol   Answer(d)

3-6. Calculate the molecular weight of the oxygen-carrier protein hemoglobin if this protein is 0.335% Fe by weight and each protein molecule contains 4 Fe atoms (AW : Fe = 55.85 amu).

   (a) Less than 1000 g/mol   (b) Between 1000 and 10000 g/mol

   (c) Between 10000 and 50000 g/mol   (d) Between 50000 and 100000 g/mol

   (e) More than 100000 g/mol

Molecular weight of hemoglobin : $(4×55.85×100)/(0.335) = 66687$ g/mol   Answer(d)

3-7. What is the mass of a molecule of CO_2? (AW : C = 12.01, O = 16.00 amu)

   (a) $4.6×10^{-23}$ g   (b) $7.31×10^{-23}$ g   (c) 28 g   (d) 44 g

Mass of $6.022×10^{23}$ molecules of CO_2 = $12.01+(2×16.00) = 44.01$ g

Mass of one molecule of CO_2 = $44.01/(6.022×10^{23}) = 7.31×10^{-23}$ g   Answer(b)

3-8. Calculate the molecular weight of sildenafil citrate (trade name viagra) $C_{22}H_{32}N_6O_4S$ (AW : H = 1.01, C = 12.01, N = 14.01, O = 16.00, and S = 32.07 amu).

   (a) Between 100 and 200 g/mol   (b) Between 200 and 300 g/mol

   (c) Between 300 and 400 g/mol   (d) Between 400 and 500 g/mol

   (e) More than 500 g/mol

Molecular weight of sildenafil citrate

= $(22×12.01)+(32×1.01)+(6×14.01)+(4×16.00)+(32.07) = 476.67$ g/mol   Answer(d)

3-9. Calculate the molecular weight of $C_{29}H_{35}NO_5$ (AW : H = 1.01, C = 12.01, N = 14.01, and O = 16.00 amu).

   (a) Between 100 and 200 g/mol   (b) Between 200 and 300 g/mol

(c) Between 300 and 400 g/mol   (d) Between 400 and 500 g/mol

(e) More than 500 g/mol

Molecular weight of $C_{29}H_{35}NO_5$ = $(29\times12.01)+(35\times1.01)+(14.01)+(5\times16.00)$ = 477.65 g/mol

Answer(d)

3-10. Elemental analysis suggests that a compound is 72.96% C, 9.36% H, 12.67% Cl, and

5.01% N by weight. What is the total number of C,H,N and Cl atoms in the molecuar formula

of this compound if the molecular weight is 279.85 g/mol? (AW : H = 1.01, C = 12.01,

N = 14.01, and Cl = 35.45 amu)

(a) 10 or less   (b) Between 11 and 20   (c) Between 21 and 30

(d) Between 31 and 40   (e) Between 41 and 50

Number of carbon atoms = $(279.85/12.01)\times(72.96/100)$ = 17

Number of hydrogen atoms = $(279.85/1.01)\times(9.36/100)$ = 26

Number of chlorine atoms = $(279.85/35.45)\times(12.67/100)$ = 1

Number of nitrogen atoms = $(279.85/14.01)\times(5.01/100)$ = 1

Total   = 45   Answer(e)

3-11. Lysine is 19.16% nitrogen by weight and each molecule contains two nitrogen atoms.

Calculate the molecular weight of lysine (AW : N = 14.01 amu).

(a) Less than100 g/mol   (b) Between 100 and 150 g/mol (c) Between 150 and 250 g/mol

(d) Between 250 and 350 g/mol   (e) More than 350 g/mol

Molecular weight of lysine = $(2\times14.01)\times(100/19.16)$ = 146 g/mol   Answer(b)

3-12. Calculate the number of carbon atoms in a mole of sucrose, $C_{12}H_{22}O_{11}$.

(a) 12   (b) 6.022×10^{23} (c) 6.62×10^{24} (d) 7.226×10^{24} (e) 1.3×10^{25}

One molecule of sucrose contains 12 carbon atoms.

One mole of sucrose will contain 12 moles of carbon atoms.

Number of carbon atoms in a mole of sucrose = $12\times6.022\times10^{23}$ = 7.226×10^{24} Answer(d)

3-13. Calculate the number of chlorine atoms in 10.4 grams of chloroform,

$CHCl_3$ (AW : H = 1.01, C = 12.01, Cl = 35.453 amu).

(a) 5.25×10^{22} (b) 1.05×10^{23} (c) 1.57×10^{23} (d) 2.10×10^{23} (e) 2.07×10^{25}

Molecular weight of chloroform = 12.01+1.01+(3×35.453) = 119.38

Number of moles of chloroform = 10.4/119.38 = 0.087 mol

1 mol of $CHCl_3$ contains 3 moles of chlorine.

0.087 mol of $CHCl_3$ will contain $3×6.022×10^{23}×0.087 = 1.57×10^{23}$ atoms of chlorine.

Answer(c)

3-14. Sodium tripolyphospahte, $Na_5P_3O_{10}$, is added to detergents to increase their cleaning power. Calculate the number of phosphorus atoms in 0.325 mol of this compound.

(a) $6.52×10^{22}$ (b) $1.96×10^{23}$ (c) $5.87×10^{23}$ (d) $6.02×10^{23}$ (e) None of the above

1 mol of sodium tripolyphosphate contains 3 moles of P.

1 mol of sodium tripolyphosphate contains $3×6.022×10^{23}$ atoms of P.

0.325 mol of sodium tripolyphosphate will contain $0.325×3×6.022×10^{23} = 5.87×10^{23}$

atoms of P.

Answer(c)

3-15. Calculate the number of electrons in 34.0 g of OH^- ions.(AW : H = 1.01, and

O = 16.00 amu)

(a) 8   (b) 10   (c) 20   (d) $1.2×10^{-24}$ (e) $1.20×10^{25}$

$1\ OH^-$ contains 8+1+1 = 10 electrons.

1 mole of OH^- contains $10×6.022×10^{23}$ electrons

Molecular weight of $OH^- = 1.01+16.00 = 17.01$ g/mol

17.01 g of OH^- contain $10×6.022×10^{23}$ electrons

34.00 g of OH^- will contain $(34.00/17.01)×10×6.022×10^{23} = 1.20×10^{25}$ electrons.

Answer(e)

3-16. Which pair of samples contain the same number of hydrogen atoms?

(a) 1 mole of NH_3 and 1 mole of N_2H_4
(b) 2 moles of NH_3 and 1 mole of N_2H_4
(c) 1 mole of NH_3 and 2 moles of N_2H_4
(d) 2 moles of NH_3 and 3 moles of N_2H_4
(e) 4 moles of NH_3 and 3 moles of N_2H_4

	NH_3	N_2H_4
	Number of H atoms	Number of H atoms
a :	$3×6.022×10^{23}$ = $1.8×10^{24}$	$4×6.022×10^{23}$ = $2.4×10^{24}$

b : $2 \times 3 \times 6.022 \times 10^{23}$ $= 3.6 \times 10^{24}$ $4 \times 6.022 \times 10^{23}$ $= 2.4 \times 10^{24}$

c : $3 \times 6.022 \times 10^{23}$ $= 1.8 \times 10^{24}$ $2 \times 4 \times 6.022 \times 10^{23}$ $= 4.8 \times 10^{24}$

d : $2 \times 3 \times 6.022 \times 10^{23}$ $= 3.6 \times 10^{24}$ $4 \times 3 \times 6.022 \times 10^{23}$ $= 7.2 \times 10^{24}$

e : $4 \times 3 \times 6.022 \times 10^{23}$ $= 7.2 \times 10^{24}$ $4 \times 3 \times 6.022 \times 10^{23}$ $= 7.2 \times 10^{24}$

Answer(e)

3-17. How many platinum atoms does 1.00 g of pure platinum contain? (AW :

Pt = 195.08 g/mol)

   (a) 195   (b) 3.09×10^{21} (c) 6.17×10^{21} (d) 1.95×10^{23} (e) 6.02×10^{23}

Number of Pt atoms in 1 g of pure platinum = $\{(6.022 \times 10^{23})/195.08\} \times 1 = 3.09 \times 10^{21}$ Answer(b)

3-18. If the atomic weight of platinum is 195.08 g/mol, what is the mass of a single platinum

atom?

   (a) 1.62×10^{-22} g   (b) 3.24×10^{-22} g   (c) 5.13×10^{-3} g   (d) 195.08 g   (e) None of the above

Mass of a single platinum atom = $195.08/6.022 \times 10^{23} = 3.24 \times 10^{-22}$ g   Answer(b)

3-19. The molecular formula of *iso*-octane is C_8H_{18}. It burns in the presence of oxygen to form

a mixture of CO_2 and H_2O. What is the ratio of moles of water to moles of CO_2 produced in

this reaction?

   (a) 1:1   (b) 2:1   (c) 9:4   (d) 9:8   (e) None of the above

$2C_8H_{18} + 25O_2 \rightarrow 16CO_2 + 18H_2O$

The ratio of moles of H_2O to moles of CO_2 = 18:16 = 9:8   Answer(d)

3-20. How many grams of CO_2 gas are produced from the combustion of 10.0 grams of *iso*-

octane? (AW : H = 1.01, C =12.01, and O = 16.00 g/mol)

   (a) 1.58 g   (b) 3.85 g   (c) 14.19 g   (d) 30.81 g   (e) 35.02 g

$2C_8H_{18} + 25O_2 \rightarrow 16CO_2 + 18H_2O$

$2\{(8 \times 12.01)+(18 \times 1.01)\} = 228.52$ grams of *iso*-octane produce

$16 \times \{12.01+(2 \times 16.00)\}$ = 704.16 grams of CO_2.

10.00 grams of *iso*-octane will produce $(704.16/228.52) \times 10 = 30.81$ g of CO_2.   Answer(d)

3-21. What is the percent by weight of fluorine in fluoroapatite, $Ca_5(PO_4)_3F$?

(AW : O = 16.00, F = 19.00, P = 30.97, and Ca = 40.08 amu)

(a) Between 0% and 0.1%   (b) Between 0.1% and 1%   (c) Between 1% and 3%

(d) Between 3% and 5%   (e) More than 5%

Percent of fluorine in $Ca_5(PO_4)_3F$: $(19\times100)/\{(5\times40.08)+(3\times30.97)+(12\times16.00)+19\} = 3.77\%$

Answer(d)

3-22. Calculate the number of C atoms in the empirical formula of nicotine if this compound is 74% C, 8.7% H, 17.3% N by weight (AW: H = 1.01, C = 12.01, and N =14.01 amu).

(a) 3   (b) 4   (c) 5   (d) 7   (e) 10

Empirical formula:

N :   17.3/14.01 = 1.2348   C : 74/12.01 = 6.17   H: 8.7/1.01 = 8.61

Divide all by 1.2348 :   N : 1   C: 5   H : 7   Answer(c)

3-23. Glucose (blood sugar), acetic acid, and formaldehyde are all 40.0% C, 6.76% H, and 53.3% O by weight. Calculate the total number of carbon,hydrogen and oxygen atoms in the empirical formula for these compounds (AW: H = 1.01, C = 12.01, and O = 16.00 amu).

(a) 3   (b) 4   (c) 5   (d) 6   (e) 9

Empirical formula:

H : 6.76/1.01 = 6.69   O : 53.3/16.00 = 3.33   C : 40.0/12.01 = 3.33

Divide all by 3.33 :

H : 2   C : 1   O : 1   Total number of atoms in the empirical formula = 4   Answer(b)

3-24. A compound that is 43.64% P and 56.36% O by weight has a molecular weight of 283.88 g/mol. What is the molecular formula of this compound?(AW : O = 16.00, and P = 30.97 amu)

(a) PO_4 (b) P_2O_3 (c) P_2O_5 (d) P_4O_{10} (e) None of the above

Number of P atoms in a molecule = $(283.88\times43.64)/(100\times30.97) = 4$

Number of O atoms in a molecule = $(283.88\times56.36)/(100\times16.00) = 10$

Molecular formula : P_4O_{10} Answer(d)

3-25. A mixed oxide of potassium and vanadium is 28.3% by weight of K and 37.0% by weight of V. What is the empirical formula? (AW : O = 16.00, K = 39.10, and V = 50.94 amu)

(a) KV   (b) KVO   (c) K_2V_3O (d) K_3V_3O (e) KVO_3

% Oxygen by weight = 100 − 28.3 − 37.0 = 34.7

Empirical formula :

K : 28.3/39.10 = 0.724   V : 37.0/50.94 = 0.726   O : 34.7/16.00 = 2.17

Divide all by 0.726 :   K : 1   V : 1   O : 3   Empirical formula : KVO_3 Answer(e)

3-26. The chief ore of manganese is an oxide known as pyrolusite, which is 36.8% O and 63.2% Mn by weight. What is the empirical formula for pyrolusite? (AW : O = 16.00, and Mn = 54.94 amu)

(a) MnO   (b) MnO_2 (c) Mn_2O_3 (d) MnO_3 (e) Mn_2O_7

Empirical formula:

O : 36.8/16.00 = 2.3   Mn : 63.2/54.94 = 1.15

Divide by 1.15 :   Mn : 1   O : 2   Empirical formula : MnO_2 Answer(b)

3-27. What is the empirical formula of nitrous oxide if this compound is 63.65% N and 36.35% by weight? (AW : O = 16.00, and N = 14.01 amu)

(a) N_2O (b) NO   (c) NO_2 (d) N_2O_3 (e) N_2O_5

Empirical formula:

N : 63.5/14.01 = 4.54   O : 36.35/16.00 = 2.27

Divide by 2.27 :   N : 2   O : 1   Empirical formula : N_2O Answer(a)

3-28. What is the atomic weight of a metal that forms an oxide, M_2O_3, that is 17.29% oxygen by weight? (AW : O = 16.00 amu)

(a) 38.3 amu   (b) 57.4 amu   (c) 76.6 amu   (d) 115 amu   (e) 229 amu

Let the atomic weight of metal M be Y amu.

$(48 \times 100)/(48+2Y) = 17.29$

$2Y+48 = 4800/17.29 = 277.62$ Y = 115 amu   Answer(d)

3-29. CISPLATIN is a commercial antitumor drug that contains 65.0% Pt, 23.7% Cl, 9.3% N and 2.0% H by weight. What is the ratio of H to Pt atoms in the empirical formula of this compound?

(AW : H = 1.01, N = 14.01, Cl = 35.45, and Pt = 195.08 amu)

(a) 1:1   (b) 2:1   (c) 3:1   (d) 4:1   (e) 6:1

Empirical formula:

Pt : 65/195.08 = 0.333   N : 9.3/14.01 = 0.664   Cl : 23.7/35.45 = 0.667   H : 2/1.01 = 1.981

Divide all by 0.333   Pt : 1   Cl : 2   N : 2   H : 6

The ratio of H atoms to Pt atoms = 6:1   Answer(e)

3-30. What is the empirical formula of a compound of N and O if a sample of the compound contains 0.483 g of N and 1.104 g of O? (AW: N = 14.01, and O =16.00 amu)

(a) N_2O (b) NO (c) NO_2 (d) N_2O_3 (e) N_2O_5

Empirical formula:

N : 0.483/14.01 = 0.0345   O : 1.104/16.00 = 0.069

Divide by 0.0345 :   N : 1   O : 2   Empirical formula : NO_2 Answer(c)

3-31. What is the atomic weight of the element that forms a fluoride , M_2F_4, that is 73.1% fluoride by weight? (AW : F = 19.00 amu)

(a) 14.0 amu   (b) 16.0 amu   (c) 28.0 amu   (d) 32.0 amu   (e) None of these

Let the atomic weight of element M be Y amu.

$\{(4\times19.00)\times100\}/\{(2Y+(4\times19)\} = 73.1$ Y = 14 amu   Answer(a)

3-32. What is the empirical formula of a compound that contains C,H and O if the compound is 40.00% C, and 6.71% H by weight? (AW: H = 1.01, C = 12.01, and O = 16.00 amu)

(a) CHO (b) CH_2O (c) CH_3O (d) CH_4O (e) None of the above

% O by weight = 100 − 40.0 − 6.71 = 53.29

Empirical formula :

C : 40/12.01 = 3.33   O : 53.29/16.00 = 3.33   H : 6.71/1.01 = 6.64

Divide by 3.33 :   C : 1   O : 1   H : 2   Empirical formula : CH_2O Answer(b)

3-33. Calculate the empirical formula of naphthalene if this compound is 93.71% C and 6.29% H by weight.(AW: H = 1.01, and C = 12.01 amu)

(a) CH (b) CH_2 (c) C_3H_2 (d) C_4H_5 (e) C_5H_4

Empirical formula:

C : 93.71/12.01 = 7.803   H : 6.29/1.01 = 6.23

Divide by 6.23:   C : 1.25   H : 1

Multiply by 4 :   C : 5   H : 4   Empirical formula : C_5H_4 Answer(e)

3-34. The molecular weight of naphthalene is 128.17 g/mol. Use the results of the previous question to calculate the total number of atoms in a naphthalene molecule.

    (a) 9   (b) 12   (c) 18   (d) 27   (e) 38

Empirical formula : C_5H_4

Empirical formula weight = $(5\times12.01)+(4\times1.01) = 64.09$

Molecular weight /Empirical formula weight = $128.17/64.09 = 2.00$

Molecular formula : $C_{10}H_8$

Total number of atoms = 18   Answer(c)

3-35. The term 'carbohydrate' once meant a compound with the empirical formula CH_2O. What is the molecular formula of glucose (blood sugar) if the molecular weight of this carbohydrate is 180 g/mol? (AW: H = 1.01, C = 12.01, and O = 16.00 amu)

    (a) CH_2O (b) $C_4H_8O_4$ (c) $C_6H_{12}O_6$ (d) $C_{12}H_{22}O_{11}$ (e) None of the above

Empirical formula : CH_2O

Empirical formula weight = $12.0 +(2\times1.01)+16.00 = 30.03$

Molecular weight /Empirical formula weight = $180/30.03 = 6.0$

Molecular formula : $C_6H_{12}O_6$ Answer(c)

3-36. A 1.854 g sample of D-ribose contains 0.01236 mol of this sugar. What is the ratio of the molecular weight to the empirical formula weight of this compound? (Empirical formula : CH_2O and AW: H = 1.01, C = 12.01, and O = 16.00 amu)

    (a) 1   (b) 2   (c) 3   (d) 5   (e) None of these

Empirical formula : CH_2O

Empirical formula weight = $12.01+(2\times1.01)+16.00 = 30.03$

Molecular weight = $1.854/0.01236 = 150$ g/mol

Molecular weight /Empirical formula weight = $150/30.03 = 5$ Answer(d)

3-37. Caffeine is a central nervous system stimulant found in coffee,tea and cola nuts. Calculate the molecular formula if caffeine is 49.48% C, 5.19% H, 28.85% N, and 16.48% O by weight and the molecular weight is 194.2 g/mol (AW: H = 1.01, C = 12.01, O =16.00, and N = 14.01 amu).

Number of C atoms = (194.2×49.48)/(12.01×100) = 8

Number of H atoms = (194.2×5.19)/(1.01×100) = 10

Number of N atoms = (194.2×28.85)/(14.01×100) = 4

Number of O atoms = (194.2×16.48)/(16.00×100) = 2

Molecular formula : $C_8H_{10}N_4O_2$

3-38. Aspartane is an artificial sweetener 160 times sweeter than cane sugar, which is 57.14% C, 6.16% H, 9.52% N, and 27.18% O by weight. Calculate the molecular formula of this compound if the molecular weight is 294.30 g/mol (AW: H = 1.01, C = 12.01, O =16.00, and N = 14.01 amu).

Number of C atoms = (294.3×57.14)/(12.01×100) = 14

Number of H atoms = (294.3×6.16)/(1.01×100) = 18

Number of N atoms = (294.3×9.52)/(14.01×100) = 2

Number of O atoms = (294.3×27.18)/(16.00×100) = 5

Molecular formula : $C_{14}H_{18}N_2O_5$

3-39. Halothane is an anesthetic that is 12.17% C, 0.51% H, 40.48 % Br, 17.96% Cl and 28.876% F by weight. Calculate the molecular formula of this compound if this molecule contains one hydrogen atom. (AW: H = 1.01, C = 12.01, Br = 79.90, Cl = 35.45, and F = 19.00 amu).

Empirical formula:

C : 12.17/12.01 = 1.01   H : 0.51/1.01 = 0.50   Br : 40.48/79.90 = 0.51

Cl : 17.96/35.45 = 0.51   F : 28.87/19.00 = 1.52

Divide by 0.5 :   C : 2 H : 1 Br : 1   Cl : 1   F : 3   Empirical formula : $C_2HBrClF_3$

Molecular formula contains one H atom. So the molecular formula is : $C_2HBrClF_3$

3-40. Elemental analysis of sarin gives the following result: 34.29% C, 7.19% H, 22.84% O, 13.56% F, and 22.1% P. How many carbon atoms does the empirical formula of this compound contain? (AW: H = 1.01, C = 12.01, O = 16.00, F = 19.00, and P = 30.97 amu).

   (a) 2 or less   (b) 3 to 6   (c) 7 to 10   (d) 11 to 15   (e) More than 15

Empirical formula :

H : 7.19/1.01 = 7.12   C : 34.29/12.01 = 2.86   O : 22.84/16.00 = 1.43   F : 13.56/19.00 = 0.71

P : 22.10/30.97 = 0.71

Divide all by 0.71: H : 10   C : 4 O : 2 F : 1 P : 1   Empirical formula : $C_4H_{10}FPO_2$

Number of C atoms in the empirical formula = 4   Answer(b)

3-41. The molecular weight of sarin is the same as the empirical weight of this compound.

What is the molecular weight of sarin in g/mol? (AW: H = 1.01, C = 12.01, O = 16.00,

 F = 19.00, and P = 30.97 amu).

   (a) Less than 100 g/mol (b) Between 100 and 150 g/mol   (c) Between 150 and 200 g/mol

   (d) Between 200 and 250 g/mol   (e) More than 250 g/mol

Empirical formula : $C_4H_{10}FPO_2$

Molecular formula : $C_4H_{10}FPO_2$

Molecular weight = (4×12.01)+(10×1.01)+(19.00)+(30.97)+(2×16.00) = 140.11 g/mol

                                                                                       Answer(b)

3-42. The balanced equation for the decomposition of ammonium dichromate

$A(NH_4)_2Cr_2O_7(s) \rightarrow BCr_2O_3$ (s) + CN_2 (g) + DH_2O (g)

has which of the following sets of coefficients?

   (a) A = 2, B = 2, C = 2, D = 4 (b) A = 1, B = 1, C = 1, D = 4 (c) A = 1, B = 1, C = 1, D = 2

   (d) A = 1, B = 1, C = 2, D = 2   (e) A = 1, B = 1, C = 1, D = 1

$(NH_4)_2Cr_2O_7(s) \rightarrow Cr_2O_3$ (s) + N_2 (g) + $4H_2O$ (g)   Answer(b)

3-43. Calculate the sum of the coefficients in the balanced chemical equation for the following

reaction:

aH_2S (g) + bO_2 (g) $\rightarrow$ cSO_2 (g) + dH_2O (g)

   (a) 6   (b) 8   (c) 9   (d) 11   (e) 17

$2H_2S$ (g) + $3O_2$ (g) $\rightarrow$ $2SO_2$ (g) + $2H_2O$ (g)   a+b+c+d = 9   Answer(c)

3-44. What is the sum of the coefficients when the following chemical equation is balanced?

$uCa_3(PO_4)_2$ (s) + vC (s) $\rightarrow$ xCa_3P_2 (s) + yCO (g)

   (a) 6   (b) 12   (c) 18   (d) 20   (e) None of the above

$Ca_3(PO_4)_2$ (s) + 8C (s) $\rightarrow$ Ca_3P_2 (s) + 8CO (g)   u+v+x+y = 18   Answer(c)

3-45. Thermal decomposition of an unknown carbonate led to a 35.09% weight loss. The unknown was which of the following compounds? (AW: C = 12.01, O = 16.00, Li = 6.94, Mg = 24.31, Ca = 40.08, Zn = 65.38, and Ba = 137.33 amu).

    (a) Li_2CO_3 (b) $MgCO_3$ (c) $CaCO_3$ (d) $ZnCO_3$ (e) $BaCO_3$

Weight loss is due to loss of CO_2.

Molecular weight of CO_2 = 12.01+(2×16.00) = 44.01

%CO_2 in Li_2CO_3 = (44.01×100)/{(2×6.94)+12.01+(3×16.00)} = 59.55

%CO_2 in $MgCO_3$ = (44.01×100)/{(24.31)+12.01+(3×16.00)} = 52.18

%CO_2 in $CaCO_3$ = (44.01×100)/{(40.08)+12.01+(3×16.00)} = 43.96

%CO_2 in $ZnCO_3$ = (44.01×100)/{(65.38)+12.01+(3×16.00)} = 35.09

%CO_2 in $BaCO_3$ = (44.01×100)/{(137.33)+12.01+(3×16.00)} = 22.30   Answer(d)

3-46. A crucible weighed 35.351 g. This crucible and a sample of $CaCO_3$, weighing 42.670 g, was heated until red hot to decompose the sample.

$CaCO_3$ (s) → CaO (s) + CO_2 (g)

What is the theoretical weight of the crucible and residue after the decomposition is complete? (AW: C = 12.01, O = 16.00, and Ca = 40.08 amu).

    (a) 3.21 g   (b) 4.100 g   (c) 38.570 g   (d) 39.451 g   (e) 49.989 g

Weight of $CaCO_3$ = 42.670 − 35.351 = 7.319 g

$CaCO_3$ (s) → CaO (s) + CO_2 (g)

40.08+12.01+48.00 = 100.09 g of $CaCO_3$ will have 40.08+16.00 = 56.08 g of residue (CaO).

Residue weight = (56.08/100.09)×7.319 = 4.101 g

The theoretical weight of the crucible and the residue = 35.351+4.101 = 39.451 g   Answer(d)

3-47. Nitrogen reacts with hydrogen to form ammonia,

N_2 (g) + 3H_2 (g) → 2NH_3 (g)

which burns in the presence of oxygen to form nitrogen oxide,

4NH_3 (g) + 5O_2 (g) → 4NO (g) + 6 H_2O (g)

which reacts with excess oxygen to form nitrogen dioxide.

2NO (g) + O_2 (g) → 2NO_2 (g)

How much nitrogen would we have to start with to make 10 moles of NO_2 ?

(a) 2.5 mol   (b) 5 mol   (c) 10 mol   (d) 15 mol   (e) 20 mol

N_2 (g) + $3H_2$ (g) → $2NH_3$ (g)

$2NH_3$ (g) + 5/2 O_2 (g) → 2NO (g) + 3 H_2O (g)

2NO (g) + O_2 (g) → $2NO_2$ (g)

1 mol of N_2 (g) produces 2 moles of NO_2 (g)

To produce 10 moles of NO_2, 5 moles of N_2 will be required.   Answer(b)

3-48. How many grams of hydrogen peroxide, H_2O_2, must decompose by the following

reaction to produce 0.400 mol of O_2? (AW: H = 1.01, and O = 16.00 amu).

$2H_2O_2$ (aq) → $2H_2O$ (l) + O_2 (g)

(a) Less than 25 g   (b) Between 25 and 40 g   (c) Between 40 and 60 g

(d) Between 60 and 80 g   (e) More than 80 g

$2H_2O_2$ (aq) → $2H_2O$ (l) + O_2 (g)

To produce 1 mole of O_2, 2{(2×1.01)+(2×16)} = 68.04 g of H_2O_2 are required.

H_2O_2 required to produce 0.400 mol of O_2 = 68.04×0.400 = 27.22 g   Answer(b)

3-49. Which of the following nitrogen containing compounds would supply the most nitrogen

per gram of fertilizer? (AW: N = 14.01 amu).

(a) Iron azide, $Fe(N_3)_2$, MW = 140 g/mol
(b) Sodium azide, NaN_3, MW = 83 g/mol
(c) Potassium nitrite, KNO_2, MW = 85.1 g/mol
(d) Potassium azide, KN_3, MW = 81.1 g/mol
(e) There is not enough information to answer this problem

		N per gram of fetilizer (g)
a: $Fe(N_3)_2$	(2×3×14.01)/140 =	0.60
b: NaN_3	(3×14.01)/83 =	0.51
c : KNO_2	14.01/85.1 =	0.16
d : KN_3	(3×14.01)/81.1 =	0.52

Answer(a)

3-50. How many oxygen molecules are required to consume 15.5 g of phosphorus when P_4 is

burned in oxygen to make tetraphosphorus decaoxide? (AW: P = 30.97 amu).

(a) 2.50   (b) 5.00   (c) 10.0   (d) $3.77×10^{23}$ (e) $3.01×10^{24}$

$P_4 + 5O_2 \rightarrow P_4O_{10}$

$5 \times 6.022 \times 10^{23}$ molecules of oxygen are required for $4 \times 30.97 = 123.88$ g of P

Number of molecules of oxygen required for 15.5 g of P = $\{(5 \times 6.022 \times 10^{23}) \times 15.5\}/123.88$

$$= 3.77 \times 10^{23} \qquad \text{Answer(d)}$$

3-51. The chemical formula of ethanol is C_2H_6O. It burns in excess oxygen to form CO_2 and H_2O. How many grams of H_2O are produced from the combustion of 25.0 g of ethanol? (AW: H = 1.01, C = 12.01, and O = 16.00 amu).

    (a) 9.78 g   (b) 18.0 g   (c) 25.0 g   (d) 29.3 g   (e) 54.1 g

$C_2H_6O + 3O_2 \rightarrow 2CO_2 + 3H_2O$

$(2 \times 12.01) + (6 \times 1.01) + 16.00 = 46.08$ g of C_2H_6O produce $3\{(2 \times 1.01) + 16\} = 54.06$ g of water.

Amount of water produced by 25 g of $C_2H_6O = (54.06 \times 25)/46.08 = 29.3$ g   Answer(d)

3-52. PF_3 reacts with XeF_4 to give PF_5. In theory, how many moles of PF_5 can be produced from 100.0 g of PF_3 and 50.0 g of XeF_4? (AW: F = 19.00, P = 30.97, and Xe = 131.29 amu).

$2PF_3 \text{ (g)} + XeF_4 \text{ (s)} \rightarrow 2PF_5 \text{ (g)} + Xe \text{ (g)}$

    (a) 0.121 mol   (b) 0.241 mol   (c) 0.482 mol   (d) 1.14 mol   (e) 2.28 mol

Molecular weight of $PF_3 = 30.97 + (3 \times 19.00) = 87.97$ g/mol

Number of moles of $PF_3 = 100/87.97 = 1.14$

Molecular weight of $XeF_4 = 131.29 + (4 \times 19.00) = 207.29$ g/mol

Number of moles of $XeF_4 = 50.0/207.29 = 0.241$

2 moles of PF_3 require 1 mole of XeF_4.

1.14 moles of PF_3 will require $1.14/2 = 0.57$ mole of XeF_4.

But only 0.241 mole of XeF_4 is available.

So XeF_4 is the limiting reagent.

1 mol of XeF_4 will produce 2 moles of PF_5

0.241 mol of XeF_4 will produce $2 \times 0.241 = 0.482$ mol of PF_5 Answer(c)

3-53. The portable stoves campers use for cooking burn propane, C_3H_8.

$AC_3H_8 \text{ (g)} + BO_2 \text{ (g)} \rightarrow CCO_2 \text{ (g)} + DH_2O \text{ (g)}$

What weight of propane would have to be burned to form 7.26 g of CO_2?

(AW: H = 1.01, C = 12.01, and O = 16.00 amu).

   (a) 2.43 g   (b) 4.84 g   (c) 7.26 g   (d) 21.8 g   (e) None of the above

C_3H_8 (g) + $5O_2$ (g) → $3CO_2$ (g) + $4H_2O$ (g)

To form $3\{(12.01+(2\times16)\} = 132.03$ g of CO_2 $(3\times12.01)+(8\times1.01) = 44.11$ g of propane have

to be burned.

Weight of propane required to form 7.26 g of CO_2 = $(44.11\times7.26)/132.03 = 2.43$ g   Answer(a)

3-54. Calculate the weight of O_2 needed to burn 10.0 g of H_2S. (AW: H = 1.01, S = 32.07,

and O = 16.00 amu).

$AH_2S + BO_2 → CSO_2 + DH_2O$

   (a) 4.70 g   (b) 6.26 g   (c) 7.04 g   (d) 9.39 g   (e) 14.08 g

$2H_2S + 3O_2 → 2SO_2 + 2H_2O$

$3\times(2\times16) = 96$ g of O_2 will burn $2\{(2\times1.01)+32.07\} = 68.18$ g of H_2S.

Weight of O_2 required to burn 10.0 g of H_2S = $(96\times10.0)/68.18 = 14.08$ g   Answer(e)

3-55. What would happen to the potential yield of SO_2 in the previous question if the amount of

O_2 was doubled?

   (a) It would decrease by a factor of 2
   (b) It would decrease by a factor of 5
   (c) It would remain constant
   (d) It would increase by a factor of 1.5
   (e) It would increase by a factor of 2

$2H_2S + 3O_2 → 2SO_2 + 2H_2O$

Molecular weight of H_2S = $(2\times1.01)+32.07 = 34.09$ g/mol

Number of mole of H_2S = 10/34.09 = 0.293

Molecular weight of O_2 = 32.00 g/mol

Number of mole of O_2 = $2\times14.08/32 = 0.88$

3 moles of O_2 will burn 2 moles of H_2S.

0.88 mole of O_2 will burn $2\times0.88/3 = 0.59$ mole of H_2S.

But only 0.293 mole of H_2S is available.

So H_2S is the limiting reagent.

The yield of SO_2 will remain constant.   Answer(c)

3-56. A sample of copper (II) sulfate, $CuSO_4$, that weighs 2.47 g picks up water from the atmosphere to form a hydrate with the formula, $CuSO_4.YH_2O$. If the sample weighs 3.86 g after it picks up water, what is the value of Y? (AW : H = 1.01, O = 16.00, S = 32.07 and Cu = 63.55 amu)

    (a) 2   (b) 2.5   (c) 3   (d) 4   (e) 5

Molecular weight of $CuSO_4$ = 63.55+32.07+(4×16.00) = 159.62 g/mol

Molecular weight of H_2O = (2×1.01)+16 = 18.02 g/mol

Weight of $CuSO_4$ = 2.47 g

Weight of water = 3.86 – 2.47 = 1.39 g

2.47 g of $CuSO_4$ pick up 1.39 g of H_2O

159.62 g (1 mole) of $CuSO_4$ will pick up (1.39×159.62)/(2.47×18.02) = 5 moles of H_2O.

                                                      Answer(e)

3-57. Nitrogen reacts with red-hot magnesium to form Mg_3N_2,

$3Mg (s) + N_2 (g) \rightarrow Mg_3N_2 (s)$

Which reacts with water to form $Mg (OH)_2$ and NH_3 :

$Mg_3N_2 (s) + 6H_2O (l) \rightarrow 3Mg(OH)_2 (aq) + 2NH_3 (aq)$

How many grams of magnesium would you have to start with to prepare 15.0 g of ammonia? (AW : H = 1.01, N = 14.01, and Mg = 24.31 amu)

    (a) 13.3 g   (b) 15.0 g   (c) 20.0 g   (d) 32.1 g   (e) None of the above

To prepare 2{(14.01+(3×1.01)} = 34.08 g of NH_3, 3×24.31 = 72.93 g of Mg are required.

Amount of Mg required to prepare 15 g of NH_3 = (72.93×15)/34.08 = 32.1 g   Answer(d)

3-58. What weight of CO_2 can be produced from 10.0 g of sucrose ($C_{12}H_{22}O_{11}$) and 10.0 g of oxygen? (AW : H = 1.01, O = 16.00, and C = 12.01 amu)

$C_{12}H_{22}O_{11} (s) + 12 O_2 (g) \rightarrow 12 CO_2 (g) + 11 H_2O (g)$

    (a) Less than 5 g   (b) Between 5 and 10 g   (c) Between 10 and 15 g

    (d) Between 15 and 20 g   (e) More than 20 g

Molecular weight of $C_{12}H_{22}O_{11}$ = (12×12.01)+(22×1.01)+(11×16.00) = 342.34 g/mol

Number of mole of $C_{12}H_{22}O_{11}$ = 10.0/342.34 = 0.0292

Number of moles of O_2 = 10/32 = 0.3125

1 mole of $C_{12}H_{22}O_{11}$ requires 12 moles of O_2.

0.0292 mole of $C_{12}H_{22}O_{11}$ will require 12×0.0292 = 0.3504 mole of O_2.

But only 0.3125 mole of O_2 is available.

So O_2 is the limiting reagent.

12 moles of O_2 will produce 12×{(12.01+(2×16.00)} = 528.12 g of CO_2

0.3125 mole of O_2 will produce (528.12×0.3125)/12 = 13.75 g of CO_2 Answer(c)

3-59. What is the limiting reagent when 10.0 g of NO react with 10.0 g of O_2 to form NO_2?

(AW : O = 16.00, and N = 14.01 amu)

$2NO\ (g) + O_2\ (g) \rightarrow 2NO_2\ (g)$

    (a) NO   (b) O_2 (c) NO_2 (d) None of these

Molecular weight of O_2 = 2×16.00 = 32.00

Molecular weight of NO = 14.01+16.00 = 30.01

2×30.01 = 60.02 g of NO will react with 32 g of O_2.

10.0 g of O_2 will require 60.02×10/32 = 18.76 g of NO. Only 10.00 g of NO are available.

So NO is the limiting reagent.   Answer(a)

3-60. How many moles of the limiting reagent are present in the previous problem?

    (a) 0.313   (b) 0.333   (c) 0.435   (d) 0.625   (e) None of these

Number of mole of NO = 10/30.01 = 0.333   Answer(b)

3-61. How many grams of HCl are produced when 10.0 g of Cl_2 and 1.00 g of H_2 react? (AW :

H = 1.01, and Cl = 35.45 amu)

$H_2\ (g) + Cl_2\ (g) \rightarrow 2HCl\ (g)$

    (a) 5.14 g   (b) 9.04 g   (c) 10.28 g   (d) 11.0 g   (e) 18.1 g

$H_2\ (g) + Cl_2\ (g) \rightarrow 2HCl\ (g)$

2×1.01 = 2.02 g of H_2 react with 2×35.45 = 70.90 g of Cl_2 to produce

2×(1.01+35.45) = 72.92 g of HCl

1.00 g of H_2 will react with 70.90/2.02 = 35.10 g of Cl_2.

But only 10 g of Cl_2 are available.

So Cl_2 is the limiting reagent.

70.90 g of Cl_2 will produce 72.92 g of HCl.

Amount of HCl produced by 10.0 g of Cl_2 = (72.92×10)/70.90 = 10.28 g   Answer(c)

3-62. What would happen to the amount of HCl produced in the previous question if the

amount of H_2 is doubled?

    (a) It would decrease by more than a factor of two
    (b) It would decrease by a factor of two
    (c) It would remain the same
    (d) It would increase by a factor of two
    (e) It would increase by more than a factor of two

10.0 g of Cl_2 will react with only 10/35.10 = 0.285 g of H_2. Increasing the amount of H_2 to

2.00 g will not change the amount of HCl produced; Cl_2 will still be the limiting reagent.

Answer(c)

3-63. How many grams of MgO can be produced by burning 10.0 g of Mg in the presence of

10.0 g of O_2? (AW : O = 16.00, and Mg = 24.31 amu)

    (a) 10.0 g   (b) 12.6 g   (c) 16.6 g   (d) 20.0 g   (e) 25.2 g

$2Mg + O_2 \rightarrow 2MgO$

2×24.31 = 48.62 g of Mg react with 32.00 g of O_2 to produce 2×(24.31+16.00) = 80.62 g of

MgO 10 g of O_2 will react with 48.62×10/32.00 = 15.19 g of Mg.

Only 10 g of Mg are available.

So Mg is the limiting reagent.

48.62 g of Mg will produce 80.62 g of MgO.

Amount of MgO produced by 10.0 g of Mg = 80.62×10/48.62 = 16.6 g   Answer(c)

3-64. When glucose reacts with O_2 in living systems, CO_2 and H_2O are produced, and a great

deal of energy is liberated.

$C_6H_{12}O_6$ (s) + $6O_2$ (g) $\rightarrow$ $6CO_2$ (g) + $6H_2O$ (g)

What weight of CO_2 can be produced from the reaction of 10.0 g of glucose and 10.0 g of O_2?

(AW : O = 16.00, H = 1.01, and C = 12.01 amu)

    (a) 2.29 g   (b) 2.44 g   (c) 13.75 g   (d) 14.7 g   (e) None of these

Molecular weight of glucose = (6×12.01)+(12×1.01)+(6×16.0) = 180.18 g/mol

Molecular weight of O_2 = 2×16 = 32 g/mol

Molecular weight of CO_2 = 12.01+(2×16.00) = 44.01 g/mol

180.18 g of glucose react with 6×32 = 192 g of O_2 to produce 6×44.01 = 264.06 g of CO_2.

10.0 g of glucose will require 192×10/180.18 = 10.66 g of O_2 . Only 10.00 g of O_2 are available. So O_2 is the limiting reagent.   192 g of O_2 will produce 264.06 g of CO_2.

Amount of CO_2 produced by 10.0 g of O_2 = 264.06×10/192 = 13.75 g   Answer(c)

3-65. "Muriatic acid' is sold in many hardware stores for cleaning bricks and tile. What is the molarity of this solution if 125 mL of the solution contains 27.4 g of HCl? (AW : H = 1.01, and Cl = 35.45 amu)

(a) 0.0938 M   (b) 0.751 M   (c) 3.43 M   (d) 6.01 M   (e) 219 M

Molecular weight of HCl = 1.01+35.45 = 36.46 g/mol

Number of mole of HCl in 27.4 g of HCl = 27.4/36.46 = 0.751

125 mL of the solution contains 0.751 mole of HCl.

Molarity of the solution = (0.751×1000)/125 = 6.01 M   Answer(d)

3-66. How many grams of baking soda, $NaHCO_3$, would be needed to neutralize 15 mL of Muriatic acid? (AW : H = 1.01, C = 12.01, O =16.00, Na = 22.99 and Cl = 35.45 amu)

   (a) Less than 1 g   (b) Between 1 and 5 g   (c) Between 5 and 10 g

   (d) Between 10 and 20 g   (e) More than 20 g

$HCl + NaHCO_3 \rightarrow NaCl + CO_2 +H_2O$

1 millimole of HCl requires 1 millimole of $NaHCO_3$.

15×6.01 = 90.15 millimoles of HCl will require 90.15 millimoles of $NaHCO_3$.

Molecular weight of $NaHCO_3$ = 22.99+1.01+12.01+(3×16.00) = 84.01 g/mol

Amount of $NaHCO_3$ required to neutralize 15 mL of Muriatic acid = (84.01×90.15)/1000

= 7.6 g   Answer(c)

3-67. Calculate the concentration of a solution prepared by dissolving 10.0 g of $NaNO_3$ in enough water to give 250 mL of solution (AW : N = 14.01, O = 16.00, and Na = 22.99 amu).

   (a) 0.029 M   (b) 0.118 M   (c) 0.471 M   (d) 2.13 M   (e) 8.50 M

Molecular weight of $NaNO_3$ = 22.99+14.01+(3×16.00) = 85 g/mol

Number of mole of $NaNO_3$ in 10.0 g of $NaNO_3$ = 10/85 = 0.1176

250 mL of solution contains 0.1176 mol of $NaNO_3$.

Molarity = (0.1176×1000)/250 = 0.471 M   Answer(c)

3-68. What is the molarity of con. phosphoric acid if this solution is 85.5% H_3PO_4 by weight and it has a density of 1.70 g/ml? (AW : H = 1.01, O = 16.00, and P = 30.97 amu)

 (a) Less than 5 M   (b) 14.8 M   (c) 17.3 M   (d) 20.3 M   (e) None of the above

Weight of H_3PO_4 in 1 L of solution = (1.7×1000×85.5)/100 = 1453.5 g

Molecular weight of H_3PO_4 = (3×1.01)+30.97+(4×16.00) = 98 g/mol

Molarity of phosphoric acid solution = 1453.5/98 = 14.8 M   Answer(b)

3-69. How many mL of 15 M NH_3 would you need to make 15.0 mL of 3.6 M NH_3 solution?

 (a) 0.036 mL   (b) 0.63 mL   (c) 3.6 mL   (d) 8.1 mL   (e) 360 mL

Number of mL of 15M NH_3 needed = (15×3.6)/15 = 3.6 mL   Answer(c)

3-70. Calculate the volume of 0.0100 M sulfuric acid (H_2SO_4) that would be needed to neutralize 10.00 mL of 0.0100 M aqueous ammonia (NH_3) solution

H_2SO_4 (aq) + 2NH_3 (aq) → $(NH_4)_2SO_4$ (aq)

 (a) 2.50 mL   (b) 5.00 mL   (c) 10.0 mL   (d) 15.0 mL   (e) 20.0 mL

2 millimoles of NH_3 are neutralized by 1 millimole of H_2SO_4.

10×0.01 = 0.1 millimole of NH_3 will be neutralized by (1×0.1)/2 = 0.05 millimole of H_2SO_4.

Volume of 0.01 M sulfuric acid required = 5 mL   Answer(b)

3-71. What is the molarity of an oxalic acid solution if it takes 37.55 mL of 0.1245 M sodium hydroxide to titrate 10.00 mL of the oxalic acid solution? (AW : H = 1.01, C = 12.01, O =16.00, and Na = 22.99 amu)

$H_2C_2O_4$ (aq) + 2NaOH (aq) → $Na_2C_2O_4$ (aq) + 2H_2O (l)

 (a) 0.06631 M   (b) 0.23375 M   (c) 0.4675 M   (d) 0.9350 M   (e) None of these

$H_2C_2O_4$ (aq) + 2NaOH (aq) → $Na_2C_2O_4$ (aq) + 2H_2O (l)

1 millimole of $H_2C_2O_4$ requires 2 millimoles of NaOH.

37.55×0.1245 = 4.6750 milimoles of NaOH will be required by

4.6750/2 = 2.3375 millimoles of $H_2C_2O_4$.

2.3375 millimoles of $H_2C_2O_4$ are in 10 mL of the solution.

Molarity = 2.3375/10 = 0.23375 M   Answer(b)

3-72. What mass of sodium oxalate is formed in the above titration?

(AW : H = 1.01, C = 12.01, O =16.00, and Na = 22.99 amu)

  (a) 0.0888 g   (b) 0.313 g   (c) 0.626 g   (d) 1.25 g

$H_2C_2O_4$ (aq) + 2NaOH (aq) → $Na_2C_2O_4$ (aq) + 2H_2O (l)

1 millimole of $H_2C_2O_4$ produces 1 millimole of $Na_2C_2O_4$.

2.3375 millimoles of $H_2C_2O_4$ will produce 2.3375 millimoles of $Na_2C_2O_4$.

Molecular weight of $Na_2C_2O_4$ = (2×22.99)+(2×12.01)+(4×16.00) = 134 g/mol

Mass of sodium oxalate formed = (2.3375×134)/1000 = 0.313 g   Answer(b)

3-73. Calculate the volume of 0.0985 M of H_2SO_4 that would be needed to neutralize 10.89 mL

of 0.1043 M aqueous ammonia solution.

H_2SO_4 (aq) + 2NH_3 (aq) → $(NH_4)_2SO_4$ (aq)

H_2SO_4 (aq) + 2NH_3 (aq) → $(NH_4)_2SO_4$ (aq)

2 millimoles of NH_3 require 1 millimole of H_2SO_4.

10.89×0.1043 = 1.1385 millimoles of NH_3 will require 1.1385/2 = 0.5679 millimoles of H_2SO_4.

Volume of 0.0985 M H_2SO_4 required = 0.5679/0.0985 = 5.77 mL

3-74. α-D-glucopyranose reacts with the periodate ion according to the following reaction:

$C_6H_{12}O_6$ (aq) + 5IO_4^- (aq) → 5IO_3^- (aq) + 5HCO_2H (aq) + H_2CO (aq)

Calculate the molarity of the α-D-glucopyranose solution if 25 mL of 0.750 M IO_4^- is required

to consume 10 mL of this solution.

$C_6H_{12}O_6$ (aq) + 5IO_4^- (aq) → 5IO_3^- (aq) + 5HCO_2H (aq) + H_2CO (aq)

1 millimole of α-D-glucopyranose requires 5 millimoles of IO_4^-.

25×0.750 = 18.75 millimoles of IO_4^- will be required by

18.75/5 = 3.75 millimoles of α-D-glucopyranose.

3.75 millimoles of α-D-glucopyranose are present in 10 mL.

Molarity of α-D-glucopyranose solution = 3.75/10 = 0.375 M

3-75. Oxalic acid reacts with the chromate ion in acidic solution according to the following reaction: $3H_2C_2O_4$ (aq) + $2CrO_4^{2-}$ (aq) + $10H^+$ (aq) → $6CO_2$ (g) + $2Cr^{3+}$ (aq) + $8H_2O$ (l)

If 10 mL of $H_2C_2O_4$ consumes 40.0 mL of 0.0250 M CrO_4^{2-} solution, what is the molarity of the oxalic acid solution ?

$3 H_2C_2O_4$ (aq) + $2CrO_4^{2-}$ (aq) + $10 H^+$ (aq) → $6CO_2$ (g) + $2Cr^{3+}$ (aq) + $8H_2O$ (l)

3 millimoles of $H_2C_2O_4$ consume 2 millimoles of CrO_4^{2-}.

40×0.0250 = 1 millimole of CrO_4^{2-} will be consumed by 3/2 = 1.5 millimoles of $H_2C_2O_4$.

1.5 millimoles of $H_2C_2O_4$ are present in 10 mL.

Molarity of the $H_2C_2O_4$ solution = 1.5/10 = 0.15 M

3-76. How much HgS precipitate forms when 113 mL of a 0.75 M CaS solution is mixed with 52 mL of a 1.21 M $Hg(NO_3)_2$ solution? (AW : S = 32.07, Ca = 40.08, and Hg = 200.59 amu)

  (a) 6.30 g   (b) 8.48 g   (c) 14.6 g   (d) 19.7 g   (e) None of the above

CaS + $Hg(NO_3)_2$ → $Ca(NO_3)_2$ + HgS

1 millimole of CaS reacts with 1 millimole of $Hg(NO_3)_2$ to precipitate 1 millimole of HgS.

Number of millimoles of CaS = 113×0.75 = 84.75

Number of millimoles of $Hg(NO_3)_2$ = 52×1.21 = 62.92

So $Hg(NO_3)_2$ is the limiting reagent.

62.92 milimoles of $Hg(NO_3)_2$ will react with 62.92 millimoles of CaS to precipitate

62.92 milimoles of HgS.

Molecular weight of HgS = 200.59+ 32.07 = 232.66 g/mol

Amount of HgS precipitated = (232.66×62.92)/1000 = 14.6 g   Answer(c)

3-77. If adding 25.0 mL of 0.6 M HCl to 1.00 L of a solution of $AgNO_3$ precipitated 1.95 g of AgCl, what was the concentration of the original $AgNO_3$ solution? ? (AW : Cl = 35.45, and Ag = 107.87 amu)

$AgNO_3$ (aq) + HCl (aq) → AgCl (s) + HNO_3 (aq)

  (a) $1.361×10^{-5}$ M   (b) $1.361×10^{-2}$ M   (c) 0.27 M   (d) 0.68 M   (e) 13.6 M

1 millimole of HCl reacts with 1 millimole of $AgNO_3$ to precipitate 1 millimole of AgCl.

Molecular weight of AgCl = 107.87+35.45 = 143.32 g/mol

Number of mole of AgCl precipitated = 1.95/143.32

Number of millimoles of AgCl precipitated = $(1.95 \times 1000)/143.32 = 13.61$

Number of millimoles of HCl added = $25.0 \times 0.6 = 15.0$

AgNO$_3$ is the limiting reagent.

Number of millimoles of AgNO$_3$ in 1 L = 13.61

Concentration of AgNO$_3$ solution = $13.61/1000 = 1.361 \times 10^{-2}$ M   **Answer(b)**

3-78. A 2.50 g sample of bronze was cut from a gong and dissolved in sulfuric acid. The copper sulfate produced was mixed with KI to form CuI and the triiodide ion, I_3^-. The I_3^- ion was then titrated with thiosulfate, $S_2O_3^{2-}$.

$Cu\ (s) + 2H_2SO_4\ (aq) \rightarrow CuSO_4\ (aq) + SO_2\ (g) + 2H_2O\ (l)$

$2CuSO_4\ (aq) + 5I^-\ (aq) \rightarrow 2CuI\ (s) + I_3^-\ (aq) + 2SO_4^{2-}\ (aq)$

$I_3^-\ (aq) + 2S_2O_3^{2-}\ (aq) \rightarrow 3I^-\ (aq) + 2S_4O_6^{2-}\ (aq)$

If it took 31.5 mL of 1.00 M thiosulfate for this titration, how many moles of copper were present in the original 2.5 g sample?

   (a) 0.0105 M   (b) 0.0158 M   (c) 0.0315 M   (d) 0.0630 M   (e) 0.0945 M

1 millimole of Cu will produce 1 millimole of CuSO$_4$ which will form 1/2 mole of I_3^-, which will need 1 millimole of $S_2O_3^{2-}$.

Number of millimoles of $S_2O_3^{2-}$ used = $31.5 \times 1 = 31.5$

Number of millimoles of Cu = 31.5

Number of moles of Cu in the original 2.5 g sample = $31.5/1000 = 0.0315$ **Answer(c)**

3-79. What was the percent by weight of copper (previous question) in the gong? (AW : Cu = 63.55 amu)

   (a) Less than 50 %   (b) Between 50 and 60 %   (c) Between 60 and 75 %

   (d) Between 75 and 85 %   (e) More than 85%

Weight of Cu in the original sample = $0.0315 \times 63.55 = 2$ g

% of Cu by weight = $(2 \times 100)/2.5 = 80\%$ **Answer(d)**

3-80. Use the following information to answer the questions:

Standardization of $S_2O_3^{2-}$:

   A. $IO_3^-\ (aq) + 8I^-\ (aq) + 6H^+\ (aq) \rightarrow 3I_3^-\ (aq) + 3H_2O\ (l)$

B. I_3^- (aq) + 2e$^-$ → 3I$^-$ (aq)

C. $2S_2O_3^{2-}$ (aq) → $S_4O_6^{2-}$ (aq) + 2e$^-$

D. Overall: IO_3^- (aq) + $6S_2O_3^{2-}$ (aq) + 6H$^+$ (aq) → I$^-$ (aq) + $3S_4O_6^{2-}$ (aq) + 3H$_2$O (l)

Reaction of Cu^{2+} with I$^-$:

E. $2Cu^{2+}$ (aq) + 3I$^-$ (aq) → 2Cu$^+$ (aq) + I_3^- (aq)

F. Cu$^+$ (aq) + I$^-$ (aq) → CuI (s)

G. Overall : $2Cu^{2+}$ (aq) + 5I$^-$ (aq) → 2CuI (s) + I_3^- (aq)

Reaction of I_3^- with $S_2O_3^{2-}$:

H. I_3^- (aq) + $2S_2O_3^{2-}$ (aq) → 3I$^-$ (aq) + $S_4O_6^{2-}$ (aq)

(i) Which of the above reactions does not involve oxidation-reduction?

(a) A   (b) B   (c) E   (d) F   (e) G

A : IO_3^- + 8I$^-$ + 6H$^+$ → $3I_3^-$ + 3H$_2$O

Oxidation number : +5 −2   −1   +1   −1/3   +1 −2   Oxidation – reduction

B : I_3^- (aq) + 2e$^-$ → 3I$^-$ (aq)

Oxidation number :   −1/3   −1   Reduction

E : $2Cu^{2+}$ (aq) + 3I$^-$ (aq) → 2Cu$^+$ (aq) + I_3^- (aq)

 Oxidation number :   +2   −1   +1   −1/3   Oxidation – reduction

F :   Cu$^+$ (aq) + I$^-$ (aq) →   CuI (s)

Oxidation number :   +1   −1   +1 −1   No oxidation – reduction

G: $2Cu^{2+}$ (aq) + 5I$^-$ (aq) → 2CuI (s) + I_3^- (aq)

Oxidation number :   +2   −1   +1 −1   −1/3   Oxidation – reduction

Answer(d)

(ii) If 22.5 mL of 0.125 M KIO$_3$ are required to react with 50.00 mL of the Na$_2$S$_2$O$_3$ solution prepared in this experiment, what is the molarity of the $S_2O_3^{2-}$ in the sodium thiosulfate solution?

(a) 2.78×10^{-3} M   (b) 1.67×10^{-2} M   (c) 5.56×10^{-2} M   (d) 3.34×10^{-1} M

(e) None of these is correct

IO_3^- (aq) + $6S_2O_3^{2-}$ (aq) + $6H^+$ (aq) → I^- (aq) + $3S_4O_6^{2-}$ (aq) + $3H_2O$ (l)

1 millimole of IO_3^- reacts with 6 millimoles of $S_2O_3^{2-}$.

22.5×0.125 = 2.781 millimoles of IO_3^- will react with

6×2.781 = 16.688 millimoles of $S_2O_3^{2-}$.

16.688 millimoles of $S_2O_3^{2-}$ are present in 50 mL.

Molarity of the $S_2O_3^{2-}$ solution = 16.688/50 = 0.334 M = $3.34×10^{-1}$ M   Answer(d)

3-81. How many g of Cu^{2+} are present in the unknown sample if 0.054 mole of $S_2O_3^{2-}$ is

required to react with the I_3^- ion produced by the unknown copper solution? (AW :

Cu = 63.55 amu)

    (a) 0.027 g   (b) 0.054 g   (c) 3.43 g   (d) 6.8 g   (e) 14 g

$2Cu^{2+}$ (aq) + $5I^-$ (aq) → 2CuI (s) + I_3^- (aq)

I_3^- (aq) + $2S_2O_3^{2-}$ (aq) → $3I^-$ (aq) + $S_4O_6^{2-}$ (aq)

So 2 moles of $S_2O_3^{2-}$ are required for 2 moles of Cu^{2+}.

0.054 mole of $S_2O_3^{2-}$ is required for 0.054 mole of Cu^{2+}.

Amount of Cu^{2+} present in the unknown sample = 0.054×63.55 = 3.43 g   Answer(c)

3-82. What is the molarity of con. HNO_3 (aq), if it takes 5.63 mL of the acid to consume 15.00

mL of 6.00 M NH_3 (aq)?

NH_3 (aq) + HNO_3 (aq) ⇌ NH_4NO_3 (aq)

    (a) Less than 1 M   (b) Between 1 and 5 M   (c) Between 5 and 10 M

    (d) Between 10 and 15 M   (e) More than 15 M

1 millimole of NH_3 (aq) will be consumed by 1 millimole of HNO_3 (aq).

15.00×6.00 = 90.00 millimoles of NH_3 willl be consumed by 90.00 millimoles of HNO_3.

Molarity of HNO_3 (aq) = 90.00/5.63 = 15.99 M   Answer(e)

3-83. Con.H_2SO_4 has a density of 1.84 g/cm^3 and this solution is 96% sulfuric acid by weight

(the rest is water). What is the molarity of this solution? (AW : H = 1.01, O = 16.00, and

S = 32.07 amu)

    (a) Less than 1 M   (b) Between 1 and 5 M   (c) Between 5 and 10 M

(d) Between 10 and 15 M   (e) More than 15 M

Molecular weight of $H_2SO_4 = (2\times1.01)+32.07+(4\times16.00) = 98.09$ g/mol

Molarity of H_2SO_4 solution $= (1.84\times1000\times96)/(100\times98.09) = 18.01$ M   Answer(e)

3-84. One way of determining blood alcohol levels is by titrating a sample of blood according to the following net ionic equation:

$$C_2H_5OH + 2Cr_2O_7^{2-} + 16H^+ \rightarrow 2CO_2 + 4Cr^{3+} + 11H_2O$$

If 8.76 mL of 0.04988 M $Cr_2O_7^{2-}$ are required for the titration of a 10.00 mL sample of blood, what is the molarity of the alcohol present in the blood sample?

   (a) Less than 0.01 M   (b) Between 0.01 and 0.03 M   (c) Between 0.03 and 0.06 M

   (d) Between 0.06 and 0.09 M   (e) More than 0.1 M

2 millimoles of $Cr_2O_7^{2-}$ are required for 1 millimole of C_2H_5OH.

$8.76\times0.04988 = 0.44$ millimole of $Cr_2O_7^{2-}$ is required by $0.44/2 = 0.22$ millimole of C_2H_5OH.

Molarity of the alcohol present in the blood sample $= 0.22/10 = 0.022$ M   Answer(b)

3-85. The reaction between fructose and the periodate ion is described by the following equation:

$$C_6H_{12}O_6 \text{ (aq)} + 5IO_4^- \text{ (aq)} \rightarrow 5IO_3^- \text{ (aq)} + 5HCO_2H \text{ (aq)} + H_2CO \text{ (aq)}$$

Calculate the molarity of a fructose solution if 22.35 mL of 1.050 M IO_4^- is required to consume 12.56 mL of this solution.

   (a) Less than 0.25 M   (b) Between 0.25 and 0.50 M   (c) Between 0.50 and 1.50 M

   (d) Between 1.50 and 2.00 M   (e) More than 2 M

1 millimole of $C_6H_{12}O_6$ consumes 5 millimoles of IO_4^-.

$22.35\times1.050 = 23.47$ millimoles of IO_4^- will be consumed by

$23.47/5 = 4.694$ millimoles of $C_6H_{12}O_6.$

Molarity of the fructose solution $= 4.694/12.56 = 0.374$ M   Answer(b)

3-86. Assume that 6.70 mL of a $Ca(OH)_2$ solution are added to 17.5 mL of 0.3750 M H_2SO_4. After this is done , the solution is still acidic and the excess H_2SO_4 is titrated with 0.2750 M KOH. If 15 mL of the KOH solution are needed to reach neutrality, what is the concentration of the $Ca(OH)_2$ solution?

H_2SO_4 (aq) + $Ca(OH)_2$ (aq) → Ca^{2+} (aq) + SO_4^{2-} (aq) + $2H_2O$ (l)

H_2SO_4 (aq) + 2KOH (aq) → $2K^+$ (aq) + SO_4^{2-} (aq) + $2H_2O$ (l)

(a) 0.0336 M   (b) 0.363 M   (c) 0.672 M   (d) 0.725 M   (e) 1.34 M

2 millimoles of KOH neutralize 1 millimole of H_2SO_4.

Excess H_2SO_4 = (15×0.2750)/2 = 2.0625 milimoles

Initial H_2SO_4 = 17.5×0.3750 = 6.5625 milimoles

H_2SO_4 neutralized by $Ca(OH)_2$ = 6.5625 – 2.0625 = 4.5 millimoles.

1 millimole of $Ca(OH)_2$ neutralizes 1 millimole of H_2SO_4.

4.5 millimoles of H_2SO_4 will be neutralized by 4.5 millimoles of $Ca(OH)_2$.

Concentration of $Ca(OH)_2$ solution = 4.5/6.7 = 0.672 M   Answer(c)

CHAPTER 4

STRUCTURE OF THE ATOM

INVALUABLE INFORMATION

Avogadro's number (N)	: 6.022142×10^{23} /mol
Planck's constant (h)	: 6.626069×10^{-34} J-s
Atomic mass unit (amu)	: $1.6605389 \times 10^{-24}$ g = $1.6605389 \times 10^{-27}$ kg
	: 1 g = 6.022142×10^{23} amu
Charge on electron (e)	: $- 1.6021765 \times 10^{-19}$ C
Mass of electron	: 5.485799×10^{-4} amu
	: 9.109383×10^{-28} g = 9.109383×10^{-31} kg
Charge on proton (p)	: $+ 1.6021765 \times 10^{-19}$ C
Mass of proton	: 1.0072765 amu
	: $1.6726217 \times 10^{-24}$ g = $1.6726217 \times 10^{-27}$ kg
Mass of neutron	: 1.0086649 amu
	: $1.6749273 \times 10^{-24}$ g = $1.6749273 \times 10^{-27}$ kg
Speed of light in vacuum (c)	: 2.99792458×10^{8} m/s
E_n	: $- 2.1799 \times 10^{-18} (Z/n)^2$ J $\quad$ n = 1,2,3.....
ΔE for hydrogen atom	: $R_H \{(1/n_1^2) - (1/n_2^2)\}$
Rydberg constant (R_H)	: 1.0974×10^{7} m^{-1}
	: 1.0974×10^{-2} nm^{-1} $\quad\quad$ (1nm^{-1} = 10^9 m^{-1})
	: $(2.1799 \times 10^{-18}$ J)

4-1. How many electrons are present in the $^{77}Se^{2-}$ ion?

(a) 32   (b) 34   (c) 43   (d) 77   (e) None of these

The atomic number of Se is 34. So there should be 34+2 = 36 electrons in $^{77}Se^{2-}$ ion. Answer(e)

4-2. How many electrons are present in the $^{126}Ba^{2+}$ ion?

(a) 56   (b) 58   (c) 78   (d) 134   (e) None of these

The atomic number of Ba is 56. There should be 56 − 2 = 54 electrons in the $^{126}Ba^{2+}$ ion.

Answer(e)

4-3. Calculate the number of electrons on a P^- ion if the atomic weight of phosphorus is 31 amu and the atomic number of the element is 15.

(a) 12   (b) 15   (c) 18   (d) 31   (e) 34

The atomic number is 15. So the number of electrons on P^{3-} ion is 15 + 3 =18.   Answer(c)

4-4. Calculate the number of electrons on a Ce^{4+} ion if the atomic number of this element is 58.

(a) 54   (b) 58   (c) 62   (d) 136   (e) 140

The atomic number is 58. So the number of electrons on Ce^{4+} ion is 58 − 4 = 54.   Answer(a)

4-5. Which has the largest (magnitude) charge to mass ratio?

(a) A Proton   (b) A Neutron   (c) An α particle   (d) An electron

	Charge (C)	Mass (kg)	Charge/mass (C/kg) (magnitude)
A proton	$1.6021765 \times 10^{-19}$	$1.6726217 \times 10^{-27}$	9.6×10^{7}
A neutron	0	$1.6749273 \times 10^{-27}$	0
An α particle	3.204353×10^{-19}	$6.6454767 \times 10^{-27}$	4.8×10^{7}
An electron	$-1.6021765 \times 10^{-19}$	9.109383×10^{-31}	1.76×10^{11}

Answer(d)

4-6. Which describes the difference between the $^{12}C, ^{13}C,$ and ^{14}C isotopes of carbon?

(a) They have the same number of protons but different number of electrons

(b) They have the same number of neutrons but different number of electrons

(c) They have the same number of electrons but different number of protons

(d) They have the same number of protons but different number of protons

(e) They have the same number of protons,electrons, and neutrons but different atomic masses

Isotopes have the same number of protons (and electrons) but diffent number of neutrons.

Answer(d)

4-7. What would you describe as the key result of the Rutherford experiment, in which a metal target was bombarded with α-particles?

(a) The α-particles were able to pass through the target

(b) The α-particles were found to knock electrons out of the target

(c) Some of the α-particles were deflected through large angles

(d) Flashes of light were emitted when the α-particles hit the ZnS screen after they passed through the target

(e) The experiment showed the need for better α-particle detectors

The key result was the large angle deflection of α-particles indicating a concentrated

positively charged region in the atom (nucleus). $\qquad$ Answer(c)

4-8. What do we mean when we say that "The energy of the electron in an atom is quantized"?

(a) The electron has a very small energy

(b) The energy of the electron is proportional to the mass of the nucleus

(c) When an electron changes its energy it emits a quantum of light

(d) The energy of an electron can have certain fixed values and not others

(e) The electron must be a wave

We mean that the energy of an electron can have certain fixed values and not others. Answer(d)

4-9. Which of the following statements is inconsistent with modern quantum mechanics?

(a) The hydrogen atom emits light at only a limited number of discreet frequencies

(b) In the ground state of the hydrogen atom the electron is constrained to a circular orbit

(c) The energy of the hydrogen atom is quantized

(d) No two electrons in an atom can have the same set of four quantum numbers

(e) The spin quantum number has a half integral value

In modern quantum mechanics electron is allowed to occupy 3-dimensional space and electrons

are not particles that can be restricted to a circular orbit. $\qquad$ Answer(b)

4-10. What is the difference between ^{14}C and ^{14}N atoms?

(a) They have the same number of protons but different number of neutrons

(b) They have the same number of neutrons but different number of protons

(c) They have the same number of neutrons and protons but different number of electrons

(d) ^{14}N has one more proton and one less neutron than ^{14}C

(e) ^{14}N has one more neutron and one less proton than ^{14}C

	Number of protons	Number of neutrons	Number of electrons
^{14}N	7	7	7
^{14}C	6	8	6

^{14}N has one more proton and one less neutron than ^{14}C.   Answer(d)

4-11. How many electrons and neutrons could be found in $^{56}Fe^{3+}$ ion formed during a

supernova explosion?

      (a)23e⁻,23p⁺, and 33n⁰   (b) 23e⁻,26p⁺, and 30n⁰   (c) 26e⁻,23p⁺, and 33n⁰

      (d) 29e⁻,26p⁺, and 30n⁰   (e) 26e⁻,29p⁺, and 27n⁰

The atomic number of Fe is 26.

$^{56}Fe^{3+}$: 26 protons , 30 neutrons, and 26 – 3 = 23 electrons.   Answer(b)

4-12. What is the wavelength in centimeters of light that has a frequency of 2.33×10^{15} s^{-1}?

    (a) 6.99×10^{25} cm   (b) 7.77×10^4 cm   (c) 1290 cm   (d) 1.29×10^{-5} cm

    (e) None of the above

Frequency×Wavelength = Speed of light

Frequency×Wavelength = 2.998×10^8 m/s

Wavelength = $(2.998 \times 10^8) / (2.33 \times 10^{15}) = 1.29 \times 10^{-7}$ m

        = 1.29×10^{-5} cm   Answer(d)

4-13. A He-Ne laser emits light with a wavelength of 6328 Å. What is the energy of a 6328 Å

photon in joules?

    (a) 2.179×10^{-19} J   (b) 3.139×10^{-19} J   (c) 5.448×10^{-19} J   (d) 6.328×10^{-19} J

    (e) 2.179×10^{-18} J

1 Å = 10^{-10} m

Wavelength = 6328 Å = 6328×10^{-10} m

Frequency×Wavelength = 2.998×10^8 m/s

Frequency = $(2.998 \times 10^8)/(6328 \times 10^{-10})$ s^{-1} = 4.738×10^{14} s^{-1}

E = hν = $(6.626 \times 10^{-34}) \times (4.738 \times 10^{14})$ = 3.139×10^{-19} J   Answer(b)

4-14. If green light has a frequency of 5.0×10^{14} s^{-1}, what is the wavelength (in meters) of this light?

(a) 6.7×10^{-24} (b) 2.0×10^{-15} (c) 6.0×10^{-7} (d) 3.0×10^8 (e) None of the above

Frequency×Wavelength = 2.998×10^8 m/s

Wavelength = $(2.998 \times 10^8)/(5.0 \times 10^{14})$ = 6.0×10^{-7} m   Answer(c)

4-15. If X-rays have shorter wavelength than UV rays which of the following statements is true?

(a) X-rays have smaller frequencies than UV rays
(b) X-rays travel faster than UV rays
(c) X-rays have more energy than UV rays
(d) X-rays have a larger amplitude than UV rays
(e) None of the above statements are true

Energy is inversely proportional to the wavelength. So X-rays have more energy than UV rays.

X-rays and UV rays have the same speed.   Answer(c)

4-16. Which of the following could convert a nonionizing form of electromagnetic radiation

into a form of ionizing radiation?

(a) Decreasing the wavelength of the radiation
(b) Decreasing the energy of the radiation
(c) Decreasing the frequency of the radiation
(d) All of the above are true
(e) None of the above are true

Nonionizing electromagnetic radiation will become ionizing form if the frequency is

increased to above a certain level. Frequency is inversely proportional to the wavelength.

So decreasing the wavelength will convert a nonionizing form of electromagnetic radiation

into ionizing form of radiation.   Answer(a)

4-17. In a 23.490 kilogauss magnetic field, ^{13}C nuclei absorb electromagnetic radiation in RF

(radio) portion of the spectrum at a frequency of 25.147 MHz (25.147×10^6 s^{-1}). Calculate the

wavelength of this radiation.

(a) Longer than 10 meters   (b) Between 10 and 0.10 meters

(c) Between 0.10 and 10^{-4} meters   (d) Between 10^{-4} and 10^{-6} meters

(e) shorter than 10^{-6} meters

Frequency×Wavelength = 2.998×10^8 m/s

Wavelength = $(2.998 \times 10^8)/(25.147 \times 10^6)$ = 11.92 m   Answer(a)

4-18. O_2 molecules can dissociate to form O atoms by absorbing electromagnetic radiation. If it takes 498 kJ to dissociate one mole of O_2 molecules to form two moles of O atoms, what would be the wavelength of the radiation that would have just energy to decompose O_2 molecules to O atoms? In what portion of the electromagnetic spectrum would this wavelength be found?

(a) Radiowave   (b) Microwave   (c) Infrared   (d) Visible   (e) UV

Energy per mole = 498 kJ = 498×10^3 J

Energy per molecule = $(498 \times 10^3)/(6.022 \times 10^{23}) = 8.27 \times 10^{-21}$ J

$E = h\nu = hc/\lambda$

λ (Wave length) = hc/E = $\{(6.626 \times 10^{-34})$ J-s $\times (2.998 \times 10^8)$ m/s$\}/(8.27 \times 10^{-21})$ J

$\qquad\qquad = 2.4 \times 10^{-7}$ m = 240 nm (UV region) $\qquad\qquad$ Answer(e)

4-19. The wavelength of the characteristic yellow-orange light emitted by sodium ions in a burner flame is 589.5923 nm. What is the energy of this light, in units of kJ/mol?

(a) Less than 1 kJ/mol   (b) Between 1 and 10 kJ/mol   (c) Between 10 and 100 kJ/mol

(d) Between 100 and 1000 kJ/mol   (e) More than 1000 kJ/mol

Wavelength = 589.5923 nm = 589.5923×10^{-9} m

$E = h\nu = hc/\lambda$

Energy per mole = $(hc/\lambda) \times (6.022 \times 10^{23})$

$\qquad\qquad = \{(6.626 \times 10^{-34}) \times (2.998 \times 10^8) \times (6.022 \times 10^{23})\}/ (589.5923 \times 10^{-9})$

$\qquad\qquad = 2.029 \times 10^5$ J/mol = 202.9 kJ/mol $\qquad\qquad$ Answer(d)

4-20. There is a fundamental difference between the long wavelength IR radiation given off by the toy ovens sold by Toys 'R Us and the shorter wavelength UV radiation emitted by the tanning booths. Which of the following statements is true?

(a) IR radiation has a higher frequency than UV radiation
(b) IR radiation carries more energy per photon than UV radiation
(c) IR radiation carries a larger amplitude than UV radiation
(d) IR radiation travels slower than UV radiation
(e) None of the above statements is true

IR radiation has a lower frequency and lower energy than UV radiation. Amplitude is determined by intensity and not by frequency. IR radiation and UV radiation have the same speed. $\qquad\qquad$ Answer(e)

4-21. A cheap spectrophotometer can be made using a light-emitting diode as the source of light. Assume that you build one of these spectrometers using an LED that gives off green light with a wavelength of 520 nm. What is the energy of a photon of this green light?

(a) 1.15×10^{-34} J   (b) 1.03×10^{-22} J   (c) 3.82×10^{-21} J   (d) 3.82×10^{-19} J

(e) None of the above

Wavelength = 520 nm = 520×10^{-9} m

$E = h\nu = hc/\lambda = \{(6.626 \times 10^{-34}) \times (2.998 \times 10^{8})\}/(520 \times 10^{-9}) = 3.82 \times 10^{-19}$ J   Answer(d)

4-22. The radiation becomes particularly dangerous to life when it carries enough energy to ionize a water molecule when it is absorbed.

$$H_2O \text{ (l)} + h\nu \rightarrow H_2O^+ + e^- \qquad \Delta H = 1200 \text{ kJ/mol}$$

Use Avogadro's number and Planck's constant to calculate the frequency of this radiation to one significant figure.

(a) 3×10^{9} s^{-1} (b) 3×10^{12} s^{-1} (c) 3×10^{15} s^{-1} (d) 3×10^{18} s^{-1} (e) 3×10^{21} s^{-1}

Energy per mole = $(h\nu) \times (6.022 \times 10^{23}) = 1200$ kJ/mol = 1200×10^{3} J/mol

Frequency = $(1200 \times 10^{3})/\{(6.022 \times 10^{23}) \times (6.626 \times 10^{-34})\} = 3 \times 10^{15}$ s^{-1} Answer(c)

4-23. In what portion of the electromagnetic spectrum are you likely to find the radiation that carries just energy to be dangerous to life because it can ionize the water that is important to living organisms?

(a) Radio/TV waves, $\lambda = 10 - 0.1$ m
(b) Microwaves, $\lambda = 0.01 - 10^{-4}$ m
(c) Ultraviolet, $\lambda = 10^{-7} - 10^{-9}$ m
(d) X-rays, $\lambda = 10^{-10} - 10^{-12}$ m
(e) γ-rays, $\lambda = 10^{-12} - 10^{-14}$ m

Energy required to ionize one mole of water = 1200 kJ/mol = 1200×10^{3} J/mol

Energy per mole = $(hc/\lambda) \times (6.022 \times 10^{23})$

Wavelength $(\lambda) = \{(6.626 \times 10^{-34}) \times (2.998 \times 10^{8}) \times (6.022 \times 10^{23})\}/(1200 \times 10^{3})$

$$= 1 \times 10^{-7} \text{ m \quad (Ultraviolet)} \qquad \text{Answer(c)}$$

4-24. According to Bohr model the band structure of the emission spectrum of the hydrogen atom suggests that :

(a) The atom is composed of a small positive nucleus and electrons
(b) There are many electrons in a hydrogen atom
(c) The position and energy of an electron can not be determined simultaneously

(d) The energies of the electron are quantized

According to Bohr model only a limited number of orbits with certain energies are allowed.

In other words energies of the electron are quantized.   Answer(d)

4-25. Which of the following statements about Bohr model is wrong?

    (a) The energy of the atom is quantized
    (b) All possible wavelengths can occur in the emission spectrum since n runs from 1 to infinity
    (c) The angular momentum of the electron is quantized
    (d) All quantized energies are negative

According to Bohr model only a limited number of orbits with certain energies are allowed

(the orbits are quantized). So all possible wavelengths can not occur in the emission spectrum.

Answer(b)

4-26. Which of the following transitions in the spectrum of the hydrogen atom results in the

absorption of a photon with the largest energy?

    (a) $n = 2$ to $n = 3$ (b) $n = 2$ to $n = 4$ (c) $n = 1$ to $n = 4$ (d) $n = 3$ to $n = 1$

    (e) $n = 7$ to $n = 1$

ΔE for hydrogen atom $= R_H \{(1/n_1^2)-(1/n_2^2)\}$ $n_2 > n_1$ $R_H = 2.1799 \times 10^{-18} J$

n_1	n_2	$\Delta E(J)$
2	3	3.0×10^{-19}
2	4	4.1×10^{-19}
1	4	2.04×10^{-18}
3	1	Emission
7	1	Emission

Answer(c)

4-27. Which transition in spectrum of the hydrogen atom results in the emission of the light

with the longest wavelength?

    (a) $n = 3$ to $n = 2$ (b) $n = 3$ to $n = 1$ (c) $n = 5$ to $n = 4$ (d) $n = 2$ to $n = 3$

    (e) $n = 1$ to $n = 3$

ΔE for hydrogen atom $= R_H \{(1/n_1^2)-(1/n_2^2)\}$ $n_2 > n_1$ $R_H = 2.1799 \times 10^{-18} J$

Transition	ΔE (J)
(a) $n = 3$ to $n = 2$	3.0×10^{-19}

(b) n = 3 to n = 1   1.94×10^{-18}

(c) n = 5 to n = 4   4.90×10^{-20}

(d) n = 2 to n = 3   Absorption

(e) n = 1 to n = 3   Absorption

ΔE is inversely proportional to the wavelength. The transition with the lowest ΔE will have the

longest wavelength.   Answer(c)

4-28. What is the energy of the light emitted when an electron falls from the n = 4 to n = 2

orbit in the hydrogen atom?

   (a) 4.09×10^{-19} J   (b) 5.45×10^{-18} J   (c) 4.36×10^{-18} J   (d) 2.62×10^{-17} J

   (e) None of the above

 ΔE for hydrogen atom = $R_H \{(1/n_1^2)-(1/n_2^2)\}$ $n_2>n_1$ R_H = 2.1799×10^{-18}J

ΔE = (2.1799×10^{-18})×(0.1875) J = 4.09×10^{-19} J   Answer(a)

4-29. If a laser is operated on the n = 5 to n = 2 transition of a H atom, what wavelength photon

would be emitted?

   (a) 380 nm   (b) 434 nm   (c) 1950 nm   (d) 3910 nm   (e) None of the above

ΔE for hydrogen atom = $R_H \{(1/n_1^2)-(1/n_2^2)\}$ $n_2>n_1$ R_H = 2.1799×10^{-18}J

ΔE = (2.1799×10^{-18})×(0.21) J

$E = h\nu = hc/\lambda$

$\lambda = hc/E$ = {(6.626×10^{-34})×(2.998×10^{8})}/ {(2.1799×10^{-18})×(0.21)} = 4.339×10^{-7}m

        = 434 nm   Answer(b)

4-30. Which of the following quantum numbers is used to describe the orientation of an orbital

in space?

   (a) n   (b) l   (c) m   (d) s   (e) None of these quantum numbers

Shape of the orbital is determined by the quantum number l, but orientation is determined

by the quantum number m.   Answer(c)

4-31. Which of the following quantum numbers can have a value that is not an integer?

   (a) n   (b) l   (c) m   (d) s

   (e) none of the quntum numbers can have any value that is not an integer n : only integer values 1,2,3 …..

l : only integer values 0,1,2,3……….(n-1)

m : only integer values -1 to $+1$

Only s can have +1/2 or −1/2 values (non integer values).   Answer(d)

4-32. What kind of orbital is described by the quantum numbers n = 4, l = 2, and m = −1?

(a) 4s   (b) 4p   (c) 4d   (d) 4f   (e) None of these

n is the principal quantum number.

Value of l	Subshell notation	number of orbitals
0	s	1
1	p	3
2	d	5
3	f	7
4	g	9
5	h	11

m : only integer values -1 to $+1$

So n = 4, l = 2, and m = −1 describes a 4d orbital.   Answer(c)

4-33. Orbital for which l = 1 is described by which of the following symbols?

(a) s   (b) p   (c) d   (d) f   (e) f

Value of l	Subshell notation	number of orbitals
0	s	1
1	p	3
2	d	5
3	f	7
4	g	9
5	h	11

So l = 1 describes a p orbital.   Answer(b)

4-34. Pauli exclusion priciple states that

(a) No two electrons in an atom can have the same spin
(b) No two electrons in an atom can occupy the same orbital
(c) Two electrons in the same orbital have identical values of spin quantum number
(d) No two electrons in an atom can have the same set of four quantum numbers
(e) No two electrons in an atom can have the same principal quantum number

According to Pauli's exclusion priciple, no two electrons in an atom can have the same set

of four quantum numbers.   Answer(d)

4-35. In the absence of an external magnetic field, the energy of an electron in an orbital in a lithium atom is determined by which pair of quantum numbers?

    (a) n and l   (b) n and m   (c) n and s   (d) l and m   (e) m and s

In the absence of an external magnetic field, the energy of an electron in an orbital in a lithium atom is determined by n and l.   Answer(a)

4-36. Two orbitals are said to be degenerate if they:

    (a) Contain the same number of electrons
    (b) Have the same value for the angular quantum number l, and different values of principal quantum number, n
    (c) Have the same set of quantum numbers, n and m, but have different values of the s quantum number
    (d) Have the same energy
    (e) Contain the same number of unpaired electrons

Two orbitals are said to be degenerate if they have the same energy.   Answer(d)

4-37. What is the maximum number of electrons that can be accomodated in the subshell for which n = 3 and l = 2?

    (a) 2   (b) 6   (c) 10   (d) 14   (e) 18

n = 3 and l = 2 . So five d orbitals will be available. Each orbital will accommodate two electrons.

Maximum number of electrons that can be accomodated $5 \times 2 = 10$.   Answer(c)

4-38. Calculate the maximum number of electrons that can fit into the n = 4 shell of orbitals.

    (a) 6   (b) 8   (c) 18   (d) 16   (e) 32

n = 4; l : integer values 0,1,2,3

l = 0 (m = 0)	One s orbital	2 electrons
l = 1 (m = −1,0,1)	Three p orbitals	6 electrons
l = 2 (m = −2,−1,0,1,2)	Five d orbitals	10 electrons
l = 3 (m = −3,−2,−1,0,1,2,3)	Seven f orbitals	14 electrons
	Total	32 electrons Answer(e)

4-39. A 25 watt bulb emits monochromatic yellow light of wavelength of 0.57μm. Calculate the rate of emission of quanta per second.

Wavelength = 0.57 μm = 0.57×10^{-6} m = 570×10^{-9} m

Energy of one photon = $hv = hc/\lambda = (6.626\times10^{-34}\times2.998\times10^8)/(570\times10^{-9})$ J $= 3.485\times10^{-19}$ J

Power of bulb = 25 watt = 25 J s^{-1}

Rate of emission of quanta = 25 J s^{-1}/3.485$\times10^{-19}$ J = 7.174$\times10^{19}$ s^{-1}

4-40. The energy associated with first orbit in the hydrogen atom is -2.1799×10^{-18} J/atom. What is the energy associated with the fifth orbit? Calculate the radius of Bohr's fifth orbit for hydrogen atom.

(1eV = 1.6022$\times10^{-19}$ J)

$E_n = - (2.1799\times10^{-18}) (Z/n)^2$ J

For hydrogen atom $E_n = - (2.1799\times10^{-18})) (1/n)^2$ J

E for the first orbit = $- (2.1799\times10^{-18})$ J = $- 13.606$ eV

E for the fifth orbit = $- (2.1799\times10^{-18})) (1/5)^2 = - (2.1799\times10^{-18}))/25 = - 8.72\times10^{-20}$ J

$$= - 0.544 \text{ eV}$$

Radius r = $a_0 n^2/Z$ $\qquad\qquad$ $a_0 = 0.0529$ nm = Bohr radius

For hydrogen atom r = $a_0 n^2$

Radius for the first orbit = $a_0 (1)^2 = a_0 = 0.0529$ nm

Radius for the fifth orbit = $a_0 (5)^2 = a_0 (25) = 1.3225$ nm

4-41. The mass of an electron is 9.11×10^{-31} kg. If its kinetic energy is 4.0×10^{-25} J (kg-m^2/s^2), calculate its wavelength.

Kinetic energy = $1/2$ mv^2 = 4.0×10^{-25} kg-m^2/s^2

$v^2 = (2\times4.0\times10^{-25})/(9.11\times10^{-31})$ m^2/s^2

v = 937.10 m/s

Wavelength = λ = h/mv = $(6.626\times10^{-34})/\{(9.11\times10^{-31})\times(937.10)\}$ m = 7.76×10^{-7} m = 776 nm

4-42. Calculate the energy required for the process:

He$^+$ (g) $\rightarrow$ He^{2+} (g) + e$^-$

The ionization energy for the H atom in the ground state is -2.1799×10^{-18} J/atom.

He$^+$ is hydrogen-like species.

$E_n = - (2.1799\times10^{-18}) (Z/n)^2$ J

Energy required for the above process = $(2.1799\times10^{-18})\times (2/1)^2$ J/atom

$$= (2.1799 \times 10^{-18}) \times (2)^2 \times (6.022 \times 10^{23}) \text{ J/mol}$$

$$= 5251 \text{ kJ/mol}$$

4-43. The ejection of photo electron from silver metal in the photoelectric effect experiment can be stopped by applying a voltage of 0.53 V, when the radiation 256.7 nm is used. Calculate the work function for the silver metal.

$(1 eV = 1.6022 \times 10^{-19} J)$

Wavelength = 256.7 nm = 2.567×10^{-7} m

$1/2 \ m_e v^2 = 0.53$ eV

$h\nu = hc/\lambda = (6.626 \times 10^{-34} \times 2.998 \times 10^8)/(2.567 \times 10^{-7})$ J = 4.83 eV

$h\nu$ = Work function for silver metal (W_o) + $1/2 \ m_e v^2$

Work function for silver metal (W_o) = $h\nu - 1/2 \ m_e v^2$ = 4.83 − 0.53 = 4.3 eV

4-44. Calculate the velocity of the electron in Bohr's first orbit of hydrogen atom.

$2\pi r = n\lambda = nh/p = nh/mv$

Radius $r = a_0 n^2/Z$ a_0 = 0.0529 nm = 0.0529×10^{-9} m = Bohr radius

$mv = nh/2\pi r = nhZ/2\pi a_0 n^2 = hZ/2\pi a_0 n$

$v = (h/2\pi a_0 m)(Z/n)$

For hydrogen atom the velocity of the electron in Bohr's first orbit = $h/2\pi a_0 m$

$= \{(6.626 \times 10^{-34}) \text{ kg-m}^2/\text{s}\}/\{2\pi \times (0.0529 \times 10^{-9} \text{ m}) \times 9.11 \times 10^{-31}\}$

$= 2.19 \times 10^6$ m/s

4-45. If the velocity of the electron in Bohr's first orbit of hydrogen atom is 2.19×10^6 m/s, calculate the de Broglie wavelength associated with it.

$\lambda = h/mv = \{(6.626 \times 10^{-34}) \text{ kg-m}^2/\text{s}\}/\{(9.11 \times 10^{-31} \text{ kg}) \times (2.19 \times 10^6 \text{ m/s})\} = 3.32 \times 10^{-10}$ m

$= 0.332$ nm

4-46. The kinetic energy of an electron in the 4th Bohr orbit of a hydrogen atom is:

   (a) $h^2/32\pi^2 m a_0^2$ (b) $h^2/64\pi^2 m a_0^2$ (c) $h^2/128\pi^2 m a_0^2$ (d) $h^2/16\pi^2 a_0^2$ (e) $h^2/8\pi^2 m a_0^2$

Radius $r = a_0 n^2/Z$ a_0 = 0.0529 nm = 0.0529×10^{-9} m = Bohr radius

$2\pi r = n\lambda = nh/p = nh/(mv)$

$mv = nh/(2\pi r) = nhZ/2\pi a_0 n^2 = hZ/(2\pi a_0 n)$

For hydrogen atom the velocity of an electron in the 4th Bohr's orbit $v = h/(8\pi a_0 m)$

The kinetic energy $= 1/2\ mv^2 = 1/2\ m\ \{h/(8\pi a_0 m)\}^2 = 1/2\ m(h^2/64\pi^2 a_0^2 m^2) = h^2/(128\pi^2 a_0^2 m)$

$\qquad\qquad\qquad\qquad\qquad\qquad\qquad\qquad\qquad\qquad\qquad\qquad$ Answer(c)

4-47. Similar to electron diffraction, neutron diffraction microscope is also used for the determination of the structure of molecules. If the wavelength used here is 800 pm, calculate the characteristic velocity associated with the neutron. (Mass of neutron $= 1.67493\times10^{-27}$ kg)

$\lambda = 800$ pm $= 800\times10^{-12}$ m $= 8.0\times10^{-10}$ m

$\lambda = h/mv$

$v = h/m\lambda$

$v = \{(6.626\times10^{-34})\ \text{kg-m}^2/\text{s}\}/\{(1.67493\times10^{-27}\ \text{kg})\times(8.0\times10^{-10}\ \text{m})\} = 494.5$ m/s

4-48. The velocity associated with a proton moving in a potential difference of 1000 V is 4.37×10^5 m/s. If a hockey ball of mass 0.1 kg is moving with this velocity, calculate the wavelength associated with this velocity.

$\lambda = h/mv = \{(6.626\times10^{-34})\ \text{kg-m}^2/\text{s}\}/\{(0.1\text{kg})\times(4.37\times10^5\ \text{m/s})\} = 1.516\times10^{-38}$ m

4-49. If a particle of mass m kg is moving with kinetic energy E J, its de Broglie wavlength would be:

$\qquad$ (a) h/2mE $\qquad$ (b) $h/\sqrt{2mE}$ $\qquad$ (c) $h/2\sqrt{mE}$ $\qquad$ (d) $h/2m\sqrt{E}$

$1/2\ mv^2 = E$

$v^2 = 2E/m$

$v = \sqrt{2E/m}$

$mv = \sqrt{2Em}$

$\lambda = h/mv = h/\sqrt{2Em}$ $\qquad\qquad\qquad\qquad\qquad\qquad\qquad\qquad$ Answer(b)

4-50. If the position of an electron is measured within an accuracy of $\pm\ 0.002$ nm, calculate the uncertainty in the veocity and the momentum of electron.

$m = 9.11\times10^{-31}$ kg

$\Delta x = 0.002\times10^{-9}$ m

$\Delta p \Delta x \geq h/4\pi$

$\Delta p \geq h/4\pi\Delta x \geq \{(6.626\times10^{-34})\ \text{kg-m}^2/\text{s}\}/\{(4\pi)\times(0.002\times10^{-9}\ \text{m})\} \geq 2.64\times10^{-23}$ kg m s^{-1}

$m\Delta v \geq h/4\pi\Delta x$

$\Delta v \geq h/4\pi m\Delta x \geq \{(6.626\times10^{-34})\ kg\text{-}m^2/s\}/\{(4\pi)\times(9.11\times10^{-31})\times(0.002\times10^{-9}\ m)\}$

$\geq 2.90\times10^7\ m\ s^{-1}$

4-51. A state in hydrogen-like species He^+ has one radial node and its energy is equal to the ground state of the hydrogen atom. This state is:

    (a) 1s   (b) 2s   (c) 3s   (d) 2p (e) 3p

$E_n = -(2.1799\times10^{-18})\ (Z/n)^2\ J$

For hydrogen $E_1 = -(2.1799\times10^{-18})\ J$

For He^+ $Z = 2$.

$E_n = -(2.1799\times10^{-18})\ (2/n)^2 = -(2.1799\times10^{-18})\ J$

So $n = 2$

Radial node $(n-1-1) = 1$. $2-1-1 = 1$ So $l = 0$

$n = 2$ and $l = 0$. The state is 2s.   Answer(b)

4-52. The Schrodinger wave equation for 2s state in hydrogen atom is

$\Psi_{2s} = \{2 - (r/a_0)\}\times\{e^{-(r/2a_0)}\}/\{(4\sqrt{2\pi})\times(a_0)^{3/2}\}$

At what value of r does the 2s radial node occur?

At radial node the probability density function reduces to zero.

Probability density function $= (\Psi_{2s})^2$

$(\Psi_{2s})^2 = \{2 - (r/a_0)\}^2\times\{e^{-(r/2a_0)}\}^2/\{(4\sqrt{2\pi})\times(a_0)^{3/2}\}^2 = 0$

    (i) $\{(2-(r/a_0)\}^2 = 0$

               $r = 2a_0$

    (ii)   Additionally, $(\Psi_{2s})^2 = 0$ if $\{e^{-(r/2a_0)}\}^2 = 0$. So there will be a node at $r = \infty$

4-53. A state in hydrogen-like species has one angular node and one radial node and its energy is equal to the ground state energy of the hydrogen atom.

The species is:

    (a) He^+ (b) Be^{3+} (c) Li^{2+} (d) B^{4+}

$E_n = -(2.1799\times10^{-18})\ (Z/n)^2\ J$

For hydrogen $E_1 = -(2.1799\times10^{-18})\ J$

Unknown hydrogen-like species has one angular node . So l = 1.

Unknown hydrogen-like species has one radial node (n − 1 − 1). So n = 3.

For unknown hydrogen-like species $E_3 = -(2.1799 \times 10^{-18})(Z/3)^2$ J $= -(2.1799 \times 10^{-18})$ J

So Z = 3   The species is Li^{2+}.   Answer(c)

4-54. Which of the following selection rules for quantum numbers is incorrectly stated?

    (a) n is an integer greater than or equal to zero
    (b) l is any integer from 0 to n−1
    (c) m is any integer from − 1 to + 1
    (d) s is either +1/2 or −1/2
    (e) All of the above selection rules are correctly stated

n : Only integer values 1,2,3…..

l : Only integer values 0,1,2, ……. n−1

m : Only integer values − 1 to + 1

s: Either +1/2 or −1/2   Answer(a)

4-55. Which of the following atomic orbitals doesn't exist?

    (a) 3f   (b) 3p   (c) 5f   (d) 5d   (e) 6s

3f : For f orbital l = 3; n should be atleast 4, so 3f doesn't exist.

3p : For p orbital l = 1; n should be atleast 2, so 3p exists

5f : For f orbital l = 3; n should be atleast 4, so 5f exists.

5d : For d orbital l = 2; n should be atleast 3, so 5d exists.

6s : For s orbital l = 0; n should be atleast 1, so 6s exists.   Answer(a)

4-56. Which of the following is a legitimate set of n,l,m, and s quantum numbers ?

    (a) 0,0,0,1/2   (b) 8,4,−3,−1/2   (c) 3,3,2,1/2   (d)2,1,−2,−1/2   (e) 5,3,3,−1

0,0,0,1/2 : n can not be zero. Not legitimate.

8,4,−3,−1/2 : n =8, l should be from 0 to 7, m should be from −4 to +4, s should be either

                1/2 or −1/2. Legitimate.

 3,3,2,1/2 : n =3, l can not be more than 2. Not legitimate.

2,1,−2,−1/2 : n =2, l =1 is allowed, for m only −1,0,+1 are allowed. Not legitimate.

5,3,3,−1 : n = 5, l = 3 allowed, m = 3 allowed, s can be only 1/2 or −1/2. Not legitimate.

                                                     Answer(b)

4-57. Which set of n,l,m,and s quantum numbers is allowed?

(a) 4,−2,−1,1/2   (b) 4,2,3,1/2   (c) 4,3,0,1   (d) 4,0,0,−1/2

4,−2,−1,1/2: n =4, l can not be negative. Not allowed

4,2,3,1/2 : m can not be more than l. Not allowed.

4,3,0,1 : s can be only 1/2 or −1/2. Not allowed.

4,0,0,−1/2 : n =4, l = 0 is allowed, m = 0 is allowed, s = −1/2 is allowed. Allowed.   Answer(d)

4-58. Which of the following sets of n,l,m, and s quantum numbers is not allowed?

(a) 1,1,0,1/2   (b) 2,0,0,1/2   (c) 3,2,−1,−1/2   (d) 4,1,−1,1/2   (e) 5,3,2,−1/2

1,1,0,1/2 : l can not be equal to n.   Not allowed.

2,0,0,1/2 : n =2, l = 0 is allowed, m = 0 is allowed, s = 1/2 is allowed. Allowed.

3,2,−1,−1/2 : n = 3, l = 2 is allowed, m = −1 is allowed, s = −1/2 is allowed. Allowed.

4,1,−1,1/2: n =4, l = 1 is allowed, m = −1 is allowed, s = 1/2 is allowed. Allowed.

5,3,2,−1/2: n = 5, l = 3 is allowed, m = 2 is allowed, s = −1/2 is allowed. Allowed.   Answer(a)

4-59. Which set of quantum numbers can be used to describe a 2p electron?

(a) 2,1,0,−1/2   (b) 2,0,0,1/2   (c) 2,2,1,1/2   (d) 3,2,1,−1/2   (e) 3,1,0,1/2

2p : n should be 2 and l should be 1.   Answer(a)

4-60. How many electrons can be placed in a 3d orbital?

(a) 2   (b) 6   (c) 8   (d) 10   (e) 18

Each 3d orbital can have only 2 electrons.   Answer(a)

4-61. What is the maximum number of unpaired electrons that can be accomodated in a 5d subshell?

(a) 4   (b) 5   (c) 10   (d) 14   (e) 6

5d has 5 orbitals. So the maximum number of unpaired electrons that can be accomodated in a 5d subshell is equal to 5. Above 5, pairing will reduce the number of unpaired electrons.

Answer(b)

4-62. Which of the following is an incorrect order of increasing energy of the atomic orbitals?

(a) 3s<4s<5s   (b) 5s<5p<5d   (c) 5s<4d<5p   (d) 5p<6s<4f

(e) None of these are incorrect

Correct order of increasing energy of the atomic orbitals :

1s<2s<2p<3s<3p<4s<3d<4p<5s<4d<5p<6s<4f<5d<6p

3s<4s<5s : Correct order   5s<5p<5d : Correct order   5s<4d<5p : Correct order

5p<6s<4f : Correct order   Answer(e)

4-63. Which set of orbitals is arranged in increasing order of energy?

    (a) 3d<4s<4p<5s<4d   (b) 3d<4s<4p<4d<5s   (c) 4s<3d<4p<5s<4d

    (d) 4s<3d<4p<4d<5s

Correct order of increasing energy of the atomic orbitals :

1s<2s<2p<3s<3p<4s<3d<4p<5s<4d<5p<6s<4f<5d<6p

3d<4s<4p<5s<4d :   3d<4s Incorrect   3d<4s<4p<4d<5s :   3d<4s Incorrect

4s<3d<4p<5s<4d :   Correct   4s<3d<4p<4d<5s :   4d<5s Incorrect Answer(c)

4-64. Which of the following sets of atomic orbitals are degenerate for a lithium atom?

    (a) $3s, 3p_x, 3d_{z2}$ (b) $2p_x, 2p_y, 2p_z$ (c) $2p_x, 3p_x, 4p_x$

    (d) None of these sets of orbitals are degenerate

    (e) All of these sets of orbitals are degenerate

Li : Electron configuration : $1s^2, 2s^1$

Lithium has 3 empty 2p orbitals and these three orbitals have the same energy.   Answer(b)

4-65. When atomic orbitals are filled according to the *aufbau* principle, the 6p orbitals are filled immediately after the

    (a) 4f   (b) 5d   (c) 6s   (d) 7s

Correct order of increasing energy of the atomic orbitals :

1s<2s<2p<3s<3p<4s<3d<4p<5s<4d<5p<6s<4f<5d<6p

6p energy level is just above 5d energy level.   Answer(b)

4-66. Which of the following orbitals is filled first when electrons are added to atomic orbitals on a xenon atom?

    (a) 4d   (b) 4f   (c) 5s   (d) 5p   (e) 5d

Correct order of increasing energy of the atomic orbitals :

1s<2s<2p<3s<3p<4s<3d<4p<5s<4d<5p<6s<4f<5d<6p

So 5s will be filled first. Electron configuration for xenon : $[Kr] 5s^24d^{10}5p^6$ Answer(c)

4-67. Which of the following orbitals would be filled first when electrons are added to a gold atom?

   (a) 4f   (b) 5d   (c) 6s   (d) 6p   (e) 6d

Correct order of increasing energy of the atomic orbitals :

1s<2s<2p<3s<3p<4s<3d<4p<5s<4d<5p<6s<4f<5d<6p<7s<5f<6d

So 6s will be filled first. Electron configuration for gold : $[Xe]6s^14f^{14}5d^{10}$ Answer(c)

4-68. In what group of the periodic table would an element with the following electron configuration belong? $1s^22s^22p^63s^23p^64s^24d^{10}4p^1$

   (a) Group 1   (b) Group 13   (c) Group 15   (d) Group 17   (e) None of the above

The electron configuration : $1s^22s^22p^63s^23p^64s^24d^{10}4p^1$: $[Ar]3d^{10}4s^24p^1$

3 valence electrons. The element belongs to group 13. The element is gallium.   Answer(b)

4-69. A single atom of element 109 has been synthesized. Use the *aufbau* principle to predict the electron configuration of this element. Which of the following elements would 109 most resemble?

   (a) Ta   (b) Re   (c) Ir   (d) Au   (e) Tl

Correct order of increasing energy of the atomic orbitals :

1s<2s<2p<3s<3p<4s<3d<4p<5s<4d<5p<6s<4f<5d<6p<7s<5f<6d<7p

The electron configuration for element 109 :

$1s^22s^22p^63s^23p^64s^23d^{10}4p^65s^24d^{10}5p^66s^24f^{14}5d^{10}6p^67s^25f^{14}6d^7$: Group 9 : $[Rn]7s^25f^{14}6d^7$

Ta : Group 5

Re : Group 7

Ir : Group 9 $1s^22s^22p^63s^23p^64s^23d^{10}4p^65s^24d^{10}5p^66s^24f^{14}5d^7$: $[Xe]6s^24f^{14}5d^7$

Au : Group 11

Tl : Group 13   Answer(c)

4-70. A single atom of element 114 was synthesized. Use the *aufbau* priciple to predict the electron configuration of this element. Which of the following elements would 114 most resemble?

   (a) Au   (b) Hg   (c) Tl   (d) Pb   (e) Po

Correct order of increasing energy of the atomic orbitals :

1s<2s<2p<3s<3p<4s<3d<4p<5s<4d<5p<6s<4f<5d<6p<7s<5f<6d<7p

The electron configuration for element 114 :

$1s^22s^22p^63s^23p^64s^23d^{10}4p^65s^24d^{10}5p^66s^24f^{14}5d^{10}6p^67s^25f^{14}6d^{10}7p^2$: Group 14

$: [Rn]7s^25f^{14}6d^{10}7p^2$

Au : Group 11

Hg : Group 12

Tl : Group 13

Pb : Group 14 $1s^22s^22p^63s^23p^64s^23d^{10}4p^65s^24d^{10}5p^66s^24f^{14}5d^{10}6p^2$: $[Xe]6s^24f^{14}5d^{10}6p^2$

Po : Group 16   Answer(d)

4-71. Theoreticians predict that the element with the atomic number 120 will be more stable than the elements recently discovered with atomic numbers between 103 and 109. On the basis of the *aufbau* principle and the order of filling of atomic orbitals, the chemistry of this element should most closely resemble the chemistry of which of the following?

   (a) Ra, group 2   (b) Pb, Group 13   (c) Po, group 16   (d) Rn, group 18

   (e) One of the transition metals between La and Hf

Correct order of increasing energy of the atomic orbitals :

1s<2s<2p<3s<3p<4s<3d<4p<5s<4d<5p<6s<4f<5d<6p<7s<5f<6d<7p<8s

The electron configuration for element 120:

$1s^22s^22p^63s^23p^64s^23d^{10}4p^65s^24d^{10}5p^66s^24f^{14}5d^{10}6p^67s^25f^{14}6d^{10}7p^68s^2$: Group 2 $[118]8s^2$

Ra : Group 2 $[Rn]7s^2$ Answer(a)

4-72. Which neutral atom has the most unpaired electrons?

   (a) Na   (b) Al   (c) Si   (d) P   (e) S

Na : $[Ne]3s^1$ One unpaired electron.

Al : $[Ne]3s^23p^1$ One unpaired electron.

Si : $[Ne]3s^23p^2$ Two unpaired electrons.

P : $[Ne]3s^23p^3$ Three unpaired electrons.

S : $[Ne]3s^23p^4$ Two unpaired electrons.   Answer(d)

4-73. Which atom contains the largest number of unpaired electrons?

   (a) B   (b) N   (c) F   (d) Ti   (e) Cu

B : $[He]2s^22p^1$ One unpaired electron.

N : $[He]2s^22p^3$ Three unpaired electrons.

F : $[He]2s^22p^5$ One unpaired electron.

Ti : $[Ar]4s^23d^2$ Two unpaired electrons.

Cu : $[Ar]4s^13d^{10}$ One unpaired electron.   Answer(b)

4-74. Which ion or atom has the largest number of unpaired electrons?

   (a) Li^+ (b) B   (c) C^{4-} (d) N   (e) O^{2-}

Li^+ : $1s^2$ No unpaired electron.

B : $[He]2s^22p^1$ One unpaired electron.

C^{4-} : $[He]2s^22p^6$ No unpaired electron.

N : $[He]2s^22p^3$ Three unpaired electrons.

O^{2-} : $[He]2s^22p^6$ No unpaired electron.   Answer(d)

4-75. Which element has the largest number of electrons for which the angular quantum

number is equal to 1?

   (a) He   (b) F   (c) S   (d) As   (e) Zn

l =1 refers to p orbital.

He : $1s^2$ No electron in p orbital

F : $1s^22s^22p^5$ 5 electrons in p orbitals

S : $1s^22s^22p^63s^23p^4$ 10 electrons in p orbitals

As : $1s^22s^22p^63s^23p^64s^23d^{10}4p^3$ 15 electrons in p orbitals

Zn : $1s^22s^22p^63s^23p^64s^23d^{10}$ 12 electrons in porbitals   Answer(d)

4-76. Possible set of quantum numbers for the last electron added to form a gallium atom

(Z = 31) in its ground state is:

   (a) 3,1,0,−1/2   (b) 3,2,1,1/2   (c) 4,0,0,1/2   (d) 4,1,1,1/2   (e) 4,2,2,1/2

The electron configuration of neutral Ga : $[Ar]4s^23d^{10}4p^1$

Last electron is added to the 4p orbital. The n should be 4 and the l should be 1.   Answer(d)

4-77. Possible set of quantum numbers for the last electron added to form an As^{3+} ion is:

   (a) 3,1,−1,1/2   (b) 4,0,0,−1/2   (c) 3,2,0,1/2   (d) 4,1,−1,1/2   (e) 5,0,0,1/2

The electron configuration of neutral As: $[Ar]4s^23d^{10}4p^3$

The electron configuration of As^{3+} : $[Ar]4s^23d^{10}$

Last electron is added to the 3d orbital. The n should be 3 and the l should be 2.   Answer(c)

4-78. The highest energy electron in element 105 could be characterized by which of the

following sets of n,l,m, and s quantum numbers?

   (a) 7,3,−3,−1/2   (b) 6,3,−1,−1/2   (c) 6,2,0,1/2   (d) 5,3,−1,−1/2

The electron configuration of element 105 : $[Rn]7s^25f^{14}6d^3$

The highest energy electron is in 6d orbital. The n should be 6 and l should be 2.   Answer(c)

4-79. Which of the following describes a possible set of quantum numbers for the last electron

added to form an Al when atomic orbitals are filled?

   (a) 1,0,0,1/2   (b) 2,0,0,1/2   (c) 2,1,1,1/2   (d) 3,0,0,1/2   (e) 3,1,1,1/2

The electron configuration of Al : $[He]2s^22p^63s^23p^1$

Last electron is in a 3p orbital. The n should be 3 and l should be 1.   Answer(e)

4-80. Which of the following sets of n,l,m, and s quantum numbers could describe the electron

removed from a neutral Al atom when the first ionization energy of Al is measured?

   (a) 1,0,0,1/2   (b) 2,0,0,1/2   (c) 2,1,1,1/2   (d) 3,0,0,1/2   (e) 3,1,1,1/2

Electron configuration of Al : $[He]2s^22p^63s^23p^1$

An electron from the 3p orbital will be removed first. The n should be 3 and l should be 1.

    Answer(e)

4-81. An element with electron configuration $1s^22s^22p^63s^23p^64s^23d^{10}4p^65s^24d^{10}5p^3$ would

belong in which group of the periodic table?

   (a) Group 1   (b) Group 13   (c) Group 14   (d) Group 15   (e) Group 17

$1s^22s^22p^63s^23p^64s^23d^{10}4p^65s^24d^{10}5p^3$ is the electron configuration of Sb (Z = 51). Group = 15.

    Answer(d)

4-82. Which electron configuration for carbon would satisfy Hund's rules?

   (a) $1s^22s^22p_x^{0}2p_y^{0}2p_z^{0}$ (b) $1s^22s^22p_x^{1}2p_y^{1}2p_z^{0}$ (c) $1s^22s^22p_x^{1}2p_y^{1}2p_z^{1}$

(d) $1s^22s^22p_x^2 2p_y^1 2p_z^1$

The electron configuration for neutral C : $1s^22s^22p^2$

According to Hund's rule there should be one electron in each degenerate orbital before pairing starts. Carbon has 3 p orbitals which are degenerate. Carbon has two electrons in p orbitals. According to Hund's rule there should be only one empty p orbital and each of the other two p orbitals should have one electron.   Answer(b)

4-83. Which is the ground state electron configuration of F atom?

   (a) $1s^22s^22p^5$ (b) $1s^22s^22p^6$ (c) $1s^22s^22p^7$ (d) $1s^22s^22p^63s^1$ (e) None of the above

Correct order of increasing energy of the atomic orbitals :

1s<2s<2p<3s<3p<4s<3d

The atomic number for F is 9. So the electron confuguration for F is : $1s^22s^22p^5$ Answer(a)

4-84. What is the value of x in the following electron configuration for silicon?

$1s^22s^22p^63s^23p^x$

   (a) 1   (b) 2   (c) 3   (d) 4   (e) 6

Correct order of increasing energy of the atomic orbitals :

1s<2s<2p<3s<3p<4s<3d

The atomic number for silicon is 14. So the electron configuration for silicon is :

$1s^22s^22p^63s^23p^2$ x = 2.   Answer(b)

4-85. What is the correct electron configuration for the Ti atom?

   (a) $1s^22s^22p^63s^23p^63d^4$ (b) $[Ar]4s^24d^2$ (c) $[Ar]4d^4$ (d) $1s^22s^22p^63s^23d^2$

   (e) $[Ar]4s^23d^2$

Correct order of increasing energy of the atomic orbitals :

1s<2s<2p<3s<3p<4s<3d

The atomic number of Ti is 22. So the electron configuration of Ti is : $1s^22s^22p^63s^23p^64s^23d^2$

The electron configuration of Ar : $1s^22s^22p^63s^23p^6$

So the electron configuration of Ti is : $[Ar]4s^23d^2$ Answer(e)

4-86. Which of the following would have the electron configuration: $1s^22s^22p^63s^23p^63d^4$

   (a) Ca^{2+} (b) Cr^{2+} (c) Fe^{2+} (d) Ti^{2+} (e) None of the above

The electron configuration of Ca : $1s^2 2s^2 2p^6 3s^2 3p^6 4s^2$

The electron configuration of Ca^{2+} : $1s^2 2s^2 2p^6 3s^2 3p^6$

The electron configuration of Cr : $1s^2 2s^2 2p^6 3s^2 3p^6 4s^1 3d^5$

4s electron will be removed first. So the electron configuration of Cr^{2+} : $1s^2 2s^2 2p^6 3s^2 3p^6 3d^4$

The electron configuration of Fe : $1s^2 2s^2 2p^6 3s^2 3p^6 4s^2 3d^6$

4s electrons will be removed first. So the electron configuration of Fe^{2+} : $1s^2 2s^2 2p^6 3s^2 3p^6 3d^6$

The electron configuration of Ti : $1s^2 2s^2 2p^6 3s^2 3p^6 4s^2 3d^2$

4s electrons will be removed first. So the electron configuration of Ti^{2+} : $1s^2 2s^2 2p^6 3s^2 3p^6 3d^2$

Answer(b)

4-87. What is the electron configuration for the P^{3+} ion?

(a) [Ne] (b) [Ne]$3s^2$ (c) [Ne]$3s^2 3p^3$ (d) [Ne]$3s^2 3p^6$

Correct order of increasing energy of the atomic orbitals :

1s<2s<2p<3s<3p<4s<3d

The atomic number of P is 15. The electron configuration of P : $1s^2 2s^2 2p^6 3s^2 3p^3$

The electron configuration of P^{3+} ion : $1s^2 2s^2 2p^6 3s^2$

The electron configuration of Ne : $1s^2 2s^2 2p^6$

So the electron configuration of P^{3+} ion is : [Ne]$3s^2$ Answer(b)

4-88. What is the electron configuration for the bromide ion, Br^-?

(a) [Ar]$4s^2 4p^5$ (b) [Ar]$4s^2 3d^{10} 4p^7$ (c) [Ar]$4s^2 3d^{10} 4p^5$ (d) [Ar]$4s^2 3d^{10} 4p^6$

(e) [Ar]$4s^2 3d^{10} 3p^6$

Correct order of increasing energy of the atomic orbitals :

1s<2s<2p<3s<3p<4s<3d<4p

The atomic number of Br is 35. The electron configuration of Br : $1s^2 2s^2 2p^6 3s^2 3p^6 4s^2 3d^{10} 4p^5$

The electron configuration of Br^- : $1s^2 2s^2 2p^6 3s^2 3p^6 4s^2 3d^{10} 4p^6$

The electron configuration of Ar : $1s^2 2s^2 2p^6 3s^2 3p^6$

The electron configuration of Br^- : [Ar]$4s^2 3d^{10} 4p^6$ Answer(d)

4-89. Which of the following describes the electron configuration for the Sn^{2+} ion?

(a) [Kr]$4d^{10}$ (b) [Kr]$5s^2$ (c) [Kr]$5s^2 5p^2$ (d) [Kr]$5s^2 4d^{10}$ (e) [Kr]$5s^2 4d^{10} 5p^2$

Correct order of increasing energy of the atomic orbitals :

1s<2s<2p<3s<3p<4s<3d<4p<5s<4d<5p<6s<4f<5d<6p<7s<5f<6d<7p<8s

The atomic number of Sn is 50.

The electron configuratuion of Sn : $1s^2 2s^2 2p^6 3s^2 3p^6 4s^2 3d^{10} 4p^6 5s^2 4d^{10} 5p^2$

The electron configuration of Sn^{2+} : $1s^2 2s^2 2p^6 3s^2 3p^6 4s^2 3d^{10} 4p^6 5s^2 4d^{10}$

The electron configuration of Kr : $1s^2 2s^2 2p^6 3s^2 3p^6 4s^2 3d^{10} 4p^6$

The electron configuration of Sn^{2+} : $[Kr]5s^2 4d^{10}$ Answer(d)

4-90. What is the electron configuration of Al^{x+} ion in the ionic compound Al_2O_3?

   (a) $1s^2 2s^2 2p^6$ (b) $1s^2 2s^2 2p^6 3s^2$ (c) $1s^2 2s^2 2p^6 3s^2 3p^1$ (d) $1s^2 2s^2 2p^6 3s^2 3p^4$

   (e) $1s^2 2s^2 2p^6 3s^2 3p^6$

The oxidation number of Al in Al_2O_3 = +3.   The atomic number of Al is 13.

The electron configuration of Al : $1s^2 2s^2 2p^6 3s^2 3p^1$

The electron configuration of Al^{3+} : $1s^2 2s^2 2p^6$ Answer(a)

4-91. If the X^{2-} has no unpaired electrons, in what group does element X belong?

   (a) Group 1   (b) Group 2   (c) Group 14   (d) Group 16   (e) Group 17

Consider period 3 of the periodic table.

Group 1 element : Na

The electron configuration of Na : $1s^2 2s^2 2p^6 3s^1$

The elctron configuartion of Na^{2-} : $1s^2 2s^2 2p^6 3s^2 3p^1$ One unpaired electron.

Group 2 element : Mg

The electron configuration of Mg : $1s^2 2s^2 2p^6 3s^2$

The electron configuratuion of Mg^{2-} : $1s^2 2s^2 2p^6 3s^2 3p^2$ Two unpaired electrons.

Group 14 element : Si

The electron configuration of Si : $1s^2 2s^2 2p^6 3s^2 3p^2$

The electron configuration of Si^{2-} : $1s^2 2s^2 2p^6 3s^2 3p^4$ Two unpaired electrons.

Group 16 element : S

The electron configuration of S : $1s^2 2s^2 2p^6 3s^2 3p^4$

The electron configuration of S^{2-} : $1s^2 2s^2 2p^6 3s^2 3p^6$ No unpaired electron.

Group 17 element : Cl

The electron configuration of Cl : $1s^2 2s^2 2p^6 3s^2 3p^5$

The electron configuration of Cl^{2-} : $1s^2 2s^2 2p^6 3s^2 3p^6 4s^1$ One unpaired electron.   Answer(d)

4-92. The results of photoelectron spectroscopy studies of the last four elements in the 2nd row of the periodic table are given below.

Element	1s	2s	2p
N	39.6	2.45	1.40 (MJ/mol)
O	52.6	3.12	1.31 (MJ/mol)
F	67.2	3.88	1.68 (MJ/mol)
Ne	84.0	4.68	2.08 (MJ/mol)

Which of the following statements is correct?

(a) The average valence electron energy of N is larger than that for O,F or Ne
(b) The average valence electron energy of O is larger than that for N,F or Ne
(c) The average valence electron energy of F is larger than that for O,N or Ne
(d) The average valence electron energy of Ne is larger than that for O,F or N

The electron configuration of N : $1s^2 2s^2 2p^3$

The electron configuration of O : $1s^2 2s^2 2p^4$

The electron configuration of F : $1s^2 2s^2 2p^5$

The electron configuration of Ne : $1s^2 2s^2 2p^6$

Average valence electron energy for O,N,F and Ne (AVEE) = $(aI_{2s} + bI_{2p})/(a + b)$

a = Number of 2s electrons I_{2s} = Ionization energy of electron from the 2s subshell

b = Number of 2p electrons I_{2p} = Ionization energy of electron from the 2p subshell

Element	AVEE (MJ/mol)
N	1.82
O	1.91
F	2.31
Ne	2.73

Answer(d)

4-93. If you believe that the average valence electron energy measures the ability of an atom to hold onto its valence electrons and therefore the ability of the atom to compete for electrons in any bond it might form, which of these atoms should be the most electronegative?

(a) N   (b) O   (c) F   (d) Ne   (e) These elements all have the same electronegativity

The average valence electron energy for Ne is larger than that for N,O,or F. Thus Ne should

be the most electronegative.   Answer(d)

4-94. The first ionization energy of the six elements in the periodic table are given below:

H   1312.0 kJ/mol   He   2372.3 kJ/mol   Li   520.2 kJ/mol

Be   899.4 kJ/mol   B   800.6 kJ/mol   C   1086.4 kJ/mol

Why is the first IE (ionization energy) of B smaller than that of Be?

   (a) Because the first IE decreases as we go across a row of the periodic table from left to right
   (b) Because the nucleus of a B (Z = 5) atom contains fewer protons than the nucleus of a Be (Z = 4) atom
   (c) Because the outer most electron on B is coming from a 2p, not a 2s orbital.
   (d) Because the atomic number of B is odd, whereas the atomic number of Be is even
   (e) For the same reason that the first IE of He is larger than that of H

The electron configuration of Be : $1s^22s^2$

The electron configuration of B : $1s^22s^22p^1$

B has one more proton than Be. So we expect that the 1st IE of B to be larger than that

of Be. But actually the 1st IE of B (800.6) < 1st IE of Be (899.4). This is because the extra

proton effect is overpowered by the fact that the outer most electron in B is a 2p

electron which is better shielded from the nucleus by the inner core of electrons than the

2s electron of Be.   Answer(c)

4-95. If a powerful enough source of energy is used in the PES experiment, it is possible to

measure the energy required to remove 2nd,3rd,4th,and so on electrons from an atom. Which of

the following would have the largest fourth ionization energy (IE)?

(a) Al   (b) Si   (c) P   (d) S   (e) Cl

The electron configuration of Al : $1s^22s^22p^63s^23p^1$

The electron configuration of Al^{3+} : $1s^22s^22p^6$

The electron configuration of Si : $1s^22s^22p^63s^23p^2$

The electron configuration of Si^{3+} : $1s^22s^22p^63s^1$

The electron configuration of P : $1s^22s^22p^63s^23p^3$

The electron configuration of P^{3+} : $1s^22s^22p^63s^2$

The electron configuration of S : $1s^22s^22p^63s^23p^4$

The electron configuration of S^{3+} : $1s^22s^22p^63s^23p^1$

The electron configuration of Cl : $1s^22s^22p^63s^23p^5$

The electron configuration of Cl^{3+} : $1s^22s^22p^63s^23p^2$

Compared to Al, other atoms have more protons. But this effect will be overpowered by the fact that in the case of Al the fourth electron has to be removed from a very stable completely filled 2p subshell (Al^{3+} has the electron configuration of Ne). So Al will have the largest fourth IE.

	4th IE (kJ/mol)
Al	11577
Si	4355.5
P	4963.6
S	4556
Cl	5158.6

Answer(a)

CHAPTER 5

THE COVALENT BOND

INVALUABLE INFORMATION

Number of Electron Groups[*]	Electron-group Geometry[**]	Hybridization	Number of Lone Pairs	VSEPR Notation	Molecular Geometry[***]
2	Linear	sp	0	AX_2	Linear
3	Trigonal Planar	sp^2	0	AX_3	Trigonal planar
3	Trigonal Planar	sp^2	1	AX_2E	Bent
4	Tetrahedral	sp^3	0	AX_4	Tetrahedral
4	Tetrahedral	sp^3	1	AX_3E	Trigonal pyramidal
4	Tetrahedral	sp^3	2	AX_2E_2	Bent
5	Trigonal bipyramidal	sp^3d	0	AX_5	Trigonal bipyramidal
5	Trigonal bipyramidal	sp^3d	1	AX_4E	Seesaw
5	Trigonal bipyramidal	sp^3d	2	AX_3E_2	T – shaped
5	Trigonal bipyramidal	sp^3d	3	AX_2E_3	Linear
6	Octahedral	sp^3d^2	0	AX_6	Octahedral
6	Octahedral	sp^3d^2	1	AX_5E	Square pyramidal
6	Octahedral	sp^3d^2	2	AX_4E_2	Square planar

[*]The number of places in the valence shell of an atom where electrons can be found, not the number of pairs of valence electrons.

[**]Distribution of electrons

[***]Repulsions between electrons is minimum in the selected geometry.

VALENCE ELECTRONS (VE)*

Element	VE	Element	VE	Element	VE	Element	VE	Element	VE
H	1								
He	2								
Li	1	Na	1	K	1	Rb	1	Cs	1
Be	2	Mg	2	Ca	2	Sr	2	Ba	2
B	3	Al	3	Sc	3	Y	3	Lu	3
C	4	Si	4	Ti	4	Zr	4	Hf	4
N	5	P	5	V	5	Nb	5	Ta	5
O	6	S	6	Cr	6	Mo	6	W	6
F	7	Cl	7	Mn	7	Tc	7	Re	7
Ne	8	Ar	8	Fe	8	Ru	8	Os	8
				Co	9	Rh	9	Ir	9
				Ni	10	Pd	0	Pt	10
				Cu	1	Ag	1	Au	1
				Zn	2	Cd	2	Hg	2
				Ga	3	In	3	Tl	3
				Ge	4	Sn	4	Pb	4
				As	5	Sb	5	Bi	5
				Se	6	Te	6	Po	6
				Br	7	I	7	At	7
				Kr	8	Xe	8	Rn	8

*The electrons on an atom that are not present in the previous rare gas, ignoring filled d or f subshells.

ELECTRONEGATIVITY (EN)

Element	EN	Element	EN	Element	EN	Element	EN	Element	EN
H	2.2								
Li	0.98	Na	0.93	K	0.82	Rb	0.82	Cs	0.79
Be	1.57	Mg	1.31	Ca	1.00	Sr	0.95	Ba	0.89
B	2.04	Al	1.61	Ga	1.81	In	1.78	Tl	1.62
C	2.55	Si	1.90	Ge	2.01	Sn	1.96	Pb	1.87
N	3.04	P	2.19	As	2.18	Sb	2.05	Bi	2.02
O	3.44	S	2.58	Se	2.55	Te	2.1	Po	2.0
F	3.98	Cl	3.16	Br	2.96	I	2.66	At	2.20

5-1. Which of the following have 5 valence electrons?

(a) Ca   (b) Ga   (c) Ge   (d) As   (e) Se

Valence electrons are the electrons on an atom that are not present in the previous rare gas, ignoring filled d and f subshells.

	Atomic electron configuration	Valence electrons	
Ca :	$[Ar]\ 4s^2$	2	
Ga:	$[Ar]\ 3d^{10}\ 4s^2\ 4p^1$	3	
Ge :	$[Ar]\ 3d^{10}\ 4s^2\ 4p^2$	4	
As:	$[Ar]\ 3d^{10}\ 4s^2\ 4p^3$	5	
Se:	$[Ar]\ 3d^{10}\ 4s^2\ 4p^4$	6	Answer(d)

5-2. How many valence electrons are in the IF_2^- ion?

(a) 7   (b) 14   (c) 20   (d) 21   (e) 22

I : [Kr] $4d^{10}5s^25p^5$ 7 valence electrons

F : [He] $2s^22p^5$ 7 valence electrons

Valence electrons (VE) in IF_2^- $= 7 + (2 \times 7) + 1 = 22$ Answer(e)

5-3. Which of the following elements is the most electronegative?

(a) S   (b) As   (c) P   (d) Se   (e) Cl

	Electronegativity
S	2.58
As	2.18
P	2.19
Se	2.55
Cl	3.16

In general electronegativity increases from bottom to top in a group and from left to right in a

period.   Answer(e)

5-4. Which of the following lists the elements in the correct order of increasing

electronegativity?

(a) Si<S<O   (b) S<Si<O   (c) S<O<Si   (d) Si<O<S   (e) O<Si<S

In general electronegativity increases from bottom to top in a group and from left to right

in a period.

	Electronegativity
S	2.58
Si	1.90
O	3.44

Answer(a)

5-5. Which series of elements is arranged in order of decreasing electronegativity?

(a) C>Si>P>As>Se   (b) O>P>Al>Mg>K   (c) Na>Li>B>N>F

(d) K>Mg>Be>O>N   (e) Li>Be>B>C>N

In general electronegativity increases from bottom to top in a group and from left to right in a

period.

	Electronegativity		Electronegativity		Electronegativity		Electronegativity
B :	2.04	C :	2.55	N :	3.04	O :	3.44
Li :	0.98	Be :	1.57	P :	2.19	F :	3.98
Na :	0.93	Mg :	1.31	Al :	1.61	Si :	1.90
K :	0.82	As :	2.18	Se :	2.55		Answer(b)

5-6. Given the following electronegativities: K = 0.8, H = 2.2, and I = 2.5, which of the following statements is true?

    (a) H_2 , I_2, and HI are non-polar covalent compounds
    (b) KI is non-polar covalent and HI is ionic
    (c) HI is non-polar covalent and I_2 is polar covalent
    (d) H_2 is non-polar covalent, while HI and KI are ionic
    (e) None of the above is true

	ΔEN
H_2	0
I_2	0
HI	0.3
KI	1.7

In general, for covalent compounds $\Delta EN < 1.2$,

           for ionic compounds $\Delta EN > 1.8$,

           and for polar compounds $\Delta EN = 1.2$ to 1.8 Answer(a)

5-7. In which of the following the bond most ionic?

    (a) BeF_2 (b) $CaCl_2$ (c) $SrBr_2$ (d) BaF_2 (e) CaI_2

	Electronegativity		Electronegativity		Electronegativity		Electronegativity
F :	3.98	Cl :	3.16	Br :	2.96	I :	2.66
Be :	1.57	Ca :	1.00	Sr :	0.95	Ba:	0.89

	ΔEN
BeF_2 :	2.41
$CaCl_2$:	2.16
$SrBr_2$:	2.01
BaF_2:	3.09

CaI$_2$: 1.66

In general, for covalent compounds ΔEN< 1.2,

for ionic compounds ΔEN> 1.8,

and for polar compounds ΔEN = 1.2 to 1.8

Higher the value of ΔEN, the more ionic the compound will be.   Answer(d)

5-8. In which of the following compounds is the bond least polar?

(a) NaCl   (b) CH$_4$ (c) CO   (d) Cl$_2$ (e) HCl

Only for Cl$_2$ ΔEN = 0. So Cl – Cl bond will be the least polar.   Answer(d)

5-9. Which of the following would form compounds with fluorine in which the bond is the least ionic?

(a) P   (b) Ca   (c) Al   (d) O   (e) Se

	Electronegativity		ΔEN
F :	3.98		
P :	2.19	PF$_5$:	1.79
Ca :	1.00	CaF$_2$:	2.98
Al :	1.61	AlF$_3$:	2.37
O :	3.44	OF$_2$:	0.54
Se :	2.55	SeF$_2$:	1.43

In general, for covalent compounds ΔEN< 1.2,

for ionic compounds ΔEN> 1.8,

and for polar compounds ΔEN = 1.2 to 1.8

Higher the value of ΔEN, the more ionic the compound will be. So The compound between O and F will be the least ionic.   Answer(d)

5-10. Which of the following compounds is most likely to be on the borderline between ionic and covalent compounds?

(a) MgO   (b) AlF$_3$ (c) AlBr$_3$

	Electronegativity		ΔEN
Mg :	1.31	MgO :	2.13

O :	3.44		
Al :	1.61		
F :	3.98	AlF_3 :	2.37
Br :	2.96	$AlBr_3$:	1.35

In general, for covalent compounds $\Delta EN < 1.2$,

for ionic compounds $\Delta EN > 1.8$,

and for polar compounds $\Delta EN = 1.2$ to 1.8 Answer(c)

5-11. Which of the following atoms, ions, or molecules could combine to form a stable covalent

bond?

(a) CH_4 and H^- (b) BF_3 and F^- (c) H and F^- (d) Ar and BF_3 (e) Ar and CH_4

C and H can not expand their valence shells. Ar is not a Lewis base. BF_3 is a Lewis acid and F^-

can act as a Lewis base. So BF_3 and F^- will combine to form a stable covalent bond. Answer(b)

5-12. The correct Lewis structure for CH_2Cl_2 contains:

(a) 2 single bonds, 2 double bonds and no non-bonding electrons
(b) 4 single bonds, and no non-bonding electrons
(c) 4 single bonds and four non-bonding electrons
(d) 2 single bonds 2 double bonds and eight non-bonding electrons
(e) 4 single bonds and 12 non-bonding electrons

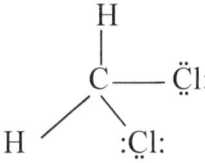

VE (Valence electrons) = $(2\times1)+4+(2\times7) = 20$

Valence electrons = Bonding electrons + Non-bonding electrons

BE (Bonding electrons) = 8   NBE (Non-bonding electrons) = 12   Answer(e)

5-13. The correct Lewis structure for NO_2^+ ion has:

(a) 2 double bonds and 8 non-bonding electrons
(b) 1 double bond, 1 single bond, and 12 non-bonding electrons
(c) 2 single bonds and 12 non-bonding electrons
(d) 2 triple bonds and 4 non-bonding electrons
(e) 1 single bond, 1 double bond and 8 non-bonding electrons

NO_2^+ : $[:\ddot{O} :: N :: \ddot{O}:]^+$ $[:\ddot{O} = N = \ddot{O}:]^+$ 2 double bonds

VE (Valence electrons) = $5 + (2\times6) - 1 = 16$

BE (Bonding electrons) = 8

NBE (Non-bonding electrons) = 8   Answer(a)

5-14. Which of the following are exceptions to the Lewis octet rule?

   (i) CO_2 (ii) BeF_2 (iii) SF_4 (iv) SO_3 (v) PCl_3

   (a) i and ii   (b) ii and iii   (c) iii and iv   (d) i,ii and iii

   (e) None are exceptions to the Lewis octet rule

CO_2 :   :Ö:: C :: Ö: :Ö = C = Ö:

VE = 16 BE = 8 NBE = 8   NBE on C = 0 Octet rule is satisfied.

BeF_2 :   F : Be : F   :F̈ – Be – F̈:   VE = 16   BE = 4   NBE = 12   NBE on Be = 0

Exception to the Lewis octet rule. Only 4 electrons are around Be atom.

SF_4 :   :F̈:

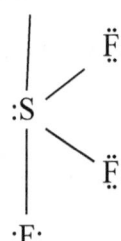

VE = 6 + (4×7) = 34   BE = 8   NBE = 26   NBE on S = 2 (1 lone pair)

1 lone pair of electrons on S atom. 10 electrons are around S atom. Exception to the Lewis octet

rule.

SO_3 :   :Ö:
                 |
        :O̤ = S – O̤:

 VE = 6 + (3×6) = 24   BE = 8   NBE = 16   NBE on S = 0   Octet rule is satisfied.

PCl_3 :

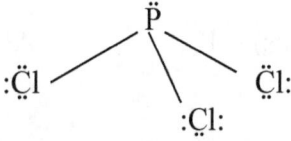

VE = 5+ (3×7) = 26   BE = 6   NBE = 20   NBE on P = 2 (1 lone pair)   Octet rule is

satisfied.   Answer(b)

5-15. Which of the following are exceptions to the Lewis octet rule?

   (i) BF_3 (ii) H_2CO (iii) XeF_4 (iv) H_3O^+ (v) ClF_3

(a) i and ii   (b) i and iii   (c) i,iii, and v   (d) ii, and iii   (e) iii, and v

BF$_3$:

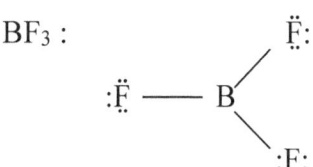

VE = 3+ (3×7) = 24   BE = 6   NBE = 18   NBE on B = 0

Only 6 electrons are around B atom. Exception to the Lewis octet rule.

H$_2$CO :

$$\text{H} \diagdown$$
$$\text{C} = \ddot{\text{O}}:$$
$$\text{H} \diagup$$

VE = (2×1) +4+6 = 12 BE = 8 NBE = 4   NBE on C = 0   Octet rule is satisfied.

XeF$_4$:

:F̈:   F̈:

        Ẍe

:F̈:   F̈:

VE = 8 + (4×7) = 36   BE = 8 NBE = 28   NBE on Xe = 4 (2 lone pairs)

12 electrons are around Xe atom. Exception to the Lewis octet rule.

H$_3$O$^+$

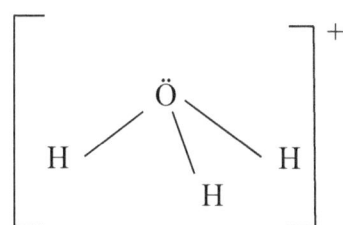

VE = (3×1) + 6 −1 = 8   BE = 6   NBE = 2   NBE on O = 2 (1 lone pair)   Octet rule is

satisfied.

ClF$_3$:

:F̈:

::Cl —— F̈:

:F̈:

$VE = 7 + (3 \times 7) = 28$ $BE = 6$ $NBE = 22$ NBE on Cl = 4 (2 lone pairs)

10 electrons are around Cl atom. Exception to the Lewis octet rule.   Answer(c)

5-16. Which of the following compounds contains only single bonds?

(a) CN^- (b) NO^+ (c) CO (d) O_2^{2-} (e) Cl_2CO

CN^- : $[:C \equiv N:]^-$ $VE = 4+5+1 = 10$ $BE = 6$ $NBE = 4$ Contains a triple bond.

NO^+ : $[:N=\ddot{O}:]^+$ $VE = 5+6-1 = 10$ $BE = 4$ $NBE = 6$ Contains a double bond.

CO : $[:C \equiv O:]$ $VE = 4+6 = 10$ $BE = 6$ $NBE = 4$ Contains a triple bond.

O_2^{2-} : $[:\ddot{O}-\ddot{O}:]^{2-}$ $VE = 6+6+2 = 14$ $BE = 2$ $NBE = 12$ Contains only a single bond.

Cl_2CO: :C̈l:

               \
                C = Ö: $VE = (2\times7)+4+6 = 24$ $BE = 8$ $NBE = 16$ NBE on C = 0 Contains a

         /\
   :C̈l: double bond.

Answer(d)

5-17. Which of the following would be the best Lewis structure for NOCl?

(a) $:\ddot{O} - N - \ddot{C}l:$ (b) $:\ddot{O} - \ddot{N} - \ddot{C}l:$ (c) $:\ddot{O} = \ddot{N} - \ddot{C}l:$ (d) $:\ddot{O} - N = \ddot{C}l:$

(e) $:\ddot{O} = \ddot{N} = \ddot{C}l:$

In a,b,d and e Lewis octet rule is violated. So c would be the best Lewis structure.   Answer(c)

5-18. Which of the following would have a Lewis structure most like that of CO_3^{2-} ion?

(a) CH_3^+ (b) NH_3 (c) NO_3^- (d) SO_3^{2-} (e) SO_4^{2-}

CO_3^{2-} :

$$\left[:\ddot{O} - \underset{\underset{:\ddot{O}:}{|}}{C} = \ddot{O}: \right]^{2-}$$

CH_3^+ :

$$\left[\underset{\underset{H}{|}}{H - C - H} \right]^{+}$$

NH_3 : $\underset{\underset{H}{|}}{H - \ddot{N} - H}$

NO_3^- :

$$\left[:\ddot{O} - \underset{\underset{:\ddot{O}:}{|}}{N} = \ddot{O}: \right]^{-}$$

SO_3^{2-} :

$$\left[:\ddot{O} - \underset{\underset{:\ddot{O}:}{|}}{\ddot{S}} = \ddot{O}: \right]^{2-}$$

$SO_4{}^{2-}$:

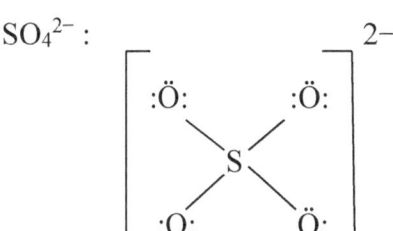

Answer(c)

5-19. Which of the following compounds would have the same electron configuration as the N_2 molecule?

(i)CaO   (ii) CO   (iii) NO   (iv) CN⁻   (v) NO⁺

(a) i   (b) ii and iv   (c) ii and iii   (d) ii,iv, and v   (e) iv and v

N_2:   :N≡N: CaO : Ionic   CO :   :C≡O: NO :   ·N̈=Ö:

CN⁻ :   [:C≡N:]⁻   NO⁺ :   [:N≡O:]⁺   Answer(d)

5-20. Which of the following is best described by two resonance structures?

(a) $NO_2{}^-$ (b) H_2O (c) OF_2 (d) ClNO   (e) BeF_2

$NO_2{}^-$:   [:Ö – N̈ = Ö:]⁻   ⟷   [:Ö = N̈ – Ö:]⁻   Two resonance structures

H_2O :   H – Ö – H   No resonance

OF_2 :   :F̈ – Ö –F̈:   No resonance

ClNO :   :Ö = N̈ – C̈l: No resonance

BeF_2 :   :F̈ – Be – F̈:   No resonance   Answer(a)

5-21. Which of the following isn't a resonance hybrid?

(a) CO_2 (b) $NO_2{}^-$ (c) N_2O (d) SO_3 (e) SCN⁻

CO_2 :   :Ö = C = Ö: No resonance

$NO_2{}^-$:   [:Ö = N̈ – Ö:]⁻   ⟷   [:Ö – N̈ = Ö:]⁻   Resonance hybrid

N_2O : :N̈ – N ≡ O: ⟷ :N ≡ N – Ö: ⟷ :N̈ = N = Ö: Resonance hybrid

SO_3: 7 resonance structures   Resonance hybrid

:Ö = S – Ö   :Ö – S = Ö: :Ö – S – Ö: :Ö = S = Ö: :Ö – S = Ö: :Ö = S – Ö: :Ö = S = Ö:
| ⟷ | ⟷ || ⟷ | ⟷ || ⟷ || ⟷ ||
:Ö: :Ö: :Ö   :Ö: :Ö   :Ö   :Ö

SCN⁻ : [:S̈ = C = N̈:]⁻ ⟷ [:S ≡ C –N̈:]⁻   ⟷   [:S̈ – C ≡ N:]⁻ Resonance hybrid

Answer(a)

5-22. Which of the following are resonance hybrids?

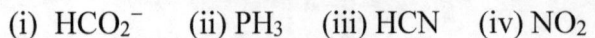
(i) HCO_2^- (ii) PH_3 (iii) HCN   (iv) NO_2

(a) i   (b) ii and iv   (c) i and iii   (d) i and iv

HCO_2^- :

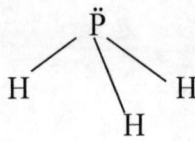

Resonance hybrid

PH_3 :

No resonance

HCN : $H - C \equiv N:$ No resonance

NO_2 :

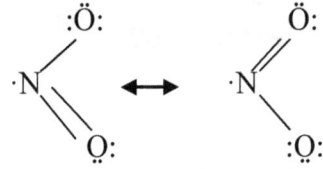

Resonance hybrid

Answer(d)

5-23. Which of the following contains the strongest C−N bond?

(a) H_2CNH (b) HCN   (c) H_3CNH_2 (d) H_3CNO

H_2CNH : $H - C = \ddot{N} - H$
                     |
                     H

HCN : $H - C \equiv \ddot{N}$

H_3CNH_2:   H
                    |
          $H - C - \ddot{N} - H$
                    | |
                    H   H

H_3CNO :   H
                    |
          $H - C - \ddot{N} = \ddot{O}:$
                    |
                    H

Triple bond is the strongest.   Answer(b)

5-24. Which of the following elements would form a compound with the following Lewis structure?

$:\ddot{O} = \ddot{X} - \ddot{C}l:$

(a) Be   (b) B   (c) C   (d) N   (e) O   (f) F

Formal charge = [Number of valence electrons in the neutral atom]− [Number of non-bonding electrons on the atom] − ½ [total number of bonding electrons on the atom]

The molecule has no charge on it. There is no formal charge on O and Cl. So neutral X should have 5 valence electrons.

	Valence electrons
Be	2
B	3
C	4
N	5
O	6
F	7

Answer(d)

5-25. If there are two pairs of non-bonding electrons in the valence shell of the central atom in the XF_4^- ion, in which group of the periodic table does the element X belong?

    (a) Group 14   (b) Group 15   (c) Group 16   (d) Group 17

Group 14 : VE = 4+(4×7)+1 = 33   BE = 8 NBE = 25

Number of non-bonding electrons in the valence shell of the central atom = 1

Group 15 : VE = 5+(4×7)+1 = 34   BE = 8 NBE = 26

Number of non-bonding electrons in the valence shell of the central atom = 2 (1 lone pair)

Group 16 : VE = 6+(4×7)+1 = 35   BE = 8 NBE = 27

Number of non-bonding electrons in the valence shell of the central atom = 3

Group 17 : VE = 7+(4×7)+1 = 36   BE = 8 NBE = 28

Number of non-bonding electrons in the valence shell of the central atom = 4 (2 lone pairs)

Answer(d)

5-26. What is the element X in the following Lewis structure?

$$H - \ddot{X} - \ddot{X} - H$$
$$\begin{array}{cc} | & | \\ H & H \end{array}$$

    (a) B   (b) C   (c) N   (d) O   (e) F

The molecule has 4 non-bonding electrons.

If X is 'B', the non-bonding electrons will be = 0

If X is 'C', the non-bonding electrons will be = 2

If X is 'N', the non-bonding electrons will be = 4

If X is 'O', the non-bonding electrons will be = 6

If X is 'F', the non-bonding electrons will be = 8 $\qquad$ Answer(c)

5-27. Which element will combine with S to form linear molecule with the formula XS_2?

(a) Li $\quad$ (b) Be $\quad$ (c) C $\quad$ (d) O

LiS_2 : $\quad$ S – Li – S $\quad$ VE = 13 $\quad$ BE = 4 $\quad$ NBE = 9 $\quad$ Not enough NBE

BeS_2: $\quad$ S – Be – S $\quad$ VE = 14 $\quad$ BE = 4 $\quad$ NBE = 10 $\quad$ Not enough NBE

CS_2: $\quad$ $:\ddot{S} = C = \ddot{S}:$ $\quad$ VE = 16 $\quad$ BE = 8 $\quad$ NBE = 8 $\quad$ NBE on C = 0

VSEPR Notation = AB_2 $\quad$ Electron-group geometry: linear $\qquad$ Molecular geometry : Linear

S_2O : $\quad$ $:\ddot{S} = \ddot{S} = \ddot{O}:$ $\quad$ VE = 18 $\quad$ BE = 8 $\quad$ NBE = 10 $\quad$ NBE on S = 2 (1 lone pair)

$\quad$ VSEPR Notation = AB_2E $\quad$ Electron-group geometry: Trigonal planar

Molecular geometry : Bent $\qquad$ Answer(c)

5-28. Which element would form an $XF_6{}^{2-}$ ion for which there are no non-bonding electrons in the Lewis structure?

(a) N $\quad$ (b) Al $\quad$ (c) Si $\quad$ (d) P $\quad$ (e) Br

$NF_6{}^{2-}$: $\quad$ VE = 5+(6×7)+2 = 49 $\quad$ BE = 12 $\quad$ NBE =37 $\quad$ NBE on 'N' = 1

$AlF_6{}^{2-}$: $\quad$ VE = 3+(6×7)+2 = 47 $\quad$ BE = 12 $\quad$ NBE =35 $\quad$ Not enough NBE

$SiF_6{}^{2-}$: $\quad$ VE = 4+(6×7)+2 = 48 $\quad$ BE = 12 $\quad$ NBE =36 $\quad$ NBE on 'Si' = 0

$PF_6{}^{2-}$: $\quad$ VE = 5+(6×7)+2 = 49 $\quad$ BE = 12 $\quad$ NBE =37 $\quad$ NBE on 'P' = 1

$BrF_6{}^{2-}$: $\quad$ VE = 7+(6×7)+2 = 51 $\quad$ BE = 12 $\quad$ NBE =39 $\quad$ NBE on 'Br' = 3 $\qquad$ Answer(c)

5-29. If X is a third row element and XO_2 has the lewis structure

$$:\ddot{O} = \ddot{X} - \ddot{O}:$$

X must be : (a) Al $\quad$ (b) Si $\quad$ (c) P $\quad$ (d) S $\quad$ (e) Cl

Al : $\quad$ VE = 15 $\quad$ BE = 6 $\quad$ NBE = 9 $\quad$ Not enough NBE

Si : $\quad$ VE = 16 $\quad$ BE = 6 $\quad$ NBE = 10 $\quad$ NBE on Si = 0

P : $\quad$ VE = 17 $\quad$ BE = 6 $\quad$ NBE = 11 $\quad$ NBE on P = 1

S : $\quad$ VE = 18 $\quad$ BE = 6 $\quad$ NBE = 12 $\quad$ NBE on S = 2 (1 lone pair)

Cl : $\quad$ VE = 19 $\quad$ BE = 6 $\quad$ NBE = 13 $\quad$ NBE on Cl = 3 $\qquad$ Answer(d)

5-30. If the XF_6^- ion has the following Lewis structure,

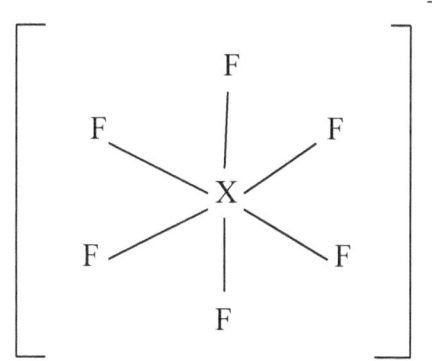

X could be which of the following elements?

(a) N   (b) Al   (c) Si   (d) P   (e) Br

Formal charge on $X = VE(X) - 0 - 6 = -1$

Valence electrons on neutral $X = 5$. So X is either N or P. 12 electrons are around X atom.

N can not expand its valence shell but P can expand its valence shell to hold 12 electrons

because it has empty 3d orbitals. So X is P.   Answer(d)

5-31. Assume that the XF_3 has the following Lewis structure :

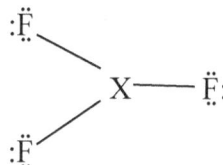

The molecule could be (a) Boron trifluoride   (b) Nitrogen trifluoride

(c) Chlorine trifluoride   (d) Phosphorus trifluoride

(e) more than one of the above

a : VE =24   BE = 6   NBE = 18 NBE on B = 0

b : VE = 26   BE = 6   NBE = 20   NBE on N = 2 (1 lone pair)

c : VE = 28   BE = 6   NBE = 22 NBE on Cl = 4 (2 lone pairs)

d : VE = 26   BE = 6   NBE = 20 NBE on P = 2 (1 lone pair)   Answer(a)

5-32. Which of the following is a correct Lewis strucuture of N_2O.

(a) :N̈ – N – Ö: (b) :N̈ = N – Ö: (c) :N ≡ N – Ö: (d) :N̈ = N̈ – Ö:

(e):N̈ = N̈ = Ö:

a : VE = 16   BE = 4   NBE = 12   NBE on central N = 0   Octet rule violation.

b : VE = 16   BE = 6   NBE = 10   NBE on central N = 0   Octet rule violation.

c : VE = 16   BE = 8   NBE = 8   NBE on central N = 0   Octet rule satisfied.

d : VE =16   BE = 6   NBE = 10   NBE on central N = 2   Octet rule violation.

e : 10 electrons are around central N atom . N atom can not expand its valence shell. Answer(c)

5-33. What are the formal charges on N (labeled 1,2,3) in arginine.

(a) N (1) = 0 , N(2) = +1, and N(3) = 0
(b) N (1) = 0 , N(2) = −1, and N(3) = 0
(c) N (1) = +1 , N(2) = 0, and N(3) = +1
(d) N (1) = −1 , N(2) = 0, and N(3) = −1
(e) N (1) = +1 , N(2) = 0, and N(3) = −1

Formal charge = [Number of valence electrons in the neutral atom]− [Number of non-bonding

electrons on the atom] − ½ [total number of bonding electrons on the atom]

	Formal charge
N^1 :	$5 - 2 - (6/2) = 0$
N^2 :	$5 - 0 - (8/2) = +1$
N^3 :	$5 - 2 - (6/2) = 0$ Answer(a)

5-34. What are the formal charges on N (labeled 1,2,3) in ciprofloxacin?

(a) N (1) = −1 , N(2) = −1, and N(3) = −1
(b) N (1) = −1 , N(2) = 0, and N(3) = 0
(c) N (1) = 0 , N(2) = 0, and N(3) = 0
(d) N (1) = 0 , N(2) = +1, and N(3) = +1
(e) N (1) = +1 , N(2) = +1 and N(3) = −1

Formal charge = [Number of valence electrons in the neutral atom]− [Number of non-bonding electrons on the atom] – ½ [total number of bonding electrons on the atom]

Formal charge

N^1 : $5 – 2 – (6/2) = 0$

N^2 : $5 – 2 – (6/2) = 0$

N^3 : $5 – 2 – (6/2) = 0$ Answer(c)

5-35. How many pairs of bonding electrons are in the valence shell of the iodine atom in the IF_2^- ion?

   (a) 0   (b) 1   (c) 2   (d) 3   (e) 4

IF_2^- : $[:\ddot{\underset{..}{F}} - \overset{..}{\underset{..}{I}} - \ddot{\underset{..}{F}}:]^-$ VE = 7 + (2×7) + 1 = 22   BE = 4   NBE = 18

NBE on central I = 6 (3 lone pairs)

Pairs of bonding electrons in the valence shell of the iodine atom = 2   Answer(c)

5-36. How many pairs of non-bonding electrons are in the valence shell of the iodine atom in the IF_2^- ion?

   (a) 0   (b) 1   (c) 2   (d) 3   (e) 4

IF_2^- : $[:\ddot{\underset{..}{F}} - \overset{..}{\underset{..}{I}} - \ddot{\underset{..}{F}}:]^-$ VE = 7 + (2×7) +1 = 22   BE = 4   NBE = 18   NBE on I = 6 (3 lone pairs)

Pairs of non-bonding electrons in the valence shell of the iodine atom = 3   Answer(d)

5-37. How many pairs of non-bonding electrons are in the valence shell of the iodine atom in the ICl_4^- ion?

   (a) 0   (b) 1   (c) 2   (d) 3   (e) 4

ICl_4^- :

VE = 7+(4×7)+1 = 36   BE = 8   NBE = 28   NBE on I = 4 (2 lone pairs)

Pairs of non-bonding electrons in the valence shell of the iodine atom = 2   Answer(c)

5-38. How many non-bonding electrons are on the central atom in the Lewis structure of XeF$_4$.

(a) 0   (b) 1   (c) 2   (d) 3   (e) 4

XeF$_4$:

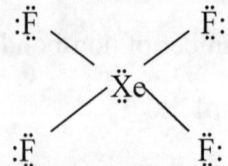

VE = 8+(4×7) = 36   BE = 8 NBE = 28   NBE on Xe = 4 (2 lone pairs)

Non-bonding electrons on the central atom = 4   Answer(e)

5-39. How many pairs of bonding electrons can be found in the valence shell of the nitrogen atom in HNO$_2$?

(a) 0   (b) 1   (c) 2   (d) 3   (e) 4

HNO$_2$:   H – Ö – N̈ = Ö

VE = 1+(2×6)+5 =18   BE = 8   NBE = 10   BE on N = 6

Pairs of bonding electrons in the valence shell of nitrogen = 3   Answer(d)

5-40. How many pairs of non-bonding electrons can be found on the chlorine atom in chlorous acid HOClO?

(a) 0   (b) 1   (c) 2   (d) 3   (e) 4

HOClO :   H – Ö – C̈l = Ö

VE = 1+6+7+6 = 20   BE = 8   NBE = 12   NBE on Cl = 4 (2 lone pairs)

Pairs of non-bonding electrons on chlorine atom = 2   Answer(c)

5-41. Which of the following elements would form an XF$_6^{2-}$ ion that has no non-bonding electrons in the valence shell of the central atom?

(a) Ca   (b) C   (c) Si   (d) S   (e) P

XF_6^{2-} :

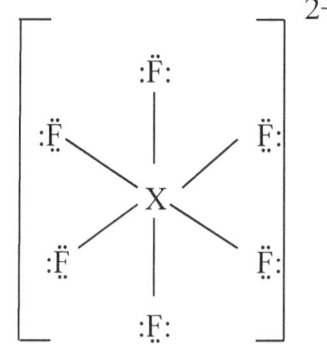

Ca : VE = 2+(6×7)+2 = 46   BE = 12   NBE = 34   Not enough non-bonding electrons

C : VE = 4+(6×7)+2 = 48   BE = 12   NBE = 36   NBE on C = 0

Si : VE = 4+(6×7)+2 = 48   BE = 12   NBE = 36   NBE on Si = 0

S : VE = 6+(6×7)+2 = 50   BE = 12   NBE = 38   NBE on S = 2 (1 lone pair)

P : VE = 5+(6×7)+2 = 49   BE = 12   NBE = 37   NBE on P = 1

X is either C or Si. 12 electrons are around the central atom. C can not expand its valence shell.

Si can expand its valence shell to hold 12 electrons because it has empty 3d orbitals. Answer(c)

5-42. Which of the following species has all its electrons paired?

   (a) Fe^{3+} (b) Zn^{2+} (c) ClO_2 (d) NO   (e) NO_2

Fe : $1s^2 2s^2 2p^6 3s^2 3p^6 4s^2 3d^6$

Fe^{3+} : $1s^2 2s^2 2p^6 3s^2 3p^6 3d^5$ All its electrons are not paired.

Zn : $1s^2 2s^2 2p^6 3s^2 3p^6 4s^2 3d^{10}$

4s electrons will be removed first. So the electron configuration of Zn^{2+} : $1s^2 2s^2 2p^6 3s^2 3p^6 3d^{10}$

All its electrons are paired.

ClO_2 : $:\ddot{O} - \cdot\ddot{Cl} = \ddot{O}:$ VE = 6+7+6 =19   BE = 6   NBE = 13   NBE on Cl = 3

All its electrons are not paired.

NO : $\cdot\ddot{N} = \ddot{O}:$ VE = 5+6 =11   BE = 4   NBE =7   NBE on N =3

  All its electrons are not paired.

NO_2 : $:\ddot{O} - \overset{\cdot}{N} = \ddot{O}:$ VE = 6+5+6 = 17   BE = 6   NBE = 11   NBE on N =1

  All its electrons are not paired.   Answer(b)

5-43. In which compound the valence electrons on the sulfur are arranged towards the corners

of a tetrahedron?

(a) SF_3^+ (b) SF_4 (c) SF_5 (d) SF_6 (e) SF_4 and SF_6

SF_3^+ :   VE = 6+7+7+7−1 = 26   BE =6   NBE = 20   NBE on S = 2 (1 lone pair)

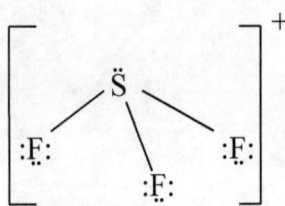

VSEPR notation : AB_3E Number of electron groups on S = 4. So the valence electron groups on the sulfur will be arranged towards the corners of a tetrahedron to reduce repulsions between electrons.   Molecular geometry: Trigonal pyramidal

SF_4: VE = 6+(4×7) = 34   BE = 8   NBE = 26   NBE on S = 2 (1 lone pair)

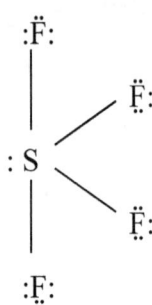

VSEPR notation : AB_4E Number of electron groups on S = 5. So the valence electron groups on sulfur will be arranged towards the corners of a trigonal bipyramid to reduce repulsions between electrons.   Molecular geometry: Seesaw

SF_5 :   VE = 6+(5×7) = 41 BE =10   NBE =31   NBE on S = 1.

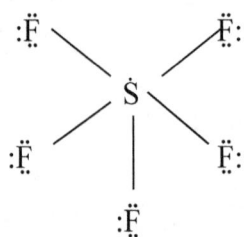

Number of electron groups on S = 6. So the valence electron groups on sulfur will be arranged towards the corners of an octahedron to reduce repulsions between electrons.

Molecular geometry: Square pyramidal

SF_6 :   VE = 6+(6×7) =48   BE = 12   NBE = 36   NBE on S = 0

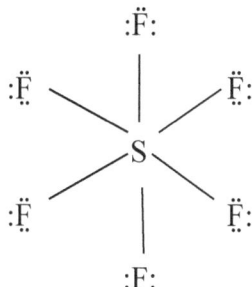

VSEPR notation : AB$_6$ Number of electron groups on S = 6. So the valence electron groups on sulfur will be arranged towards the corners of an octahedron to reduce repulsions between electrons.   Molecular geometry: Octahedral   Answer(a)

5-44. What is the distribution of electrons around the S atom in H$_2$SO$_4$?

   (a) Linear   (b) Trigonal planar   (c) Tetrahedral   (d) Trigonal bipyramidal

   (e) Octahedral

H$_2$SO$_4$:

$$H - \ddot{O} - \overset{\displaystyle :\ddot{O}:}{\underset{\displaystyle :\ddot{O}:}{S}} - \ddot{O} - H$$

VE = (2×1)+(4×6)+6 =32   BE = 12   NBE = 20   NBE on S = 0

Number of places where electrons are found in the valence shell of the S atom = 4. So the electron distribution around S atom is terahedral to reduce repulsions between electrons.

                    Answer(c)

5-45. What is the distribution of electrons around N atom in HNO$_3$?

   (a) Linear   (b) Trigonal planar   (c) Tetrahedral   (d) Trigonal bipyramidal

   (e) Octahedral

HNO$_3$:

$$H - \ddot{O} - N \Big\langle \begin{array}{c} \ddot{O}: \\[4pt] \ddot{O}: \end{array}$$

VE = 1+(3×6)+5 = 24   BE = 8   NBE = 16 NBE on N =0

Number of places where electrons are found in the valence shell of the N atom = 3. So the electron distribution around N atom is trigonal planar to reduce repulsions between electrons.

                    Answer(b)

5-46. There is a single pair of non-bonding electrons on the central atom in the Lewis structure of ICl_5. The molecule has:

(a) A trigonal pyramidal distribution of electrons and an octahedral molecular geometry
(b) An octahedral distribution of electrons and a square pyramidal geometry
(c) A trigonal bipyramidal distribution of electrons and a seesaw molecular geometry
(d) An octahedral distribution of electrons and a trigonal bipyramidal geometry
(e) A trigonal bipyramidal distribution of electrons and a trigonal bipyramidal geometry

ICl_5 :

:Çl   Çl:

      Ï

:Çl | Çl:

    :Çl:

VE = 7+(5×7) = 42   BE = 10   NBE = 32   NBE on I = 2 (1 lone pair)

VSEPR notation : AB_5E Number of electron groups = 6. So the valence electron groups on iodine will be arranged towards the corners of an octahedron to reduce repulsions between electrons.   Molecular geometry: Square pyramidal.   Answer(b)

5-47. The bond angles in most symmetric AB_x molecules are the same. Which of the following is an exception to the rule?

(a) NH_3 (b) SF_6 (c) PCl_5 (d) SO_3 (e)CO_2

NH_3 :

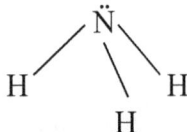

VE = 5+(3×1) = 8   BE = 6 NBE = 2   NBE on N = 2 (1 lone pair)

VSEPR notation : AB_3E Number of electron groups = 4. So the valence electron groups on nitrogen will be arranged towards the corners of a tetrahedron to reduce repulsions between electrons.   Molecular geometry: Trigonal pyramidal   The bond angles will be same.

SF_6 :

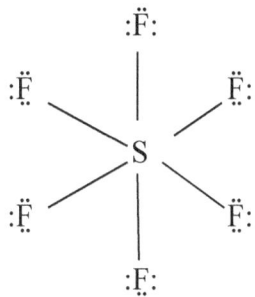

VE = 6+(6×7) =48   BE = 12   NBE = 36   NBE on S = 0

VSEPR notation : AB_6 Number of electron groups = 6. So the valence electron groups on sulfur will be arranged towards the corners of an octahedron to reduce repulsions between electrons.   Molecular geometry: Octahedral   The bond angles will be same.

PCl_5 :

VE = 5+(5×7) = 40   BE = 10   NBE = 30   NBE on P = 0

VSEPR notation : AB_5 Number of electron groups = 5. So the valence electron groups on phophorus will be arranged towards the corners of a trigonal bipyramid to reduce repulsions between electrons. Molecular geometry: Trigonal bipyramidal. The bond angles will not be same. Some Cl – P – Cl angles will be 90° and some Cl – P – Cl angles will be 120° .

SO_3 :   7 resonance structures; NBE on S = 0

Molecular geometry : Trigonal planar. The bond angles will be same.

CO_2 :   :Ö = C = Ö: VE = 4+(2×6) = 16   BE = 8   NBE = 8   NBE on C = 0

Molecular geometry : Linear   The O-C-O bond angle will be 180°.   Answer(c)

5-48. Which term best describes the shape of the I_3^- ion?

   (a) Linear   (b) Bent   (c) Trigonal planar   (d) Trigonal pyramidal

   (e) Trigonal bipyramidal

I_3^- : $[:\ddot{\underset{..}{I}} - \overset{..}{\underset{..}{I}} - \ddot{\underset{..}{I}}:]^-$

VE = (3×7)+1 = 22 BE = 4   NBE = 18 NBE on Central I = 6 (3 lone pairs)

VSEPR notation : AB_2E_3 Number of electron groups = 5. So the valence electron groups on iodine will be arranged towards the corners of a trigonal bipyramid to reduce repulsions between electrons.   Molecular geometry : Linear   Answer(a)

5-49. What is the molecular geometry of the compound formed by the reaction of nitrogen with fluorine?

   (a) Bent   (b) Trigonal planar   (c) Trigonal pyramidal   (d) trigonal bipyramidal

   (e) None of the above

N_2 reacts with F_2 to form NF_3 at high temperatures (N_2 is too inert to react with F_2 at room temperature).

NF_3 :

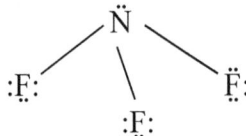

VE = 5+(3×7) = 26   BE = 6   NBE = 20   NBE on N = 2 (1 lone pair)

VSEPR notation : AB_3E Number of electron groups = 4. So the valence electron groups on nitrogen will be arranged towards the corners of a tetrahedron to reduce repulsions between electrons. Molecular geometry : Trigonal pyramidal   Answer(c)

5-50. What is the molecular geometry of the SF_5^+ ion?

   (a) Seesaw   (b) Square planar   (c) Square pyramidal   (d) Trigonal bipyramidal

   (e) Octahedral

SF_5^+ :

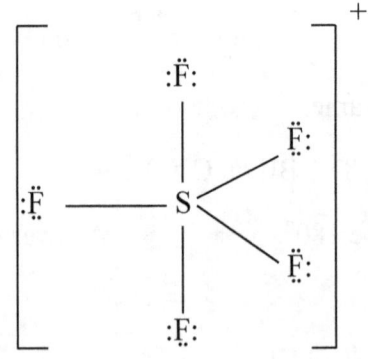

VE = 6+(5×7)−1 = 40   BE = 10 NBE = 30   NBE on S =0

VSEPR notation : AB_5 Number of electron groups = 5. So the valence electron groups on sulfur will be arranged towards the corners of a trigonal bipyramid to reduce repulsions between electrons.   Molecular geometry : Trigonal bipyramidal   Answer(d)

5-51. What is the molecular geometry of Xenon tetrafluoride?

  (a) Tetrahedral   (b) Trigonal bipyramidal   (c) Seesaw   (d) Square planar

  (e) Octahedral

XeF$_4$:

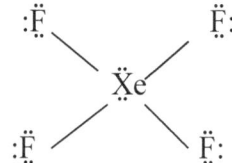

VE = 8+(4×7) = 36   BE = 8   NBE = 28   NBE on Xe = 4 (2 lone pairs)

VSEPR notation : AB$_4$E$_2$ Number of electron groups = 6. So the valence electron groups on xenon will be arranged towards the corners of an octahedron to reduce repulsions between electrons.   Molecular geometry : Square planar   Answer(d)

5-52. The shape of the NO$_2^-$ ion is best described as:

  (a) Linear   (b) T-shaped   (c) Bent   (d) Square planar   (e) Pyramidal

NO$_2^-$:

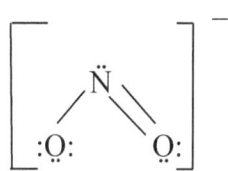

VE = 5+(2×6)+1 = 18   BE = 6 NBE = 12   NBE on N = 2 (1 lone pair)

VSEPR notation : AB$_2$E Number of electron groups (The number of places in the valence shell of Cl atom where electrons can be found, not the number of pairs of valence electrons) = 3.

Electron-group geometry : Trigonal planar. Molecular geometry : Bent   Answer(c)

5-53. What is the best description of the geometry of the SF$_3^+$ ion?

  (a) Bent or angular   (b) Trigonal planar   (c) Trigonal pyramidal   (d) tetrahedral

  (e) Trigonal bipyramidal

SF$_3^+$:

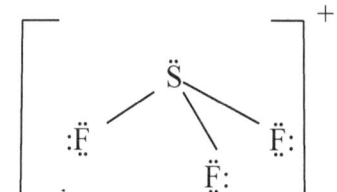

$VE = 6+(3\times7)-1 = 26$ $\qquad$ $BE = 6$ $\quad$ $NBE = 20$ $\quad$ NBE on $S = 2$ (1 lone pair)

VSEPR notation : AB_3E Number of electron groups = 4. So the valence electron groups on sulfur will be arranged towards the corners of a tetrahedron to reduce repulsions between electrons. $\qquad$ Molecular geometry : Trigonal pyramidal $\qquad$ Answer(c)

5-54. The molecular geometry of the IO_3^- ion is described as:

(a) Bent or angular $\quad$ (b) Trigonal planar $\quad$ (c) Trigonal pyramidal

(d) Trigonal bipyramidal $\quad$ (e) T-shaped

IO_3^- :

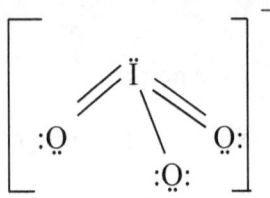

$VE = 7+(3\times6)+1 = 26$ $\quad$ $BE = 10$ $\quad$ $NBE = 16$ $\quad$ NBE on $I = 2$ (1 lone pair)

VSEPR notation : AB_3E Number of electron groups = 4. So the valence electron groups on iodine will be arranged towards the corners of a tetrahedron to reduce repulsions between electrons. $\qquad$ Molecular geometry : Trigonal pyramidal $\qquad$ Answer(c)

5-55. Which of the following is a Lewis acid that has a trigonal planar shape?

(a) $BeCl_2$ $\quad$ (b) PCl_3 $\quad$ (c) $Ca(OH)_2$ $\quad$ (d)BF_3 $\quad$ (e) H_3O^+

For trigonal planar shape the VSEPR notation should be AB_3.

$BeCl_2$: $\quad$:Cl – Be – Cl: $\quad$ $VE = 2+(2\times7) = 16$ $BE = 4$ $NBE = 12$ NBE on $BE = 0$

VSEPR notation : AB_2 Number of electron groups = 2.

Electron-group geometry: Linear $\qquad$ Molecular geometry : Linear

PCl_3 :

$$\text{:Cl} \diagup \overset{\ddot{P}}{\diagdown} \text{Cl:}$$
$$\text{:Cl:}$$

$VE = 5+(3\times7) = 26$ $\qquad$ $BE = 6$ $\quad$ $NBE = 20$ $\qquad$ NBE on $P = 2$ (1 lone pair)

VSEPR notation : AB_3E Number of electron groups = 4. So the valence electron groups on phosphorus will be arranged towards the corners of a tetrahedron to reduce repulsions between electrons.

Molecular geometry : Trigonal pyramidal

$Ca(OH)_2$: The compound is a base.

BF_3:

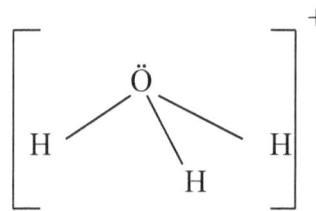

$VE = 3+(3\times7) = 24$ $\qquad$ $BE = 6$ $\qquad$ $NBE = 18$ $\qquad$ NBE on B = 0

VSEPR notation : AB_3 Number of electron groups = 3.

Electron-group geometry : Trigonal planar $\quad$ Molecular geometry : Trigonal planar

BF_3 is a Lewis acid because it accepts a pair of electrons from $:NH_3$ to form

$$
\begin{array}{ccc}
 & F & H \\
 & | & | \\
F - & B - N & -H \\
 & | & | \\
 & F & H
\end{array}
$$

H_3O^+ :

$$
\left[\begin{array}{c}
\ddot{O} \\
H \quad \diagdown \quad H \\
H
\end{array}\right]^+
$$

$VE = 6+(3\times1)-1 = 8$ $\qquad$ $BE = 6$ $\qquad$ $NBE = 2$ $\qquad$ NBE on O = 2 (1 lone pair)

VSEPR notation : AB_3E Number of electron groups = 4. So the valence electron groups on

oxygen will be arranged towards the corners of a tetrahedron to reduce repulsions between

electrons. Molecular geometry : Trigonal pyramidal $\hfill$ Answer(d)

5-56. Which of the following molecules are best described as bent or angular in shape?

$\quad$ (i) H_2S $\quad$ (ii) CO_2 $\quad$ (iii) ClNO $\quad$ (iv) NH_2^- $\quad$ (v) O_3

$\quad$ (a) i and ii $\quad$ (b) i, iii, and iv $\quad$ (c) i,iii,iv, and v $\quad$ (d) i,iii, and iv

$\quad$ (e) All of these molecules

H_2S :
$$
\begin{array}{c}
\ddot{S} \\
\diagup \ddot{} \diagdown \\
H \qquad H
\end{array}
$$

$VE = 6+(2\times1) = 8$ $\qquad$ $BE = 4$ $\quad$ $NBE = 4$ $\qquad$ NBE on S = 4 (2 lone pairs)

VSEPR notation : AB_2E_2 Number of electron groups = 4. So the valence electron groups on

sulfur will be arranged towards the corners of a tetrahedron to reduce repulsions between electrons.   Molecular geometry : Bent

CO_2 : $:\ddot{O}=C=\ddot{O}:$ VE = 4+(2×6) = 16   BE = 8   NBE = 8   NBE on C = 0

Molecular geometry : Linear   The O –C –O bond angle will be 180°.

ClNO :

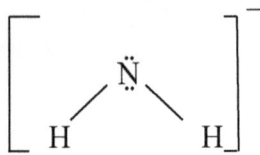

VE = 7+5+6 = 18   BE = 6   NBE = 12   NBE on N = 2 (1 lone pair)

VSEPR notation : AB_2E Number of electron groups = 3.

Electron-group geometry : Trigonal planar

Molecular geometry : Bent

NH_2^- :

VE = 5+2+1 = 8   BE = 4   NBE = 4   NBE on N = 4 (2 lone pairs)

VSEPR notation : AB_2E_2 Number of electron groups = 4. So the valence electron groups on

nitrogen will be arranged towards the corners of a tetrahedron to reduce repulsions between

electrons.

Molecular geometry : Bent

O_3 :

VE = 3×6 = 18   BE = 6   NBE = 12   NBE on Central O = 2 (1 lone pair)

VSEPR notation : AB_2E Number of electron groups = 3.

Electron-group geometry : Trigonal planar   Molecular geometry : Bent   Answer(c)

5-57. Which of the following molecules are planar?

        (i)SO_3 (ii) SO_3^{2-} (iii) NO_3^- (iv) PF_3 (v) BH_3

        (a) i and ii   (b) i and iii   (c) i,iii, and v   (d) ii and iv   (e) iii and v

SO_3 : 7 resonance structures.

$:\ddot{O}=S-\ddot{O}:$ $:\ddot{O}-S=\ddot{O}:$ $:\ddot{O}-S-\ddot{O}:$ $:\ddot{O}=S=\ddot{O}:$ $:\ddot{O}-S=\ddot{O}:$ $:\ddot{O}=S-\ddot{O}:$ $:\ddot{O}=S=\ddot{O}:$
         | | || | || || ||
     $:\ddot{O}:$ $:\ddot{O}:$ $:\ddot{O}$ $:\ddot{O}:$ $:\ddot{O}$ $:\ddot{O}$ $:\ddot{O}$

NBE on S = 0

VSEPR notation : AB$_3$ Number of electron groups = 3.

Electron-group geometry : Trigonal planar Molecular geometry : Trigonal planar

SO_3^{2-} :

$$\left[\ddot{\underset{..}{O}} - \overset{..}{\underset{|}{\underset{\ddot{\underset{..}{O}}}{S}}} = \ddot{O} \right]^{2-}$$

VE = 6+(3×6)+2 = 26   BE = 8   NBE =18   NBE on S = 2 (1 lone pair)

VSEPR notation : AB$_3$E Number of electron groups = 4.

Electron-group geometry : Tetrahedral Molecular geometry : Trigonal pyramidal

NO_3^-

$$\left[\ddot{\underset{..}{O}} - \overset{}{\underset{|}{\underset{\ddot{\underset{..}{O}}}{N}}} = \ddot{O} \right]^{-}$$

VE = 5+(3×6)+1 = 24   BE = 8   NBE =16   NBE on N = 0

VSEPR notation : AB$_3$ Number of electron groups = 3.

Electron-group geometry : Trigonal planar Molecular geometry : Trigonal planar

PF$_3$:

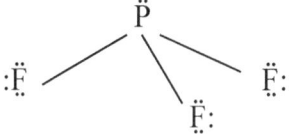

VE = 5+(3×7) =26   BE = 6   NBE = 20   NBE on P = 2 (1 lone pair)

VSEPR notation : AB$_3$E Number of electron groups = 4.

Electron-group geometry : Tetrahedral Molecular geometry : Trigonal pyramidal

BH$_3$:

$$H - B \overset{\diagup H}{\underset{\diagdown H}{}}$$

VE = 3+(3×1) = 6   BE = 6 NBE = 0   NBE on B = 0

VSEPR notation : AB$_3$ Number of electron groups = 3.

Electron-group geometry : Trigonal planar   Molecular geometry : Trigonal planar

i,iii, and v are planar.   Answer(c)

5-58. Which of the following molecules are tetrahedral?

(i) SiF_4 (ii) CH_4 (iii) NF_4^+ (iv) BF_4^- (v) TeF_4

(a) i and ii   (b) i,ii, and iii   (c) i,ii,iii, and iv   (d) ii,iii, and iv   (e) v

SiF_4 :

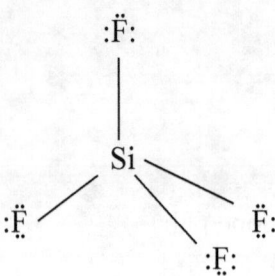

VE = 4+(4×7) = 32   BE = 8   NBE = 24   NBE on Si = 0

VSEPR notation : AB_4 Number of electron groups = 4.

Electron-group geometry : Tetrahedral   Molecular geometry : Tetrahedral

CH_4 :

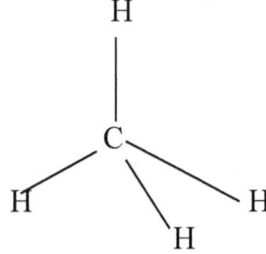

VE = 4+(4×1) = 8   BE = 8   NBE = 0   NBE on C = 0

 VSEPR notation : AB_4 Number of electron groups = 4.

Electron-group geometry : Tetrahedral   Molecular geometry : Tetrahedral

NF_4^+ :

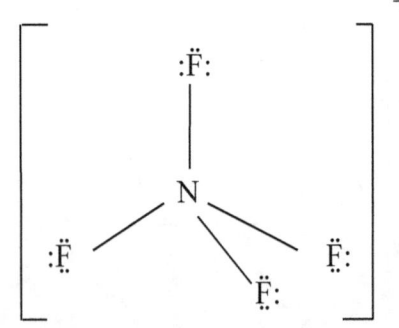

VE = 5+(4×7)-1 = 32   BE = 8   NBE = 24   NBE on N = 0

VSEPR notation : AB_4 Number of electron groups = 4.

Electron-group geometry : Tetrahedral   Molecular geometry : Tetrahedral

BF_4^- :

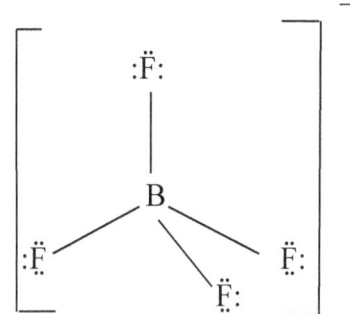

VE = 3+(4×7)+1 = 32   BE = 8   NBE = 24   NBE on B = 0

VSEPR notation : AB_4 Number of electron groups = 4.

Electron-group geometry : Tetrahedral   Molecular geometry : Tetrahedral

TeF_4 :

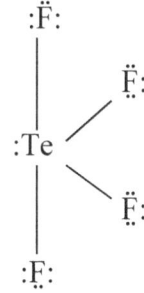

VE = 6+(4×7) = 34   BE = 8   NBE = 26   NBE on Te = 2 (1 lone pair)

VSEPR notation : AB_4E Number of electron groups = 5.

Electron-group geometry : Trigonal bipyramidal   Molecular geometry : Seesaw

i,ii,iii, and iv are tetrahedral.   Answer(c)

5-59. Which of the following molecules are linear?

   (i) C_2H_2 (ii) CO_2 (iii) NO_2^- (iv) NO_2^+ (v) H_2O

   (a) i and ii   (b) i,ii, and iii   (c) i,ii, and iv   (d) ii and iii   (e) v

C_2H_2 : $H - C \equiv C - H$ VE = (2×4)+(2×1) = 10   BE = 10   NBE = 0

 C atoms are sp hybridized.   Molecular geometry : Linear

CO_2 : $:\ddot{O} = C = \ddot{O}:$ VE = 4+(2×6) = 16   BE = 8   NBE = 8   NBE on C = 0

VSEPR notation : AB_2 Number of electron groups = 2.

Molecular geometry : Linear

NO_2^- :

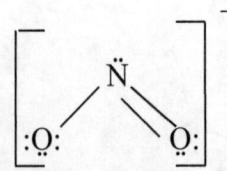

$VE = 5+(2×6)+1 = 18$ $BE = 6$ $NBE = 12$ NBE on $N = 2$ (1 lone pair)

VSEPR notation : AB_2E Number of electron groups = 3.

Electron-group geometry : Trigonal planar   Molecular geometry : Bent

NO_2^+ : $[:\ddot{O} = N - \ddot{O}:]^+$

$VE = 5+(2×6)-1 = 16$ $BE = 6$ $NBE = 10$ NBE on $N = 0$

VSEPR notation : AB_2 Number of electron groups = 2.

Electron-group geometry : Linear   Molecular geometry : Linear

H_2O :

$VE = 6+(2×1) = 8$ $BE = 4$ $NBE = 4$ NBE on $O = 4$ (2 lone pairs)

VSEPR notation : AB_2E_2 Number of electron groups = 4.

Electron-group geometry : Tetrahedral   Molecular geometry : Bent

i,ii, and iv are linear.   Answer(c)

5-60. Which of the following has a square planar shape?

   (a) SF_4 (b) ClO_4^- (c) CCl_4 (d) ICl_4^- (e) PO_4^{3-}

SF_4: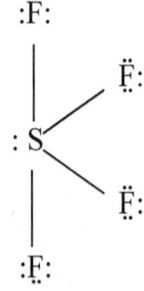

$VE = 6+(4×7) = 34$ $BE = 8$ $NBE = 26$ NBE on $S = 2$ (1 lone pair)

VSEPR notation : AB_4E Number of electron groups = 5. So the valence electron groups on sulfur will be arranged towards the corners of a trigonal bipyramid.

Molecular geometry: Seesaw

ClO_4^- :

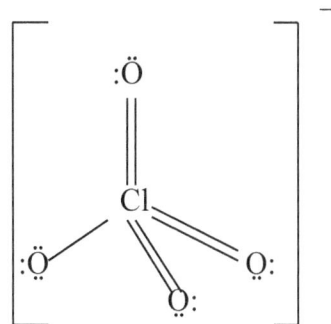

VE = 7+(4×6)+1 = 32   BE = 14   NBE = 18   NBE on Cl = 0

VSEPR notation : AB₄

Number of electron groups (The number of places in the valence shell of Cl atom

where electrons can be found, not the number of pairs of valence electrons)= 4.

Electron-group geometry : Tetrahedral Molecular geometry : Tetrahedral

CCl_4 :

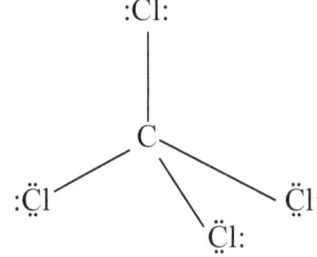

VE = 4+(4×7) = 32   BE = 8   NBE = 24   NBE on C = 0

VSEPR notation : AB₄   Number of electron groups = 4.

Electron-group geometry : Tetrahedral   Molecular geometry : Tetrahedral

ICl_4^- :

VE = 7+(4×7)+1 = 36   BE = 8   NBE = 28   NBE on I = 4 (2 lone pairs)

VSEPR notation : AB₄E₂   Number of electron groups = 6.

Electron-group geometry : Octahedral   Molecular geometry : Square planar

$PO_4{}^{3-}$:

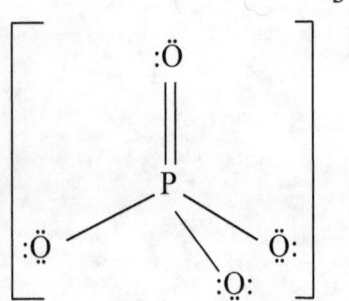

$VE = 5+(4\times6)+3 = 32$ $BE = 10$ $NBE = 22$ NBE on $P = 0$

VSEPR notation : AB_4 Number of electron groups = 4.

Electron-group geometry : Tetrahedral   Molecular geometry : Tetrahedral   Answer(d)

5-61. Which of the following compounds have the same geometry?

   (a) $NH_2{}^-$ and H_2O (b) $NH_2{}^-$ and BeH_2 (c) H_2O and BeH_2 (d) $NH_2{}^-$, H_2O and BeH_2

$NH_2{}^-$:

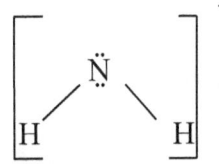

$VE = 5+(2\times1)+1 = 8$ $BE = 4$ $NBE = 4$ NBE on $N = 4$ (2 lone pairs)

VSEPR notation : AB_2E_2 Number of electron groups = 4.

Electron-group geometry : Tetrahedral Molecular geometry : Bent

H_2O :

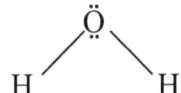

$VE = 6+(2\times1) = 8$ $BE = 4$ $NBE = 4$ NBE on $O = 4$ (2 lone pairs)

VSEPR notation : AB_2E_2 Number of electron groups = 4.

Electron-group geometry : Tetrahedral   Molecular geometry : Bent

BeH_2: $H - Be - H$

$VE = 2+(2\times1) = 4$ $BE = 4$ $NBE = 0$ NBE on $Be = 0$

VSEPR notation : AB_2 Number of electron groups = 2.

Electron-group geometry : Linear   Molecular geometry : Linear

$NH_2{}^-$ and H_2O have the same geometry.   Answer(a)

5-62. Which of the following compounds have the same shape or geometry?

(a) SF_4 and CH_4 (b) CO_2 and H_2O (c) CO_2 and BeH_2 (d) N_2O and NO_2

(e) PCl_4^+ and PCl_4^-

SF_4 :

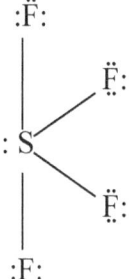

VE = 6+(4×7) = 34   BE = 8 NBE = 26 NBE on S =2 (1 lone pair)

VSEPR notation : AB_4E Number of electron groups = 5. So the valence electron groups on sulfur will be arranged towards the corners of a trigonal bipyramid.

Molecular geometry: Seesaw

CH_4:

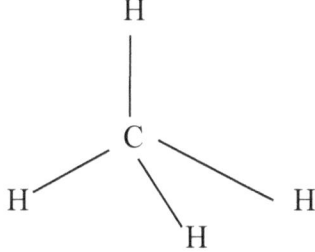

VE = 4+(4×1) = 8   BE =8   NBE = 0   NBE on C = 0

 VSEPR notation : AB_4 Number of electron groups = 4.

Electron-group geometry : Tetrahedral   Molecular geometry : Tetrahedral

CO_2 : $:\ddot{O} = C = \ddot{O}:$ VE = 4+(2×6) = 16   BE = 8 NBE = 8 NBE on C = 0

VSEPR notation : AB_2 Number of electron groups = 2.   Molecular geometry : Linear

H_2O:

VE = 6+(2×1) = 8   BE = 4   NBE = 4   NBE on O = 4 (2 lone pairs)

VSEPR notation : AB_2E_2 Number of electron groups = 4.

Electron-group geometry : Tetrahedral

Molecular geometry : Bent

BeH_2:   H – Be – H

VE = 2+(2×1) = 4   BE = 4   NBE = 0   NBE on Be = 0

VSEPR notation : AB$_2$ Number of electron groups = 2.

Electron-group geometry : Linear   Molecular geometry : Linear

N_2O :   :N ≡ N – Ö:

VE = (2×5)+6 = 16   BE = 8   NBE = 8   NBE on central N = 0

VSEPR notation : AB$_2$ Number of electron groups = 2.

Electron-group geometry : Linear   Molecular geometry : Linear

NO_2 :

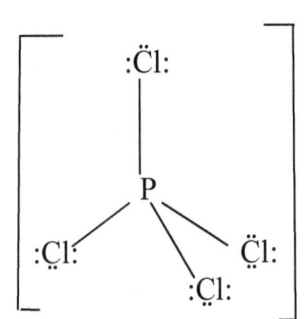

VE = 5+(2×6) = 17 BE = 6   NBE = 11   NBE on N =1

Number of electron groups = 3.

Electron-group geometry : Trigonal planar. Molecular geometry : Bent

PCl_4^+ :   +

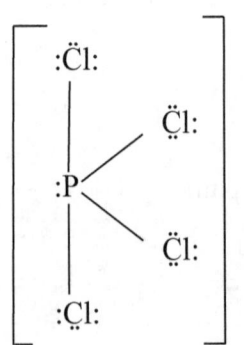

VE = 5+(4×7)−1 = 32   BE = 8   NBE = 24   NBE on P = 0

VSEPR notation : AB$_4$ Number of electron groups = 4.

Electron-group geometry : Tetrahedral   Molecular geometry : tetrahedral

PCl_4^- :   −

VE = 5+(4×7)+1 = 34   BE = 8   NBE = 26   NBE on P = 2 (1 lone pair)

VSEPR notation : AB$_4$E Number of electron groups = 5.

Electron-group geometry : Trigonal bipyramidal. Molecular geometry: Seesaw

CO_2 and BeH_2 have the same molecular geometry.   Answer(c)

5-63. Which of the following molecules or ions does not have a tetrahedral shape?

(a) SiF_4 (b) NF_4^+ (c) BF_4^- (d) SeF_4 (e) CF_4

SiF_4 :

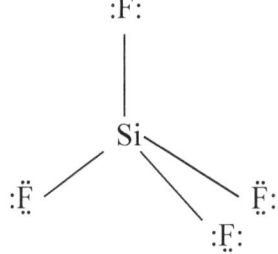

VE = 4+(4×7) = 32   BE = 8   NBE = 24   NBE on Si = 0

VSEPR notation : AB_4 Number of electron groups = 4.

Electron-group geometry : Tetrahedral Molecular geometry : Tetrahedral

NF_4^+ :

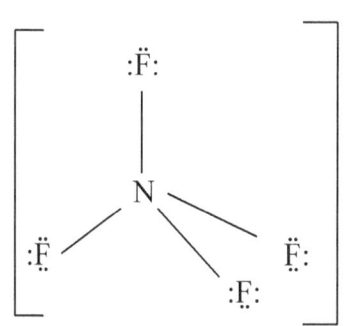

VE = 5+(4×7)−1 = 32   BE = 8   NBE = 24   NBE on N = 0

VSEPR notation : AB_4 Number of electron groups = 4.

Electron-group geometry : Tetrahedral Molecular geometry : Tetrahedral

BF_4^- :

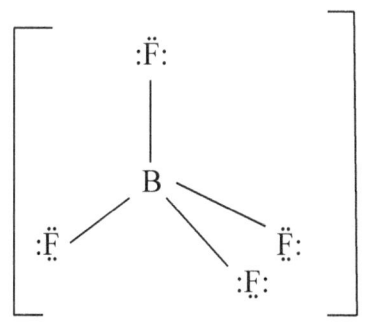

VE = 3+(4×7)+1 = 32   BE = 8   NBE = 24   NBE on B = 0

VSEPR notation : AB_4 Number of electron groups = 4.

Electron-group geometry : Tetrahedral Molecular geometry : Tetrahedral

SeF₄ :

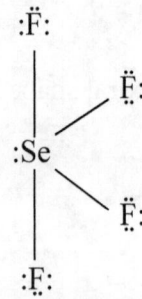

VE = 6+(4×7) = 34   BE = 8 NBE = 26   NBE on Se = 2 (1 lone pair)

VSEPR notation : AB₄E Number of electron groups = 5.

Electron-group geometry : Trigonal bipyramid. Molecular geometry: Seesaw

CF₄ :

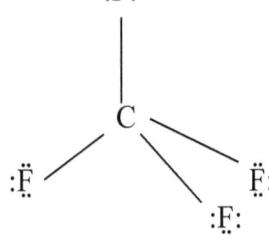

VE = 4+(4×7) = 32   BE = 8 NBE = 24   NBE on C = 0

VSEPR notation : AB₄ Number of electron groups = 4.

Electron-group geometry : Tetrahedral Molecular geometry : Tetrahedral

SeF₄ does not have a tetrahedral shape.   Answer(d)

5-64. Which of the following molecules or ions isn't planar?

   (a) NH₃   (b) BF₃   (c) CO₃²⁻   (d) NO₃⁻

NH₃ :

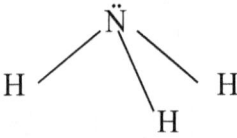

VE = 5+(3×1) = 8   BE = 6   NBE = 2   NBE on N = 2 (1 lone pair)

VSEPR notation : AB₃E   Number of electron groups = 4.

Electron-group geometry : Tetrahedral   Molecular geometry : Trigonal pyramidal

BF₃ :

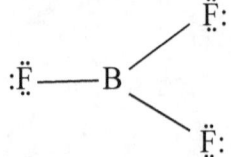

VE = 3+(3×7) = 24   BE = 6   NBE = 18   NBE on B = 0

VSEPR notation : AB$_3$ Number of electron groups = 3.

Electron-group geometry : Trigonal planar Molecular geometry : Trigonal planar

CO$_3^{2-}$:

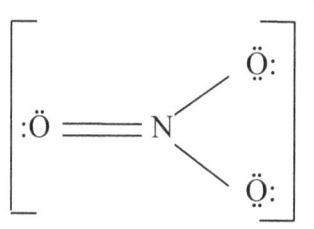

VE = 4+(3×6)+2 = 24   BE = 8   NBE = 16   NBE on C = 0

VSEPR notation : AB$_3$ Number of electron groups = 3.

Electron-group geometry : Trigonal planar Molecular geometry : Trigonal planar

NO$_3^-$:

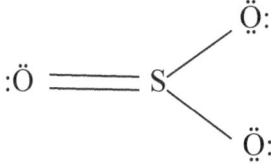

VE = 5+(3×6)+1 = 24   BE = 8   NBE = 16   NBE on N = 0

VSEPR notation : AB$_3$ Number of electron groups = 3.

Electron-group geometry : Trigonal planar Molecular geometry : Trigonal planar

NH$_3$ isn't planar.   Answer(a)

5-65. Which of the following species isn't planar?

   (a) SO$_3$ (b) SO$_3^{2-}$ (c) SO$_2$ (d) BF$_3$ (e) NO$_3^-$

SO$_3$:

VE = 6+(3×6) =24   BE = 8   NBE = 16   NBE on S = 0

VSEPR notation : AB$_3$ Number of electron groups = 3.

Electron-group geometry : Trigonal planar Molecular geometry : Trigonal planar

115

SO_3^{2-} :

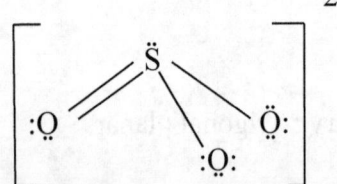

VE = 6+(3×6)+2 = 26   BE = 8   NBE = 18   NBE on S = 2 (1 lone pair)

VSEPR notation : AB₃E   Number of electron groups = 4.

Electron-group geometry : Tetrahedral Molecular geometry : Trigonal pyramidal

SO_2 :

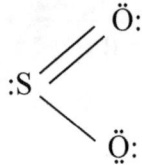

VE = 6+(2×6) = 18   BE = 6   NBE = 12   NBE on S = 2 (1 lone pair)

VSEPR notation : AB₂E Number of electron groups = 3.

Electron-group geometry : Trigonal planar   Molecular geometry : Bent

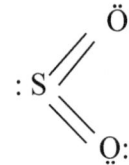

VE = 6+(2×6) = 18   BE = 8   NBE = 10   NBE on S = 2 (1 lone pair)

VSEPR notation : AB₂E Number of electron groups = 3.

Electron-group geometry : Trigonal planar   Molecular geometry : Bent

BF_3 :

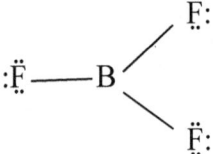

VE = 3+(3×7) = 24   BE = 6   NBE = 18   NBE on B = 0

VSEPR notation : AB₃ Number of electron groups = 3.

Electron-group geometry : Trigonal planar Molecular geometry : Trigonal planar

NO_3^- :

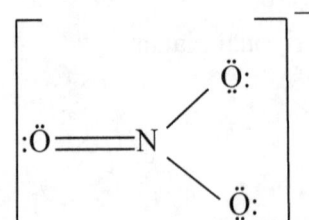

VE = 5+(3×6)+1 = 24   BE = 8   NBE = 16   NBE on N = 0

VSEPR notation : AB$_3$ Number of electron groups = 3.

Electron-group geometry : Trigonal planar Molecular geometry : Trigonal planar

SO$_3^{2-}$ is not planar.   Answer(b)

5-66. What is the distribution of electrons in the valence shell of the Xe atom in the XeF$_3^+$ ion?

   (a) Trigonal planar   (b) Tetrahedral   (c) Square planar   (d) Trigonal bipyramidal

   (e) Octahedral

XeF$_3^+$:

VE = 8+(3×7)−1 = 28   BE = 6 NBE = 22   NBE on Xe = 4 (2 lone pairs)

VSEPR notation : AB$_3$E$_2$

Number of electron groups (The number of places in the valence shell of Xe atom where

electrons can be found) = 5

Electron-group geometry (distribution of electrons): Trigonal bipyramidal   Answer(d)

5-67. What would be the best description of the shape of the XeF$_3^+$ ion?

   (a) Linear   (b) Bent   (c) Seesaw   (d) T-shaped   (e) tetrahedral

XeF$_3^+$:

VE = 8+(3×7)−1 = 28   BE = 6 NBE = 22   NBE on Xe = 4 (2 lone pairs)

VSEPR notation : AB$_3$E$_2$

 Number of electron groups (The number of places in the valence shell of Xe atom

where electrons can be found) = 5

Electron-group geometry (distribution of electrons): Trigonal bipyramidal

Molecular geometry : T – Shaped   Answer(d)

5-68. Tellurium hydride is one of the foulest smelling compounds known. How many pairs of non-bonding electrons are in the valence shell of the tellurium atom in TeH_2?

(a) 0   (b) 1   (c) 2   (d) 3   (e) 4

TeH_2 :

$$\overset{..}{\underset{..}{Te}}\diagdown\diagup$$
H   H

VE = 6+(2×1) = 8   BE = 4   NBE = 4   NBE on Te = 4 (2 lone pairs)   Answer(c)

5-69. What is the best description of the distribution of electrons in the valence shell of the Te atom in TeH_2?

(a) Linear   (b) Trigonal planar   (c) Tetrahedral   (d) Trigonal pyramidal

(e) Octahedral

TeH_2 :

$$\overset{..}{\underset{..}{Te}}\diagdown\diagup$$
H   H

VE = 6+(2×1) = 8   BE = 4   NBE = 4   NBE on Te = 4 (2 lone pairs)

VSEPR notation : AB_2E_2

 Number of electron groups (The number of places in the valence shell of Te atom where electrons can be found) = 4

Electron-group geometry (distribution of electrons): Tetrahedral   Answer(c)

5-70. What is the best description of the shape of the TeH_2 molecule?

(a) Linear   (b) Bent   (c) Trigonal planar   (d) tetrahedral   (e) Seesaw

TeH_2 :

$$\overset{..}{\underset{..}{Te}}\diagdown\diagup$$
H   H

VE = 6+(2×1) = 8   BE = 4   NBE = 4   NBE on Te = 4 (2 lone pairs)

VSEPR notation : AB_2E_2

 Number of electron groups (The number of places in the valence shell of Te atom where electrons can be found) = 4

Electron-group geometry (distribution of electrons): Tetrahedral

Molecular geometry : Bent   Answer(b)

5-71. The best way for the body to get rid of excess nitrogen from the metabolism of proteins is by excreting a planar molecule which has the following structure:

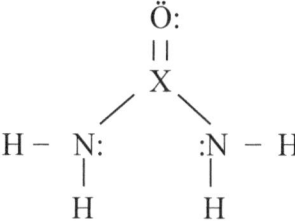

$$H - \underset{\underset{H}{|}}{N:} \qquad \underset{\underset{H}{|}}{:N} - H$$

Which element does the 'X' stand for in this Lewis structure?

   (a) C   (b) N   (c) O   (d) P   (e) S

Thye compound is a neutral molecule.   NBE on X = 0

C : VE = $6+4+(2\times5)+(4\times1) = 24$ BE = 16 NBE = 8 NBE on C = 0

N : VE = $6+(3\times5)+(4\times1) = 25$ BE = 16 NBE = 9 NBE on N = 1

O : VE = $(2\times6)+(2\times5)+(4\times1) = 26$ BE = 16 NBE = 10 NBE on O = 2 (1 lone pair)

P : VE = $6+5+(2\times5)+(4\times1) = 25$ BE = 16 NBE = 9 NBE on P = 1

S : VE = $6+6+(2\times5)+(4\times1) = 26$ BE = 16 NBE = 10 NBE on S = 2(1 lone pair) Answer(a)

5-72. If there is a single pair of NBE on the X atom in the Lewis structure of XF_4, in which group of the periodic table would you be most likely to find element 'X'?

   (a) Group 13   (b) Group 14   (c) Group 15   (d) Group 16   (e) Group 17

XF_4 :

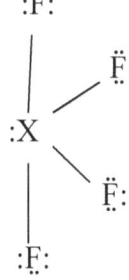

The compound is a neutral molecule.   NBE on X = 2

Group 13 : VE = $3+(4\times7) = 31$ BE = 8   NBE = 23 Not enough NBE.

Group 14 : VE = $4+(4\times7) = 32$ BE = 8   NBE = 24 NBE on X = 0

Group 15 : VE = $5+(4\times7) = 33$ BE = 8   NBE = 25 NBE on X = 1

Group 16 : VE = $6+(4\times7) = 34$ BE = 8   NBE = 26 NBE on X = 2 (1 lone pair)

Group 17 : VE = $7+(4\times7) = 35$ BE = 8   NBE = 27 NBE on X = 3   Answer(d)

5-73. What would be the distribution of electrons in the valence shell of the 'X' atom in the previous question?

    (a) Trigonal planar   (b) Tetrahedral   (c) Trigonal bipyramidal   (d) Octahedral

    (e) None of these

XF_4 :

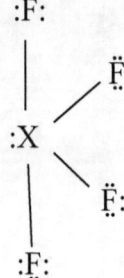

The compound is a neutral molecule.   NBE on X = 2 (1 lone pair)

VE = 6+(4×7) = 34   BE = 8 NBE = 26   NBE on X = 2 (1 lone pair)

VSEPR notation : AB_4E

 Number of electron groups (The number of places in the valence shell of X atom

where electrons can be found) = 5

Electron-group geometry (distribution of electrons): Trigonal bipyramidal   Answer(c)

5-74. What would be the shape of the XF_4 molecule in the previous question?

    (a) Octahedral   (b) Seesaw   (c) Tetrahedral   (d) Square pyramidal

    (e) Trigonal bipyramidal

XF_4 :

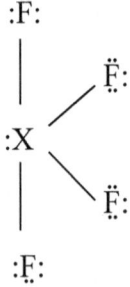

The compound is a neutral molecule.   NBE on X = 2 (1 lone pair)

VE = 6+(4×7) = 34   BE = 8 NBE = 26   NBE on X = 2 (1 lone pair)

VSEPR notation : AB_4E

Number of electron groups (The number of places in the valence shell of X atom

where electrons can be found) = 5

Electron-group geometry (distribution of electrons): Trigonal bipyramidal

Molecular geometry : Seesaw   Answer(b)

5-75. Which element would form a neutral XF_3 molecule with a trigonal planar geometry?

    (a) Be   (b) B   (c) C   (d) N   (e) Cl

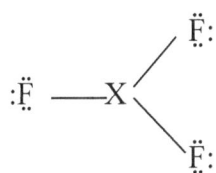

The molecule is neutral. Molecular geometry : Trigonal planar

Be : VE = 2+(3×7) = 23   BE = 6   NBE =17 Not enough NBE

B : VE = 3+(3×7) = 24   BE = 6   NBE =18   NBE on B = 0

VSEPR notation : AB_3 Number of electron groups = 3.

Electron-group geometry : Trigonal planar   Molecular geometry : Trigonal planar

C : VE = 4+(3×7) = 25   BE = 6   NBE =19   NBE on C = 1

Number of electron groups = 4.

Electron-group geometry : Tetrahedral   Molecular geometry : Trigonal pyramidal

N : VE = 5+(3×7) = 26   BE = 6   NBE = 20   NBE on N = 2 (1 lone pair)

VSEPR notation : AB_3E Number of electron groups = 4.

Electron-group geometry : Tetrahedral   Molecular geometry : Trigonal pyramidal

Cl : VE = 7+(3×7) = 28   BE = 6   NBE = 22   NBE on Cl = 4 (2 lone pairs)

VSEPR notation : AB_3E_2 Number of electron groups = 5.

Electron-group geometry : Trigonal bipyramidal   Molecular geometry : T – shaped

BF_3 has a trigonal planar shape.   Answer(b)

5-76. The compound XF_3 has a trigonal pyramidal geometry. Element X is a member of :

    (a) Group 2   (b) Group 13   (c) Group 15   (d) Group 16   (e) Group 17

XF_3 is a neutral molecule.   Molecular geometry : Trigonal pyramidal

Group 2: VE = 2+(3×7) = 23   BE = 6   NBE =17 Not enough NBE

Group 13 : VE = 3+(3×7) = 24   BE = 6   NBE =18 NBE on X = 0

VSEPR notation : AB_3 Number of electron groups = 3.

Electron-group geometry : Trigonal planar Molecular geometry : Trigonal planar

Group 15 : VE = 5+(3×7) = 26   BE = 6   NBE = 20   NBE on X = 2 (1 lone pair)

VSEPR notation : AB_3E Number of electron groups = 4.

Electron-group geometry : Tetrahedral   Molecular geometry : Trigonal pyramidal

Group 16 : VE = 6+(3×7) = 27   BE = 6   NBE = 21   NBE on X = 3

Number of electron groups = 5.

Electron-group geometry : Trigonal bipyramidal   Molecular geometry : T – shaped

Group 17 : VE = 7+(3×7) = 28   BE = 6   NBE = 22   NBE on X = 4 (2 lone pairs)

VSEPR notation : AB_3E_2 Number of electron groups = 5.

Electron-group geometry : Trigonal bipyramidal   Molecular geometry : T – shaped

Group 15 element has a trigonal pyramidal geometry.   Answer(c)

5-77. Which of the following set of hybrid orbitals would give rise to a trigonal bipyramidal geometry?

   (a) sp   (b) sp^2 (c) sp^3 (d) sp^3d (e) sp^3d^2

Number of Electron Groups*	Hybridization	Electron-group Geometry**
2	sp	Linear
3	sp^2	Trigonal Planar
4	sp^3	Tetrahedral
5	sp^3d	Trigonal bipyramidal
6	sp^3d^2	Octahedral

The number of places in the valence shell of an atom where electrons can be found, not the number of pairs of valence electrons.

**Distribution of electrons   Answer(d)

5-78. An element we will represent as X forms a compound with the empirical formula XCl_5 that undergoes an interesting reaction when it crystallizes to form a mixed salt of XCl_4^+ and XCl_6^- ions. Assume that XCl_4^+ has tetrahedral geometry and there are no NBE in the valence shell of the X atom.In which column of the periodic table are you most likeley to find element X?

(a) Group 14 (C,Se,Ge…)   (b) Group 15 (N,P,As…)   (c) Group16 (O,S,Se…)

(d) Group 17 (F,Cl,Br…)   (e) Group 18 (He,Ne,Ar…)

XCl_4^+ NBE on X = 0   Molecular geometry : Tetrahedral

Group 14 : VE = 4+(4×7)-1 = 31   BE = 8   NBE = 23   Not enough NBE

Group 15 : VE = 5+(4×7)-1 = 32   BE = 8   NBE = 24   NBE on X = 0

Group 16 : VE = 6+(4×7)-1 = 33   BE = 8   NBE = 25   NBE on X = 1

Group 17 : VE = 7+(4×7)-1 = 33   BE = 8   NBE = 26   NBE on X = 2 (1 lone pair)

Group 18 : VE = 8+(4×7)-1 = 34   BE = 8   NBE = 27   NBE on X = 3

In the case of group 15 NBE on X = 0.   Answer(b)

5-79. The Lewis structure of metformin is shown is shown below:

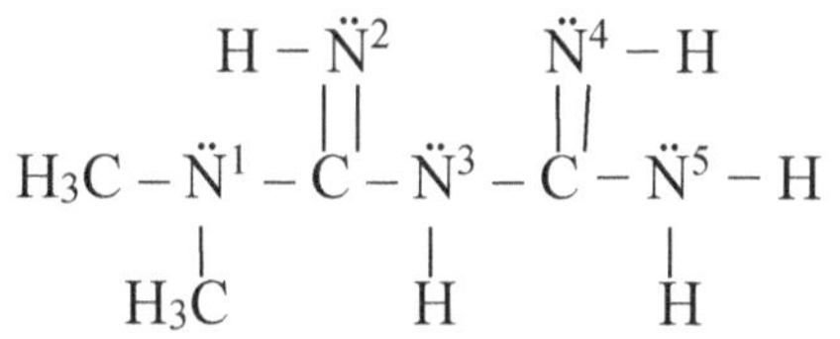

$$H - \ddot{N}^2 \qquad \ddot{N}^4 - H$$
$$\| \qquad \qquad \|$$
$$H_3C - \ddot{N}^1 - C - \ddot{N}^3 - C - \ddot{N}^5 - H$$
$$| \qquad \quad | \qquad \quad |$$
$$H_3C \qquad H \qquad H$$

Use the VSEPR or VSED theories to predict the geometry around nitrogen atom in the center

of the molecule (N^3).

   (a) Bent or angular   (b) Trigonal planar   (c) Trigonal pyramidal

   (d) Trigonal bipyramidal   (e) Tetrahedral

N^3 : NBE on N = 2 (1 lone pair)   VSEPR notation : AB_3E

Number of electron groups (The number of places in the valence shell of N atom where

electrons can be found) = 4

Electron-group geometry (distribution of electrons): Tetrahedral

Geometry around N^3 : Trigonal pyramidal   Answer(c)

5-80. The Lewis structure of terpin hydrate is shown below:

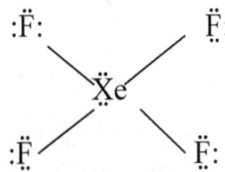

Use the VSEPR theory to predict the distribution of electrons in the valence shell of the oxygen atoms in this compound.

(a) Linear   (b) Bent or angular   (c) Trigonal planar   (d) Tetrahedral

(e) Trigonal bipyramidal

NBE on each O atom = 4 (2 lone pairs)

Number of electron groups (The number of places in the valence shell of O atom

where electrons can be found) = 4

Electron-group geometry (distribution of electrons): Tetrahedral   Answer(d)

5-81. Which of the following describes the XeF_4 molecule best?

(a) Trigonal bipyramidal distribution of VE on the Xe with T - shaped molecular geometry
(b) Trigonal bipyramidal distribution of VE on the Xe with a trigonal planar molecular geometry
(c) Trigonal bipyramidal distribution of VE on the Xe with a pyramidal molecular geometry
(d) Octahedral distribution of VE on the Xe with a square planar molecular geometry
(e) Octahedral distribution of VE on the Xe with a seesaw molecular geometry

XeF_4:   :F̈:   F̈:

                    Ẍe

        :F̈   F̈:

VE = 8+(4×7) = 36   BE = 8   NBE = 28   NBE on Xe = 4 (2 lone pairs)

VSEPR notation : AB_4E_2 Number of electron groups = 6. So the valence electron groups on

xenon will be arranged towards the corners of an octahedron to reduce repulsions

between electrons.   Molecular geometry : Square planar   Answer(d)

5-82. How many pairs of NBE are in the valence shell of the central atom in the I_3^- ion?

(a) 0   (b) 1   (c) 2   (d) 3   (e) 4

I_3^- : $[:\ddot{I} - \ddot{I} - \ddot{I}:]^-$

VE = (3×7)+1 = 22 BE = 4 NBE = 18 NBE on Central I = 6 (3 lone pairs)

VSEPR notation : AB_2E_3 Number of electron groups = 5. So the valence electron groups on iodine will be arranged towards the corners of a trigonal bipyramid to reduce repulsions between electrons.   Answer(d)

5-83. How many pairs of NBE can be found on each nitrogen atom in the hydrazine molecule?

(a) 0   (b) 1   (c) 2   (d) 3

(e) The two nitrogen atoms contain a different number of NBE

Hydrazine: $\begin{array}{c} H - \ddot{N} - \ddot{N} - H \\ | \quad\quad | \\ H \quad H \end{array}$

VE = (2×5)+(4×1) = 14   BE = 10   NBE = 4   NBE on each nitrogen atom = 2 (1 lone pair)

                                                                                                      Answer(b)

5-84. Predict the distribution of electrons in the valence shell of the nitrogen atoms in hydrazine.

(a) Linear   (b) Bent or angular   (c) Trigonal planar   (d) Trigonal pyramidal

(e) Tetrahedral

Hydrazine: $\begin{array}{c} H - \ddot{N} - \ddot{N} - H \\ | \quad\quad | \\ H \quad H \end{array}$

VE = (2×5)+(4×1) = 14   BE = 10   NBE on each nitrogen atom = 2 (1 lone pair)

Number of electron groups (The number of places in the valence shell of each N atom where electrons can be found) = 4

Electron-group geometry (distribution of electrons): Tetrahedral   Answer(e)

5-85. Use the VSEPR theory to predict the geometry around each nitrogen atom in hydrazine.

(a) Linear   (b) Bent or angular   (c) Trigonal planar   (d) Trigonal pyramidal

(e) tetrahedral

Hydrazine:
$$H - \ddot{N} - \ddot{N} - H$$
$$\quad\quad\quad | \quad\quad |$$
$$\quad\quad\quad H \quad\quad H$$

VE = (2×5)+(4×1) = 14   BE = 10   NBE on each nitrogen atom = 2 (1 lone pair)

Number of electron groups (The number of places in the valence shell of each N atom

where electrons can be found) = 4

Electron-group geometry (distribution of electrons): Tetrahedral

Geometry around each N atom : Trigonal pyramidal   Answer(d)

5-86. Which type of hybrid orbitals are used by the carbon atom in CO_2?

(a) sp   (b) sp^2 (c) sp^3 (d) dsp^3 (e) d^2sp^3

CO_2 :   :$\ddot{O}$ = C = $\ddot{O}$:   VE = 4+(2×6) = 16   BE = 8   NBE = 8   NBE on C = 0

Number of electron groups (The number of places in the valence shell of C atom

where electrons can be found, not the number of pairs of valence electrons) = 2

Electron-group geometry (distribution of electrons): Linear   Hybridization : sp   Answer(a)

5-87. In which of the following compounds is the central nitrogen atom sp^2 hybridized?

(a) NO_2^+ (b) NO_2 (c) NO_2^- (d) NO_3^- (e) NO_2, NO_2^-, and NO_3^-

NO_2^+ :   [:$\ddot{O}$ = N - $\ddot{O}$:]$^+$

VE = 5+(2×6)-1 = 16   BE = 6   NBE = 10   NBE on N = 0

Number of electron groups (The number of places in the valence shell of N atom

where electrons can be found, not the number of pairs of valence electrons) = 2

Electron-group geometry (distribution of electrons): Linear   Hybridization : sp

NO_2 :

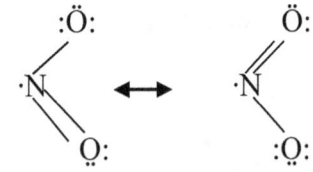

VE = 5+(2×6) = 17   BE = 6   NBE = 11   NBE on N = 1

Number of electron groups (The number of places in the valence shell of N atom

where electrons can be found, not the number of pairs of valence electrons) = 3

Electron-group geometry (distribution of electrons): Trigonal planar   Hybridization : sp^2

NO_2^- :

VE = 5+(2×6)+1 = 18   BE = 6   NBE =12   NBE on N = 2 (1 lone pair)

Number of electron groups (The number of places in the valence shell of N atom

where electrons can be found, not the number of pairs of valence electrons) = 3

Electron-group geometry (distribution of electrons): Trigonal planar Hybridization : sp^2

NO_3^- :

VE = 5+(3×6)+1 = 24   BE = 8   NBE = 16   NBE on N = 0

Number of electron groups (The number of places in the valence shell of N atom

where electrons can be found, not the number of pairs of valence electrons) = 3

Electron-group geometry (distribution of electrons): Trigonal planar Hybridization : sp^2

Answer(e)

5-88. What is the hybridization of N atom in the N_2 molecule?

   (a) sp   (b) sp^2 (c) sp^3 (d) sp^3d (e) sp^3d^2

N_2 :   :N≡N:

VE = (2×5) = 10   BE = 6   NBE = 4   NBE on each N = 2 (1 lone pair)

Number of electron groups (The number of places in the valence shell of each N atom

where electrons can be found, not the number of pairs of valence electrons) = 2

Electron-group geometry (distribution of electrons): Linear   Hybridization : sp   Answer(a)

5-89. What is the hybridization of central iodine atom in the I_3^- ion?

   (a) sp   (b) sp^2 (c) sp^3 (d) sp^3d (e) sp^3d^2

I_3^- :   [:Ï – Ï – Ï:]⁻

VE = (3×7)+1 = 22 BE = 4 NBE = 18 NBE on Central I = 6 (3 lone pairs)

Number of electron groups (The number of places in the valence shell of central iodine atom where electrons can be found) = 5

Electron-group geometry (distribution of electrons): Trigonal bipyramidal

Hybridization : sp^3d Answer(d)

5-90. What is the hybridization of S atom in the SF_3^+ ion?

   (a) sp   (b) sp^2 (c) sp^3 (d) sp^3d (e) sp^3d^2

SF_3^+ :

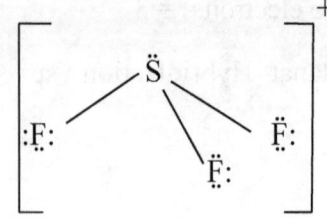

VE = 6+(3×7)−1 = 26   BE = 6   NBE = 20 NBE on S = 2 (1 lone pair)

Number of electron groups (The number of places in the valence shell of central sulfur atom where electrons can be found) = 4

Electron-group geometry (distribution of electrons): Tetrahedral

Hybridization : sp^3 Answer(c)

5-91. What is the hybridization of the beryllium atom in BeF_2?

   (a) sp   (b) sp^2 (c) sp^3 (d) sp^3d (e) sp^3d^2

BeF_2 : $:\ddot{F} - Be - \ddot{F}:$

VE = 2+(2×7) =16   BE = 4 NBE = 12   NBE on Be = 0

Number of electron groups (The number of places in the valence shell of central Be atom where electrons can be found) = 2

Electron-group geometry (distribution of electrons): Linear   Hybridization : sp   Answer(a)

5-92. What is the hybridization of the xenon atom in the XeF_3^+ ion?

   (a) sp   (b) sp^2 (c) sp^3 (d) sp^3d (e) sp^3d^2

XeF_3^+ :

$$\begin{array}{c}
:\ddot{F}: \\
| \\
::Xe - \ddot{F}: \\
| \\
:\ddot{F}:
\end{array}\Bigg]^{+}$$

VE = 8+(3×7)−1 = 28   BE = 6   NBE = 22   NBE on Xe = 4 (2 lone pairs)

Number of electron groups (The number of places in the valence shell of central xenon atom where electrons can be found) = 5

Electron-group geometry (distribution of electrons): Trigonal bipyramidal

Hybridization : sp^3d Answer(d)

5-93. What is the hybridization of the nitrogen atom in the NO_2^+ ion?

    (a) sp   (b) sp^2 (c) sp^3 (d) sp^3d (e) sp^3d^2

NO_2^+ : $[:\ddot{O} = N - \ddot{O}:]^+$

VE = 5+(2×6)-1 = 16   BE = 6   NBE = 10   NBE on N = 0

Number of electron groups (The number of places in the valence shell of N atom

where electrons can be found, not the number of pairs of valence electrons) = 2

Electron-group geometry (distribution of electrons): Linear   Hybridization : sp   Answer(a)

5-94. The central atom in all of the following compounds has the same hybridization, except

    (a) NO_2^- (b) CO_3^{2-} (c) BF_3 (d) NH_3

    (e) The hybridization is same in all the compounds

NO_2^- :

VE = 5+(2×6)+1 = 18   BE = 6 NBE =12   NBE on N = 2 (1 lone pair)

Number of electron groups (The number of places in the valence shell of N atom

where electrons can be found, not the number of pairs of valence electrons) = 3

Electron-group geometry (distribution of electrons): Trigonal planar Hybridization : sp^2

CO_3^{2-} :

VE = 4+(3×6)+2 = 24   BE = 8   NBE = 16 NBE on C = 0

Number of electron groups (The number of places in the valence shell of C atom

where electrons can be found, not the number of pairs of valence electrons) = 3

Electron-group geometry (distribution of electrons): Trigonal planar   Hybridization : sp^2

BF₃ :

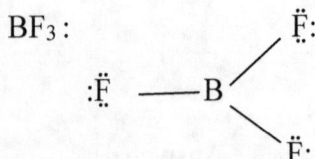

VE = 3+(3×7) = 24   BE = 6   NBE = 18   NBE on B = 0

Number of electron groups (The number of places in the valence shell of B atom

where electrons can be found) = 3

Electron-group geometry (distribution of electrons): Trigonal planar   Hybridization : sp^2

NH₃ :

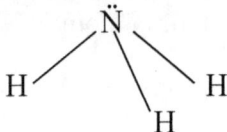

VE = 5+(3×1) = 8   BE = 6   NBE = 2   NBE on N = 2 (1 lone pair)

Number of electron groups (The number of places in the valence shell of N atom

where electrons can be found) = 4

Electron-group geometry (distribution of electrons): Tetrahedral Hybridization : sp^3 Answer(d)

5-95. Which of the following molecules should be paramagnetic?

    (i) B_2 (ii) C_2 (iii) N_2 (iv) O_2 (v) F_2

    (a) i   (b) i and ii   (c) i and iv   (d) i,ii, and iv   (e) ii,ii, and iv

For O_2 and F_2 the increasing order of energies of various molecular orbitals is:

$\sigma_{2s} < \sigma_{2s}^* < \sigma_{2p} < \pi_x = \pi_y < \pi_x^* = \pi_y^* < \sigma_{2p}^*$

O_2 : The electron configuration of O : $1s^2 2s^2 2p^4$

There are 12 valence electrons in O_2 molecule.

The valence electron configuration of O_2:

$(\sigma_{2s})^2 (\sigma_{2s}^*)^2 (\sigma_{2p})^2 \{(\pi_x)^2 = (\pi_y)^2\} \{(\pi_x^*)^1 = (\pi_y^*)^1\}$

The molecule has unpaired electrons. The molecule is paramagnetic.

F_2 : The electron configuration of O : $1s^2 2s^2 2p^5$

There are 14 valence electrons in F_2 molecule.

The valence electron configuration of F_2:

$(\sigma_{2s})^2 (\sigma_{2s}^*)^2 (\sigma_{2p})^2 \{(\pi_x)^2 = (\pi_y)^2\} \{(\pi_x^*)^2 = (\pi_y^*)^2\}$

The molecule does not have an unpaired electron. The molecule is diamagnetic.

For B_2, C_2, and N_2 (due to 2s-2p$_z$ mixing) the increasing order of energies

of various molecular orbitals is:

$\sigma_{2s} < \sigma_{2s}^* < \pi_x = \pi_y < \sigma_{2p} < \pi_x^* = \pi_y^* < \sigma_{2p}^*$

B_2 : The electron configuration of B : $1s^2 2s^2 2p^1$

There are 6 valence electrons in B_2.

The valence electron configuration of B_2:

$(\sigma_{2s})^2 (\sigma_{2s}^*)^2 \{(\pi_x)^1 = (\pi_y)^1\}$

The molecule has unpaired electrons. The molecule is paramagnetic.

C_2 : The electron configuration of C : $1s^2 2s^2 2p^2$

There are 8 valence electrons in C_2.

The valence electron configuration of C_2:

$(\sigma_{2s})^2 (\sigma_{2s}^*)^2 \{(\pi_x)^2 = (\pi_y)^2\}$

The molecule does not have an unpaired electron. The molecule is diamagnetic.

N_2 : The electron configuration of N : $1s^2 2s^2 2p^3$ There are 10 valence electrons in N_2.

The valence electron configuration of N_2:

$(\sigma_{2s})^2 (\sigma_{2s}^*)^2 \{(\pi_x)^2 = (\pi_y)^2\} (\sigma_{2p})^2$

The molecule does not have an unpaired electron. The molecule is diamagnetic.

B_2 and O_2 are paramagnetic.   Answer(c)

5-96. Which of the following has the largest bond order?

   (a) B_2 (b) C_2 (c) N_2 (d) O_2 (e) F_2

Bond order = {(Bonding electrons) – (Antibonding electrons)}/2

Molecule	Bonding electrons	Antibonding electrons	Bond order
B_2	4	2	1
C_2	6	2	2
N_2	8	2	3
O_2	8	4	2
F_2	8	6	1

N_2 has the largest bond order. Answer(c)

5-97. Use the following MO diagram to predict which of the following has the largest bond order.

(a) NO^+ (b) NO   (c) NO^-

(d) All of the above have the same bond order

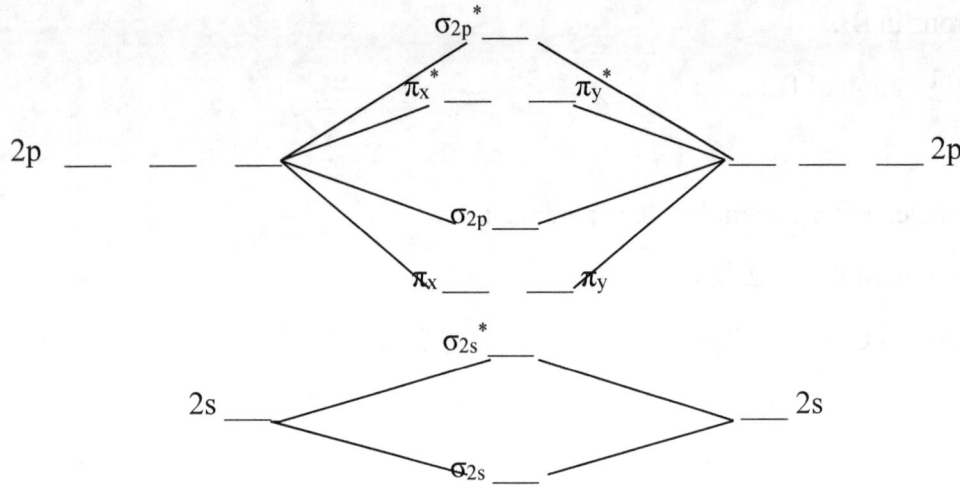

NO^+ : The ion has 10 valence electrons.

The valence electron configuration of NO^+ :

$(\sigma_{2s})^2(\sigma_{2s}{}^*)^2\{(\pi_x)^2=(\pi_y)^2\}(\sigma_{2p})^2$

NO : The molecule has 11 valence electrons.

The valence electron configuration of NO :

$(\sigma_{2s})^2(\sigma_{2s}{}^*)^2\{(\pi_x)^2=(\pi_y)^2\}(\sigma_{2p})^2(\pi_x{}^*)^1$

NO^- : The ion has 12 valence electrons.

The valence electron configuration of NO^- :

$(\sigma_{2s})^2(\sigma_{2s}{}^*)^2\{(\pi_x)^2=(\pi_y)^2\}(\sigma_{2p})^2\{(\pi_x{}^*)^1=(\pi_y{}^*)^1\}$

	Bonding electrons	Antibonding electrons	Bond order
NO^+	8	2	3.0
NO	8	3	2.5
NO^-	8	4	2.0

NO^+ has the largest bond order. Answer(a)

5-98. Which of the following lists the species in the correct order of decreasing stability?

(a) $O_2^+>O_2>O_2^->O_2^{2-}$ (b) $O_2^{2-}>O_2^->O_2>O_2^+$ (c) $O_2^->O_2^+>O_2>O_2^{2-}$

(d) $O_2>O_2^+>O_2^{2-}>O_2^-$

O_2^+ : The species has 11 valence electrons.

The valence electron configuration of O_2^+ :

$(\sigma_{2s})^2(\sigma_{2s}^*)^2(\sigma_{2p})^2\{(\pi_x)^2=(\pi_y)^2\}(\pi_x^*)^1$

O_2 : The species has 12 valence electrons.

The valence electron configuration of O_2 :

$(\sigma_{2s})^2(\sigma_{2s}^*)^2(\sigma_{2p})^2\{(\pi_x)^2=(\pi_y)^2\}\{(\pi_x^*)^1=(\pi_y^*)^1\}$

O_2^- : The species has 13 valence electrons.

The valence electron configuration of O_2^- :

$(\sigma_{2s})^2(\sigma_{2s}^*)^2(\sigma_{2p})^2\{(\pi_x)^2=(\pi_y)^2\}\{(\pi_x^*)^2=(\pi_y^*)^1\}$

O_2^{2-} : The species has 14 valence electrons.

The valence electron configuration of O_2^{2-} :

$(\sigma_{2s})^2(\sigma_{2s}^*)^2(\sigma_{2p})^2\{(\pi_x)^2=(\pi_y)^2\}\{(\pi_x^*)^2=(\pi_y^*)^2\}$

Electrons in antibonding molecular orbitals destabilize the species. Based on the number of electrons in antibonding molecular orbitals the stability order is $O_2^+>O_2>O_2^->O_2^{2-}$.

Answer(a)

CHAPTER 6

IONIC AND METALLIC BONDS

6-1. Write a balanced equation for the following reaction: $Li (s) + O_2 (g) \rightarrow$

$4Li (s) + O_2 (g) \rightarrow 2Li_2O (s)$

6-2. Magnesium reacts with hydrogen to form compound A, which is a white solid at room temperature. It also reacts with HCl to form gas B and an aqueous solution of compound C. Identify the products of these reactions and write balanced equations for each reaction.

$Mg (s) + H_2 (g) \rightarrow MgH_2 (s)$

$Mg (s) + 2HCl (aq) \rightarrow MgCl_2 (aq) + H_2 (g)$

A : $MgH_2 (s)$

B : $H_2 (g)$

C : $MgCl_2$

6-3. Determine which element is oxidised and which is reduced when lithium reacts with nitrogen to form lithium nitride.

$6Li (s) + N_2 (g) \rightarrow 2 Li_3 N (s)$

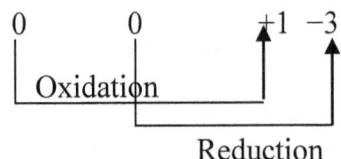

Li is oxidised and N_2 is reduced.

6-4. Determine which is oxidised and which is reduced in the following reaction:

$Sr (s) + 2H_2O (l) \rightarrow Sr_2^{2+} (aq) + 2OH^- (aq) + H_2 (g)$

| 0 | +1 −2 | +2 | | 0 |

Sr is oxidised and H_2O is reduced.

6-5. Identify the oxidising agent and reducing agent in the following reaction:

$Ca (s) + H_2 (g) \rightarrow Ca H_2 (g)$

| 0 | 0 | +2 −1 |

Ca is the reducing agent and H_2 is the oxidising agent.

6-6. Use the following reactions to determine the relative strengths of Na, Mg, and Al metal as reducing agents:

$2Na + MgCl_2 \rightarrow 2NaCl + Mg$

$Al + MgBr_2 \not\rightarrow$

Al can not reduce $MgBr_2$. So Mg is a stronger reducing agent than Al.

Na can reduce $MgCl_2$ to Mg. So Na is a stronger reducing agent than Mg.

Reducing strength: Na>Mg>Al

6-7. Compare the sizes of neutral sodium and chlorine atoms and the Na^+ and Cl^- ions.

	Covalent radius (nm)		Ionic radius (nm)
Na	0.157	Na^+	0.095
Cl	0.099	Cl^-	0.181

Na : [Ne] $3s^1$ Na^+ : [Ne]

11 protons in sodium nucleus hold the 10 electrons in Na^+ more tightly than the 11 electrons in neutral sodium atom. So neutral sodium atom is larger than Na^+ ion.

Cl : [Ne] $3s^23p^5$ Cl^- : [Ne] $3s^23p^6$

17 protons in chlorine nucleus can not hold the 18 electrons in Cl^-ion as tightly as they can the 17 electrons in neutral chlorine atom. So Cl^- ion is larger than neutral chlorine atom.

6-8. Predict which is larger in each of the following pairs of atoms or ions.

   (a) S^{2-} or O^{2-} (b) Na or Al   (c) C^{4-} or F^- (d) P^{3-} or P

(a) O : [He] $2s^22p^4$ O^{2-} : [He] $2s^22p^6$

   S : [Ne] $3s^23p^4$ S^{2-} : [Ne] $3s^23p^6$

As we proceed from O^{2-} to S^{2-} the principal quantum number increases from 2 to 3. So the outer electrons in S^{2-} will be farther from the nucleus and S^{2-} will be larger than O^{2-}.

(b) Na : [Ne]$3s^1$ 11 protons in the nucleus

   Al : [Ne] $3s^23p^1$ 13 protons in the nucleus

The outer electrons are in the same valence shell and the effective nuclear charge in Al is more than that in Na. So Na will be larger than Al.

(c) C : [He] $2s^22p^2$ C^{4-} : [He] $2s^22p^6$

F : [He] $2s^22p^5$ F^- : [He] $2s^22p^6$

F^- has nine protons in the nucleus and C^{4-} has 6 protons in the nucleus; the effective

nuclear charge will be more in the case of F^-. So C^{4-} will be larger than F^-.

(d) P : [Ne] $3s^23p^3$ P^{3-} : [Ne] $3s^23p^6$

15 protons in phosphorus nucleus can not hold the 18 electrons in P^{3-} ion as tightly as they

can the 15 electrons in neutral phosphorus atom. So P^{3-} ion will be larger than the neutral

phosphorus atom.

6-9. Use the Bohr model to calculate the wavelength of the photon that would have to be

absorbed to ionize a neutral hydrogen atom. (IE: H = 1312 kJ/mol)

Ionisation energy per mole = 1312 kJ = 1312×10^3 J

Energy per molecule = $1312 \times 10^3 / 6.022 \times 10^{23}$ = 217.9×10^{-20} J

$E = hc/\lambda$

Wavelength = $\{(6.626 \times 10^{-34}) \times (2.998 \times 10^8)\} / 217.9 \times 10^{-20}$ = 0.091×10^{-6} m

Wavelength = 91 nm

6-10. Which of the following metals is the most 'active', the metal which should react most

rapidly with oxygen or water vapor in the atmosphere?

   (a) Be   (b) Ca   (c) Ba   (d) Al   (e) Ga

Active metals are found exclusively in Group1 and Group 2 of the periodic table.

The activity increases as we proceed from right to left. The activity increases as we go down in

a column because the elements become more metallic. So Ba will be the most 'active' metal.

Answer(c)

6-11. Which of the following metals would be the most reactive towards air and water ?

   (a) Ca   (b) Al   (c) Ag   (d) Sn   (e) Pb

In the above list there are no metals from Group1 and only one metal from

Group 2 (Ca). Metals of Group 1 and Group 2 are the active metals.   Answer(a)

6-12. Which of the following compounds is least likely to be ionic?

   (a) MgF_2 (b) $MgCl_2$ (c) $MgBr_2$ (d) BaF_2

EN (Electronegativity) : Mg = 1.31, F = 3.98 , Cl = 3.16, Br = 2.96, and Ba = 0.89

ΔEN : MgF_2 = 2.67 , $MgCl_2$ = 1.85, $MgBr_2$ = 1.65, BaF_2 = 3.09

In general the compound is less likely to be ionic if ΔEN <1.8.   Answer(c)

6-13. Which of the following is the most likely product of the reaction between magnesium metal and nitrogen?

   (a) MgN   (b) Mg_2N (c) MgN_2 (d) Mg_2N_3 (e) Mg_3N_2

Mg : [Ne] $3s^2$ N : [He] $2s^22p^3$

3 Mg (s) + N_2 (g) → $Mg_3 N_2$

   0   0   +2   −3   Answer(e)

6-14. Predict the product of the following reaction: Sr (s) + P_4 (s) →

Sr : [Kr] $5s^2$ P : [Ne] $3s^23p^3$

6 Sr (s) + P_4 (s) → $2Sr_3 P_2$

   0   0   +2   −3

6-15. What would be the product of the reaction between Al metal and sulfur?

Al : [Ne] $3s^23p^1$ S : [Ne] $3s^23p^4$

16 Al (s) + 3 S_8 (s) → 8 $Al_2 S_3$

   0   0   +3   −2

6-16. What compound is formed by the reaction of Na with P_4?

Na : [Ne] $3s^1$ P : [Ne] $3s^23p^3$

12Na (s) + P_4(s) → 4Na_3 P

   0   0   +1   −3

6-17. Use the positions of silicon and fluorine in the periodic table to predict the most likely product of the reaction between these elements.

Si : [Ne] $3s^23p^2$ F : [He] $2s^22p^5$

Si + 2F_2 → SiF_4

6-18. Use the positions of gallium and oxygen in the periodic table to predict the formula for gallium oxide.

Ga : [Ar] $3d^{10}4s^24p^1$

O : [He] $2s^22p^4$

$$4Ga + 3O_2 \quad \rightarrow \quad 2Ga_2O_3$$

$$0 \qquad 0 \qquad\qquad +3 \ -2$$

6-19. An element, X, forms an ionic compound CaX. X is a member of which group?

Ca: $[Ar]4s^2$

Ca   X

+2 −2

X should be a member of group 16.

6-20. Which of the following elements would be most likely to form an oxide with the formula XO and a hydride with the formula XH_2?

  (a) Na   (b) Mg   (c) Al   (d) Si   (e) P

$O : [He]\ 2s^2 2p^4$ \qquad\qquad $H : 1s^1$

X   O \qquad\qquad\qquad X H_2

+2 −2 \qquad\qquad\qquad +2 −1

$Na : [Ne]\ 3s^1$ \qquad $Mg : [Ne]\ 3s^2$ \qquad $Al : [Ne]\ 3s^2 3p^1$ \quad $Si : [Ne]\ 3s^2 3p^2$ \quad $P : [Ne]\ 3s^2 3p^3$

The oxidation number of the element in the products = +2

The element should be Mg. \hfill Answer(b)

6-21. An element reacts with hydrogen and oxygen to form ionic compounds with the formula MH_4 and MO_2. In which group of the periodic table does this element belong?

$H : 1s^1$ \qquad\qquad $O : [He]\ 2s^2 2p^4$

M H_4 \qquad\qquad\qquad M O_2

+4 −1 \qquad\qquad\qquad +4 −2

The oxidation number of the element in the products = +4. It should belong to group 14.

6-22. What group of metals react with sulfur to form M_2S_3 sulfides, react with fluorine to form MF_3 fluorides and react with acid to form M^{3+} ions and H_2 gas?

$S : [Ne]\ 3s^2 3p^4$

$F : [He]\ 2s^2 2p^5$

M_2 S_3 \qquad\qquad\qquad M F_3

+3 −2 \qquad\qquad\qquad +3 −1

The oxidation number of the metals in the products = +3. They should belong to group 13.

6-23. Which of the following elements would be the most likely to react with hydrogen to form a compound with the formula XH that reacts with water to produce a basic solution and H_2 gas?

(a) Na   (b) Mg   (c) Al   (d) Si   (e) P

	Compound with hydrogen	
Na	NaH	$NaH + H_2O \rightarrow NaOH + H_2$ (g)
Mg	MgH_2	
Al	AlH_3	
Si	SiH_4	
P	PH_3	Answer(a)

6-24. Magnesium metal reacts with hydrogen gas to form a white solid (A) with high melting point. Compound A reacts with water to form compound B, which is a gas and an aqueous solution of compound C, which is another white solid. Magnesium metal reacts with hydrochloric acid to form gas B and an aqueous solution of compound D. Identify compounds A, B, C, and D and write balanced chemical equations for each reaction.

Mg (s) + H_2 (g) $\rightarrow$ MgH_2 (s)

MgH_2 (s) + $2H_2O$ (l) $\rightarrow$ $Mg(OH)_2$ (aq) + $2H_2$ (g)

Mg (s) + $2HCl$ (aq) $\rightarrow$ $MgCl_2$ (aq) + H_2 (g)

A: MgH_2(s)   B : H_2 (g)   C : $Mg(OH)_2$ D : $MgCl_2$

6-25. Which of the following equations represents the reaction that would be expected when Ca reacts with H_2O?

(a) Ca (s) + $2H_2O$ (l) $\rightarrow$ Ca^{2+} (aq) + $2OH^-$ (aq)
(b) Ca (s) + $2H_2O$ (l) $\rightarrow$ CaH_2 (aq) + $2OH^-$ (aq)
(c) Ca (s) + $2H_2O$ (l) $\rightarrow$ $Ca(OH)_2$ (aq) + $2H^+$ (aq)
(d) Ca (s) + $2H_2O$ (l) $\rightarrow$ $Ca(OH)_2$ (aq) + H_2 (g)
(e) None of these equations corresponds to the reaction expected

Ca reacts with water at room temperature.

$Ca \rightarrow Ca^{2+} + 2e^-$

$2H_2O + 2e^- \rightarrow H_2 + 2OH^-$

Ca (s) + $2H_2O$ (l) $\rightarrow$ Ca^{2+} (aq) + $2OH^-$ (aq) + H_2 (g)   Answer(d)

6-26. Which of the following would not be a product of the reaction between potassium metal and an aqueous solution of hydrochloric acid?

(a) K^+ (aq) (b) H_2 (g) (c) O_2 (g) (d) Cl^- (aq)

$2K$ (s) $+ 2HCl$ (aq) $\rightarrow 2K^+$ (aq) $+ 2Cl^-$ (aq) $+ H_2$ (g)   Answer(c)

6-27. Which of the following does not form OH^- ions when dissolved in water?

(a) LiH (b) Li_2O (c) Li_3N

(d) All of the above form OH^- ions when dissolved in H_2O

$LiH + H_2O$ (l) $\rightarrow Li^+$ (aq) $+ OH^-$ (aq) $+ H_2$ (g)

$Li_2O + H_2O$ (l) $\rightarrow 2Li^+$ (aq) $+ 2OH^-$ (aq)

$Li_3N + 3H_2O$ (l) $\rightarrow NH_3$ (g) $+ 3Li^+$ (aq) $+ 3OH^-$ (aq)   Answer(d)

6-28. Which of the following reactions does not occur?

(a) $Ca(s) + Cl_2$ (g) $\rightarrow CaCl_2$
(b) CaO (s) $+ H_2O$ (l) $\rightarrow Ca(OH)_2$ (aq)
(c) CaH_2 (s) $+ 2H_2O$ (l) $\rightarrow Ca(OH)_2$ (aq) $+ 2H_2$ (g)
(d) Ca (s) $+ 2H_2O$ (l) $\rightarrow Ca^{2-}$ (aq) $+ 2H_3O^+$ (aq)
(e) All of these reactions occur as written

$Ca \rightarrow Ca^{2+} + 2e^-$

$2H_2O + 2e^- \rightarrow H_2 + 2OH^-$

Ca (s) $+ 2H_2O$ (l) $\rightarrow Ca^{2+}$ (aq) $+ 2\ OH^-$ (aq) $+ H_2$ (g)   Answer(d)

6-29. Which of the following describes the products of the reaction of KH with water?

(a) K^+,H^+,OH^- (b) K^+,H_2,OH^- (c) K^+,OH^-,H^+ (d) K^- and H_3O^+ (e) K and H_2

KH (s) $+ H_2O$ (l) $\rightarrow K^+$ (aq) $+ OH^-$ (aq) $+ H_2(g)$ Answer(b)

6-30. Which of the following describes what happens when sodium hydride is added to water at room temperature?

(a) There is no discernable reaction
(b) O_2 is evolved and the solution becomes basic
(c) O_2 and H_2 are evolved and the solution remains neutral
(d) H_2 is evolved and the solution becomes basic
(e) The solution becomes basic, but no gas is evolved

NaH (s) $+ H_2O$ (l) $\rightarrow Na^+$ (aq) $+ OH^-$ (aq) $+ H_2(g)$ Answer(d)

6-31. What are the products of the reaction between sodium peroxide and water?

(a) Sodium, oxygen, and hydrogen

140

(b) Sodium hydroxide and oxygen

(c) Sodium hydroxide and hydrogen

(d) Sodium oxide and hydrogen

(e) An aqueous solution of sodium peroxide

An aqueous solution of sodium peroxide.   Answer(e)

6-32. Which of the following describes the products of the reaction of Na_2O with water?

(a) Na^+ and O^{2-} (b) Na^+, H_2, and OH^- (c) Na^+ and OH^-

(d) Na^+, H_2, and H^+ (e) Na and O_2

$Na_2O + H_2O$ (l) $\rightarrow 2Na^+$ (aq) $+ 2\,OH^-$ (aq)   Answer(c)

6-33. Which of the following ions or molecules is not formed when Li_3N dissolves in water?

(a) Li^+ (aq)   (b) H^+ (c) OH^- (aq)   (d) NH_3 (aq)

$Li_3N + 3H_2O$ (l) $\rightarrow NH_3$ (aq) $+ 3\,Li^+$ (aq) $+ 3OH^-$ (aq) H^+ is not formed.   Answer(b)

6-34. Which of the following compounds has sulfur in a +6 oxidation state?

(a) SO_2 (b) H_2S (c) H_2SO_3 (d) H_2SO_4 (e) $H_2S_2O_3$

S O₂	H₂ S	H₂ S O₃	H₂ S O₄	H₂ S₂ O₃
+4 −2	+1 −2	+1 +4 −2	+1 +6 −2	+1 +2 −2

Answer(d)

6-35. For the reaction $3S$ (s) $+ 2KClO_3$ (s) $\rightarrow 3SO_2$ (g) $+ 2KCl$ (s), which of the following is

true?

(a) S is reduced   (b) $KClO_3$ is the reducing agent   (c) S is the oxidizing agent

(d) All of the above are true   (e) None of the above are true

$3S + 2K\,Cl\,O_3 \rightarrow 3S\,O_2 + 2K\,Cl$

0   +1 +5 −2   +4 −2   +1 −1

In this reaction S is oxidized by $KClO_3$ (oxidizing agent) to SO_2.   Answer(e)

6-36. In the following oxidation-reduction reaction,

Sr (s) $+ 2H_2O$ (l) $\rightarrow Sr^{2+}$ (aq) $+ 2OH^-$ (aq) $+ H_2$ (g)

which of the following statements is true?

(a) H_2O is oxidized   (b) Sr acts as the oxidizing agent   (c) H_2 is reduced

(d) H_2O is the oxidizing agent   (e) More than one of the above is true

$$Sr\ (s)\ +\ 2H_2O\ (l)\ \rightarrow\ Sr^{2+}\ (aq)\ +\ 2OH^-\ (aq)\ +\ H_2\ (g)$$

$$0 \qquad\quad +1\ -2 \qquad\quad +2 \qquad\qquad -2\ +1 \qquad\quad 0$$

In this reaction Sr is oxidized by H_2O (oxidizing agent) to Sr^{2+}.   Answer(d)

6-37. Phosphorus is reduced in which of the following reactions?

    (a) $P_4\ (s)\ +\ 5O_2\ (g)\ \rightarrow\ P_4O_{10}\ (s)$
    (b) $P_4\ (s)\ +\ 3O_2\ (g)\ \rightarrow\ P_4O_6\ (s)$
    (c) $P_4\ (s)\ +\ 6F_2\ (g)\ \rightarrow\ 4PF_3\ (g)$
    (d) $P_4\ (s)\ +\ 6Ca\ (s)\ \rightarrow\ 2Ca_3P_2\ (s)$
    (e) $P_4\ (s)\ +\ 6Cl_2\ (g)\ +\ 12H_2O\ (l)\ \rightarrow\ 4H_3PO_3\ (aq)\ +\ 12HCl\ (aq)$

$$P_4\ (s)\ +\ 5O_2\ (g)\ \rightarrow\ P_4\ O_{10}\ (s)$$

$$0 \qquad\quad 0 \qquad\qquad +5\ -2 \qquad\qquad\qquad\qquad\qquad\text{P is oxidised}$$

$$P_4\ (s)\ +\ 3O_2\ (g)\ \rightarrow\ \ P_4\ O_6\ (s)$$

$$0 \qquad\quad 0 \qquad\qquad +3\ -2 \qquad\qquad\qquad\qquad\qquad\text{P is oxidised}$$

$$P_4\ (s)\ +\ 6F_2\ (g)\ \rightarrow\ \ 4P\ \ F_3\ (g)$$

$$0 \qquad\quad 0 \qquad\qquad +3\ -1 \qquad\qquad\qquad\qquad\qquad\text{P is oxidised}$$

$$P_4\ (s)\ +\ 6Ca\ (s)\ \rightarrow\ 2Ca_3P_2\ (s)$$

$$0 \qquad\quad 0 \qquad\qquad +2\ \ -3 \qquad\qquad\qquad\qquad\qquad\text{P is reduced}$$

$$P_4\ (s)\ +\ 6Cl_2\ (g)\ +\ 12H_2O\ (l)\ \rightarrow\ 4H_3\ P\ O_3\ (aq)\ +\ 12H\ Cl\ (aq)$$

$$0 \qquad\quad 0 \qquad\qquad +1\ -2 \qquad\quad +1\ +3\ -2 \qquad\quad +1\ -1 \qquad\text{P is oxidised} \qquad\text{Answer(d)}$$

6-38. Which is not an oxidation-reduction reaction?

    (a) $2Na\ +\ Cl_2\ \rightarrow\ 2NaCl$
    (b) $2HCl\ \rightarrow\ H_2\ +\ Cl_2$
    (c) $2NO_2\ +\ H_2O\ \rightarrow\ HNO_3\ +\ HNO_2$
    (d) $Na_2CO_3\ +\ 2HCl\ \rightarrow\ 2NaCl\ +\ CO_2\ +\ H_2O$
    (e) All of these are oxidation reduction reactions

$$2Na\ +\ Cl_2\ \rightarrow\ 2Na\ Cl$$

$$0 \qquad\quad 0 \qquad\quad +1\ -1 \qquad\qquad\qquad\qquad\text{oxidation-reduction reaction}$$

$$2HCl\ \rightarrow\ H_2\ +\ Cl_2$$

$$+1\ -1 \qquad 0 \quad\ 0 \qquad\qquad\qquad\qquad\text{oxidation-reduction reaction}$$

$$2N\ O_2\ +\ H_2\ O\ \rightarrow\ H\ N\ O_3\ +\ H\ N\ O_2$$

$$+4\ -2 \quad +1\ -2 \quad +1\ +5\ -2 \quad +1\ +3\ -2 \qquad\text{oxidation-reduction reaction}$$

$Na_2CO_3 + 2HCl \rightarrow 2NaCl + CO_2 + H_2O$

+1 +4 −2   +1 −1   +1 −1   +4 −2   +1 −2   not an oxidation-reduction reaction   Answer(d)

6-39. Which of the following reactions involves oxidation-reduction? For those that are redox reactions, identify which reagent is oxidized and which reagent is reduced.

   (a) Ca_3P_2 (s) + $6H_2O$ (l) → $3Ca(OH)_2$ (aq) + $2PH_3$ (g)
   (b) PH_3 (g) + $2O_2$ (g) → H_3PO_4 (s)
   (c) PH_3 (g) + HCl (g) → PH_4Cl (s)
   (d) P_4 (s) + $5O_2$ (g) → P_4O_{10} (s)

Ca_3P_2 (s) + $6H_2O$ (l) → $3Ca(OH)_2$ (aq) + $2PH_3$ (g)

+2 −3   +1 −2   +2 −2 +1   −3 +1   Not an oxidation-reduction reaction.

 PH_3 (g) + $2O_2$ (g) → H_3PO_4 (s)

−3 +1   0   +1 +5 −2   Oxidation-reduction reaction; PH_3 is

                                                    oxidized and O_2 is reduced.

PH_3 (g) + HCl (g) → PH_4Cl (s)

−3 +1   +1 −1   −3 +1 −1   Not an oxidation-reduction reaction.

P_4 (s) + $5O_2$ (g) → P_4O_{10} (s)

0   0   +5 −2   Oxidation-reduction reaction; P_4 is

                                                    oxidized and O_2 is reduced.   Answer(b and d)

6-40. Which of the following statements about the following reaction is false?

$2Na$ (s) + $2NH_3$ (l) → $2NaNH_2$ (s) + H_2 (g)

   (a) Na is the reducing agent
   (b) NH_3 is oxidized
   (c) Na is better reducing agent than H_2
   (d) There is no change in the oxidation state of nitrogen
   (e) This is not a redox reaction

$2Na$ (s) + $2NH_3$ (l) → $2NaNH_2$ (s) + H_2 (g)

0   −3 +1   +1 −3 +1   0

In this reaction Na is oxidized and NH_3 is the oxidizing agent.   Answer(b)

6-41. Identify the reducing agent and the oxidizing agent in the following reaction:

H_2O_2 (aq) + 2HI (aq) → $2H_2O$ (l) + I_2 (s)

H_2O_2 (aq) + 2H I (aq) → $2H_2O$ (l) + I_2 (s)

+1 −1   +1 −1   +1 −2   0

In this reaction I^- is oxidized to I_2. HI is the reducing agent and H_2O_2 is the oxidizing agent.

6-42. Identify the conjugate oxidizing agent for each of the following reducing agents.

   (a) Na   (b) Zn   (c) H^- (d) Sn^{2+}

2Na (s) + $2H_2O$ (l) → $2Na^+$ (aq) + $2OH^-$ (aq) + H_2(g) Na^+ is the conjugate oxidizing agent.

Zn (s) + 2HCl (aq) → Zn^{2+} (aq) + $2Cl^-$ (aq) + H_2 (g) Zn^{2+} is the conjugate oxidizing agent.

NaH (s) + H_2O (l) → Na^+ (aq) + OH^- (aq) + H_2 (g) H_2 is the conjugate oxidizing agent.

$2Fe^{3+} + Sn^{2+} → Sn^{4+} + 2Fe^{2+}$ Sn^{4+} is the conjugate oxidizing agent.

6-43. Use the table of relative reducing strengths to determine which (if any) of the following

reactions should occur as written.

   (a) Mg (s) + 2NaCl (s) → $MgCl_2$ (s) + 2Na (s)
   (b) 3Na (s) + $AlCl_3$ (l) → 3NaCl (s) + Al (s)
   (c) Al_2O_3 (s) + 2Fe (s) → Fe_2O_3 (s) + 2Al (s)
   (d) Pb (s) + 2NaCl (s) → $PbCl_2$ (s) + 2Na (s)
   (e) All of these reactions should occur as written
   (f) None of these reactions should occur as written

$$E^0_{red}$$

$Na^+ + e^- \rightleftharpoons$ Na   − 2.7109

$Mg^{2+} + 2e^- \rightleftharpoons$ Mg   − 2.3705

$Al^{3+} + 3e^- \rightleftharpoons$ Al   − 1.706

$Pb^{2+} + 2e^- \rightleftharpoons$ Pb   − 0.1263

$Fe^{3+} + 3e^- \rightleftharpoons$ Fe   − 0.036

Reducing strength Na>Mg>Al>Pb>Fe

Mg can not reduce Na^+ Pb can not reduce Na^+ Fe can not reduce Al^{3+}

But Na can reduce Al^{3+} to Al. Only reaction (b) can occur as written.   Answer(b)

6-44. Which of the following reducing agents are not strong enough to reduce Fe_2O_3 to Fe?

   (a) Na   (ii) Mg   (iii) Al   (iv) Ag   (v) H_2

$$E^0_{red}$$

$Na^+ + e^- \rightleftharpoons$ Na   − 2.7109

$$Mg^{2+} + 2e^- \rightleftharpoons Mg \qquad\qquad -2.3705$$

$$Al^{3+} + 3e^- \rightleftharpoons Al \qquad\qquad -1.706$$

$$Fe^{3+} + 3e^- \rightleftharpoons Fe \qquad\qquad -0.036$$

$$2H^+ + 2e^- \rightleftharpoons H_2 \qquad\qquad 0.000$$

$$Ag^+ + e^- \rightleftharpoons Ag \qquad\qquad 0.7996$$

Reducing strength: $Na>Mg>Al>Fe>H_2>Ag$

So only Na, Mg, and Al can reduce Fe_2O_3 to Fe. Ag and H_2 are not strong enough to reduce

Fe_2O_3 to Fe. Answer(iv and v)

6-45. Which of the following oxides can be reduced to the metal with H_2?

(a) Na_2O (b) MgO (c) Al_2O_3 (d) Fe_2O_3 (e) HgO

$$E^0_{red}$$

$$Na^+ + e^- \rightleftharpoons Na \qquad\qquad -2.7109$$

$$Mg^{2+} + 2e^- \rightleftharpoons Mg \qquad\qquad -2.3705$$

$$Al^{3+} + 3e^- \rightleftharpoons Al \qquad\qquad -1.706$$

$$Fe^{3+} + 3e^- \rightleftharpoons Fe \qquad\qquad -0.036$$

$$2H^+ + 2e^- \rightleftharpoons H_2 \qquad\qquad 0.000$$

$$Hg^{2+} + 2e^- \rightleftharpoons Hg \qquad\qquad 0.851$$

Reducing strength: $Na>Mg>Al>Fe>H_2>Hg$

So only HgO can be reduced to Hg by H_2. Answer(e)

6-46. Which of the following reducing agents should be able to reduce Sn^{2+} to tin metal?

(i) Na (ii) Mg (iii) Al (iv) Fe (v) Hg

$$E^0_{red}$$

$$Na^+ + e^- \rightleftharpoons Na \qquad\qquad -2.7109$$

$$Mg^{2+} + 2e^- \rightleftharpoons Mg \qquad\qquad -2.3705$$

$$Al^{3+} + 3e^- \rightleftharpoons Al \qquad\qquad -1.706$$

$$Fe^{3+} + 3e^- \rightleftharpoons Fe \qquad\qquad -0.036$$

$$Sn^{2+} + 2e^- \rightleftharpoons Sn \qquad\qquad -0.1364$$

$$Hg^{2+} + 2e^- \rightleftharpoons Hg \qquad\qquad 0.851$$

Reducing strength: Na>Mg>Al>Fe>Sn>Hg

So Na, Mg, Al and Fe will be able to reduce Sn^{2+} to Sn.   Answer(i,ii,iii, and iv)

6-47. Which element or ion is the strongest reducing agent?

(a) K^+ (b) K   (c) H_2 (d) Ag^+ (e) Ag

	E^0_{red}
$K^+ + e^- \rightleftharpoons K$	-2.924
$2H^+ + 2e^- \rightleftharpoons H_2$	0.000
$Ag^+ + e^- \rightleftharpoons Ag$	0.7996

K is the strongest reducing agent. K^+, H^+, and Ag^+ are conjugate oxidizing agents.   Answer(b)

6-48. Which of the following reactions will occur as written?

(a) $Zn\ (s) + 2NaCl \rightarrow ZnCl_2\ (s) + 2Na\ (s)$
(b) $Fe_2O_3\ (s) + 3Cu\ (s) \rightarrow 2Fe\ (s) + 3CuO\ (s)$
(c) $3K\ (s) + AlCl_3 \rightarrow 3KCl\ (s) + Al\ (s)$
(d) All of these reactions will occur as written
(e) None of these reactions will occur as written

	E^0_{red}
$K^+ + e^- \rightleftharpoons K$	-2.924
$Na^+ + e^- \rightleftharpoons Na$	-2.7109
$Al^{3+} + 3e^- \rightleftharpoons Al$	-1.706
$Zn^{2+} + 2e^- \rightleftharpoons Zn$	-0.7628
$Fe^{3+} + 3e^- \rightleftharpoons Fe$	-0.036
$Cu^{2+} + 2e^- \rightleftharpoons Cu$	0.3402
$Hg^{2+} + 2e^- \rightleftharpoons Hg$	0.851

Reducing strength: K>Na>Al>Zn>Fe>Cu

So Zn can not reduce Na^+ to Na. Cu can not reduce Fe^{3+} to Fe.

K can reduce Al^{3+} to Al.   Answer(c)

6-49. Metal ores are roasted to convert sulfides into corresponding oxides.

$2ZnS\ (s) + 3O_2\ (g) \rightarrow 2ZnO\ (s) + 2SO_2\ (g)$

Is this oxidation-reduction reaction? If yes, what is oxidized and what is reduced?

$2ZnS (s) + 3O_2 (g) \rightarrow 2ZnO (s) + 2SO_2 (g)$

+2 −2   0   +2 −2   +4 −2   Oxidation-reduction reaction.

O_2 is reduced and S^{2-} is oxidized.

6-50. Which of the following is the best explanation for the reaction between Na and Cl_2 to

form NaCl.

(a) Sodium likes to give up electrons to form Na^+ ions
(b) Enough energy is given off when neutral chlorine atom picks up an electron to form Cl^- to
    remove an electron from a neutral sodium to form a Na^+ ion
(c) It takes enormous energy to generate Na^+ and Cl^- ions, but once this is done, even more energy
    is given off when these ions come together to form an ionic lattice
(d) Chemists would agree that the above explanations are all equally valid
(e) None of the above explanations are valid

$Na (s) \rightarrow Na (g)$	$\Delta H = 107.3$ kJ/mol
$\frac{1}{2} Cl_2 (g) \rightarrow Cl (g)$	$\Delta H = 121.7$ kJ/mol
$Na (g) \rightarrow Na^+ (g) + e^-$	$\Delta H = 495.8$ kJ/mol
$Cl (g) + e^- \rightarrow Cl^- (g)$	$\Delta H = -348.8$ kJ/mol
$Na (s) + \frac{1}{2} Cl_2 (g) \rightarrow Na^+ (g) + Cl^- (g)$	$\Delta H = 376$ kJ/mol
$Na^+ (g) + Cl^- (g) \rightarrow NaCl (s)$	$\Delta H = -787.3$ kJ/mol (Lattice energy)
$Na (s) + \frac{1}{2} Cl_2 (g) \rightarrow NaCl (s)$	$\Delta H = -411.3$ kJ/mol Answer(c)

6-51. The lattice energy of NaCl is a measure of the energy given off in which of the following

reactions?

(a) $Na (s) + Cl_2 (s) \rightarrow 2NaCl (s)$
(b) $Na^+ (s) + Cl^- (s) \rightarrow NaCl (s)$
(c) $Na (g) + Cl (g) \rightarrow NaCl (g)$
(d) $Na^+ (g) + Cl^- (g) \rightarrow NaCl (g)$
(e) $Na^+ (g) + Cl^- (g) \rightarrow NaCl (s)$

Lattice energy of NaCl is a measure of the energy given off when one mole of NaCl (s) is

formed from one mole of gaseous Na^+ and one mole of gaseous Cl^- ions.   Answer(e)

6-52. Which of the following compounds will have the smallest lattice energy?

(a) CsI   (b) NaI   (c) NaF   (d) BaTe   (e) MgO

Lattice energy will be smaller when the ions are larger and will increase when the charge on

the ions increase. So CsI will have the smallest lattice energy.   Answer(a)

6-53. Which of the following ionic solids would have the highest lattice energy?

(a) NaI   (b) MgS   (c) BeO   (d) CsBr   (e) MgO

The lattice energy will be higher when the ions are smaller and will increase rapidly when the charge on the ions increase. So BeO will have the highest lattice energy.   Answer(c)

6-54. Which of the following has the largest lattice energy (assume the negative ions are all approximately the same size)?

(a) $NaClO_4$ (b) Na_2SO_4 (c) NaH_2PO_4 (d) Na_3PO_4 (e) $CsClO_4$

The lattice energy will be higher when the ions are smaller and will increase rapidly when the charge on the ions increase. So Na_3PO_4 will have the highest lattice energy.   Answer(d)

6-55. The lattice energy of Cs_2Te (interatomic distance = 0.390 nm) should be closest to that of

(a) CsI (interatomic distance = 0.385 nm)
(b) Cs_2O (interatomic distance = 0.309 nm)
(c) LiF (interatomic distance = 0.196 nm)
(d) Li_2Te (interatomic distance = 0.281 nm)
(e) SrI_2 (interatomic distance = 0.400 nm)

The lattice energy of ionic solid A will be closest to the lattice energy of B, if the interatomic distance and the charges on the ions are almost equal in both ionic solids.   Answer(e)

6-56. What is the maximum possible oxidation state of a P atom?

(a) +2   (b) +3   (c) +4   (d) +5   (e) None of the above

P : [Ne] $3s^2 3p^3$ P has 5 valence electrons.

For P, apart from 3s and 3p orbitals, 3d orbitals are available for bonding. So P can expand its valence shell to hold more than 8 electrons.

Maximum possible oxidation state is +5; e.g., PF_5, P_4O_{10} Answer(d)

6-57. What is the oxidation state of Mo atom in Li_2MoO_4?

Li_2 Mo O_4

+1 +6 −2   The oxidation state of Mo is +6.

6-58. All but one of the following species contains nitrogen in the same oxidation state. Which one is different?

(a) HNO_2 (b) NH_2^- (c) NH_3 (d) NH_4Cl

148

H N O_2	N H_2^-	N H_3	N H_4 Cl	
+1 +3 −2	− 3 +1	−3 +1	−3 +1 −1	Answer(a)

6-59. Calculate the oxidation number of the metal atom in each of the following compounds?

(a) $Re_2Cl_8^{2-}$ (b) $Cr_2Cl_9^{3-}$ (c) $Mo_2Cl_8^{4-}$

Re_2 Cl_8^{2-}	Cr_2 Cl_9^{3-}	$Mo_2Cl_8^{4-}$
+3 −1	+3 −1	+2 −1

6-60. The active ingredient in Rolaids has the formula $NaAl(OH)_2CO_3$.

Calculate the oxidation state of Al atom in this compound.

Na Al (O H)$_2$ C O$_3$

+1 +3   −2 +1 +4 −2

6-61. Arrange the following compounds in order of increasing oxidation state of the carbon

atom.

(a) CO   (b) CO_2 (c) H_2CO (d) CH_3OH (e) CH_4

C O	C O_2	H$_2$ C O	C H$_3$ O H	C H$_4$
+2 −2	+4 −2	+1 0 −2	−2 +1 −2 +1	−4 +1

$CH_4<CH_3OH<H_2CO<CO<CO_2$

6-62. Which compounds contain hydrogen in a negative oxidation state?

(i) H_2S (ii) $LiAlH_4$ (iii) CH_4 (iv) CaH_2

Hydrogen is reduced by elements that are less electronegative to form compounds in which

its oxidation number is −1.

Electronegativity		Electronegativity		
H	2.2	Li	0.98	
S	2.58	Al	1.61	
C	2.55	Ca	1.0	Answer(ii and iv)

6-63.(i) Prussian blue is a pigment with the formula Fe_4 [Fe $(CN)_6$]$_3$. If this compound

contains the [Fe $(CN)_6$]$_3^{4-}$ ion, what is the oxidation state of the other four iron atoms?

Fe_4	[Fe $(CN)_6$]$_3$
+3	−4

(ii) Turnbull's blue is a pigment with the formula $Fe_3[Fe(CN)_6]_2$. This compound contains the $[Fe(CN)_6]_2^{3-}$ ion. What is the oxidation state of the other three iron atoms?

Fe₃   [Fe(CN)₆]₂

+2   −3

6-64. Calculate the oxidation state of iron in $BaFeO_4$.

Ba Fe O₄

+2 +6 −2

6-65. What is the oxidation state of Mn in $LiMnO_4$?

Li Mn O₄

+1 +7 −2

6-66. What is the oxidation state of P in $LiPF_6$?

Li   P   F₆

+1   +5   −1

6-67. What is the name for Li_3N?   Lithium Nitride

6-68. Which of the following formula/name combinations is incorrect?

   (a) H_2SO_3/Sulfurous acid   (b) HCO_3^-/Bicarbonate ion (c) $HBrO_3$/Bromic acid

   (d) ClO_4^-/ Hypochlorite ion   (e) $CuSO_4$/Copper (II) sulfate

ClO_4^- is perchlorate ion.   Answer(d)

6-69. What is the better name for each compound?

   (a) Phosphorus pentoxide (P_2O_5)
   (b) Iron Oxide (Fe_2O_3)
   (c) Sodium carbonate ($NaHCO_3$)
   (d) Chlorine monoxide (Cl_2O)
   (e) Copper bromide ($CuBr_2$)

P_2O_5 – Diphosphorus pentoxide Fe_2O_3 – Iron (III) oxide or ferric oxide

$NaHCO_3$ – Sodium-bi-carbonate Cl_2O – Dichlorine monoxide

$CuBr_2$ – Copper (II) bromide

6-70. Explain why $CaBr_2$ is calcium bromide but $FeBr_2$ has to be called Iron (II) bromide?

The typical oxidation state of calcium is +2, but the oxidation state of iron can be +2 or +3.

So we have to specify the oxidation state of iron.

6-71. Write the formulas for

    (a) Tetraphosphorus trisulfide   (b) Silicon dioxide   (c) Carbon disulfide

    (d)Carbon tetrachloride   (e) Phosphorus pentafluoride

(a) P_4S_3 (b) SiO_2 (c) CS_2 (d) CCl_4 (e) PF_5

6-72. Write the formulas for

    (a) Silicon tetrafluoride   (b) Sulfur hexafluoride   (c) Oxygen difluoride

    (d) Dichloro heptoxide   (e) Chlorinetrifluoride

(a) SiF_4 (b) SF_6 (c) OF_2 (d) Cl_2O_7 (e) ClF_3

6-73. Write the formula for

    (a) Tin (II) chloride   (b) Mercury (II) nitrate   (c) Tin (IV) sulfide

    (d) Chromium (III) oxide   (e) Iron (II) phosphide

(a) $SnCl_2$ (b) $Hg(NO_3)_2$ (c) SnS_2 (d) Cr_2O_3 (e) Fe_3P_2

6-74. Name the following?

    (a) KNO_3 (b) Li_2CO_3 (c) $BaSO_4$ (d) Na_2SO_3 (e) PbI_2

(a) Potassium nitrate   (b) Lithium carbonate   (c) Barium sulfate   (d) Sodium sulfite

(e) Lead (II) iodide

6-75. Name the following:

    (a) $AlCl_3$ (b) Na_3N (c) Ca_3P_2 (d) Li_2S (e) MgO

(a)Aluminum chloride   (b) Sodium nitride   (c) Calcium phosphide   (d) Lithium sulfide

(e) Magnesium oxide

6-76. Name the following:

    (a) Sb_2S_3 (b) $SnCl_2$ (c) SF_4 (d) $SrBr_2$ (e) $SiCl_4$

(a)Antimony sulfide   (b) Tin (II) chloride   (c) Sulfur tetrafluoride   (d) Strontium bromide

(e) Silicon tetrachloride

6-77. Write the formulas of the following common acids.

    (a) Acetic   (b) Hydrochloric   (c) Sulfuric   (d) Phosphoric   (e) Nitric

(a)CH_3COOH (b) HCl (c) H_2SO_4 (d) H_3PO_4 (e) HNO_3

CHAPTER 7
GASES

INVALUABLE INFORMATION

1 atm = 760 mm Hg = 760 torr = 101.325 kPa

STP: 0 °C and 1 atm

$N = 6.022142 \times 10^{23}$ /mol

$R = 0.08206$ L-atm/mol-K

$\{P + a\,(n^2/V^2)\}\,\{V - nb\} = nRT$

$a = 27\,R^2 T_c^2/\,64\,P_c$ and $b = RT_c/8P_c$

7-1. The atmospheric pressure was 745.8 mm Hg. Calculate the pressure in units of atm.

760 mm Hg = 1 atm

745.8 mm Hg = 745.8/760 atm = 0.9813 atm.

7-2. Assume that the volume of a balloon filled with H_2 is 1.00 L at 25 °C. Calculate the volume of the balloon when it is cooled to – 78 °C in a low temperature bath made by adding dry ice to acetone.

$V_1/V_2 = T_1/T_2$

$T_1 = 273.15 + 25 = 298.15$ $T_2 = 273.15 - 78 = 195.15$

$V_2 = V_1\,(T_2/T_1) = (195.15/298.15) \times 1 = 0.6545$ L

7-3. Calculate the value of the ideal gas constant R, if exactly 1 mole of ideal gas occupies a volume of 22.414 liters at 0 °C and 1 atm pressure.

   (i)   PV = nRT

$R = PV/nT = (1 \times 22.414)/(1 \times 273.15) = 0.08206$ L-atm/mol-K

   (i)   PV = nRT

$1\,cm^3 = 1\,mL$ $1\,dm = 10\,cm$ $1\,dm^3 = 10^3 cm^3 = 10^3\,mL = 1\,L$

$1J\ Pa^{-1} = (kg\ m^2\ s^{-2})\ (kg\ m^{-1}\ s^{-2})^{-1} = m^3$

$1dm = 10^{-1}\ m \quad 1dm^3 = 10^{-3}\ m^3 = 10^{-3}\ J\ Pa^{-1}$

$P = 1\ atm = 101.325 \times 10^3\ Pa$

$V = 22.414\ L = 22.414\ dm^3 = 22.414 \times 10^{-3}\ m^3 = 22.414 \times 10^{-3}\ J\ Pa^{-1}$

$R = PV/nT = \{(101.325 \times 10^3) \times (22.414 \times 10^{-3})\}/(1 \times 273.15) = 8.314\ J\ K^{-1}\ mol^{-1}$

7-4. What is the mass of O_2 that can be stored at 21 ^oC and 170 atm in a cylinder with a volume of 60.0 L?

$PV = nRT$

$n = PV/RT = (170 \times 60)/(0.08206 \times 294.15) = 422.57$ moles

Molar mass of $O_2 = 32$ g/mol

Mass of O_2 that can be stored $= 422.57 \times 32 = 13522$ g

7-5. Calculate the mass of air in a hot air balloon that has a volume of 4.00×10^5 L, at temperature 30 ^oC and pressure 748 mm Hg. Average molar mass of air is 29.0 g/mol.

760 mm Hg = 1 atm

748 mm Hg = 748/760 = 0.98421 atm

$n = PV/RT = (0.98421 \times 4.00 \times 10^5)/(0.08206 \times 303.15) = 15825.5$ mol

Molar mass of air = 29.0 g/mol

Mass of the air $= 15825.5 \times 29 \times 10^{-3}$ kg $= 458.9$ kg

7-6. Calculate the molecular weight of butane if 0.5813 g of this gas fills a 250.0 mL flask at 24.4 ^oC and a pressure of 742.6 mm Hg.

$n = PV/RT = (742.6 \times 250)/(760 \times 1000 \times 0.08206 \times 297.55) = 0.01$ mol

Molecular weight $\times$ 0.01 = 0.5813 g

Molecular weight = 0.5813/0.01 = 58.13 g/mol

7-7. Calculate the density in g/L of O_2 gas at 0 ^oC and 1 atm.

$n = PV/RT = 1/(.08206 \times 273.15) = 0.0446$ mol

Density $= 0.0446 \times 32 = 1.428$ g/L

7-8. Calculate the total pressure of a mixture that contains 1.00 g H_2 and 1.00 g He in a 5 L containerat 21 ^oC.

Total pressure $P_T = P_{Hydrogen} + P_{Helium}$

$P_{Hydrogen} = (1/2.02) \times (0.08206 \times 294.15)/5 = 2.39$ atm

$P_{Helium} = (1/4.00) \times (0.08206 \times 294.15)/5 = 1.21$ atm

$P_T = 2.39 + 1.21 = 3.6$ atm

7-9. Calculate the average velocity of a H_2 molecule at 0 °C if the average velocity of an O_2 molecule at this temperature is 500 m/s.

$v_{Hydrogen}/v_{Oxygen} = \sqrt{(m_{Oxygen}/m_{Hydrogen})} = \sqrt{(32/2.02)} = 3.98$

$v_{Hydrogen} = 3.98 \times 500$ m/s $= 1990$ m/s

7-10. Calculate the number of grams of O_2 that can be collected by displacing water from a 250 ml flask at 21 °C and 746.2 mm Hg.

$P_{Total} = P_{Oxygen} + P_{water}$ (vapor pressure of water at 21 °C).

$P_{Oxygen} = P_{Total} - P_{water}$ (vapor pressure of water at 21 °C).

$= 746.2$ mm Hg $- 18.65$ mm Hg

$= 727.55$ mm Hg

$n = PV/RT = \{(727.55/760) \times (250/1000)\}/(0.08206 \times 294.15) = 9.9 \times 10^{-3}$ mole

Molar mass of oxygen $= 32$ g/mole

Mass of $O_2 = 9.9 \times 10^{-3} \times 32 = 0.317$ g

7-11. Which of the following elements or compounds is most likely to be a gas at room temperature?

(a) A metal such as sodium (Na)
(b) An alloy of two metals such as bronze, which is a mixture of copper (Cu) and Tin (Sn)
(c) A salt such as MgO
(d) A nonmetal such as chlorine (Cl_2)
(e) None of these elements or compounds will be a gas at room temperature

Elements that are gases at room temperature are all non-metals.   Answer(d)

7-12. Which graph isn't a straight line for an ideal gas?

(a) V versus T (n and P constant)
(b) T versus P (n and V constant)
(c) P versus 1/V (n and T constant)
(d) n versus 1/T (P and V constant)
(e) n versus 1/P (V and T constant)

$PV = nRT$

So n versus 1/P will not be a straight line.   Answer(e)

7-13. If equal weights of O_2 and N_2 are placed in identical containers at the same temperature,

which of the following statements is true?

(MW: N_2 = 28.01 g/mol ; O_2 = 32.0 g/mol)

   (a) Both flasks contain the same number of molecules
   (b) The pressure in the flask that contains N_2 will be greater than the pressure in the flask that
       contains O_2.
   (c) There will be more molecules in the flask that contains O_2 than in the flask that contains N_2
   (d) This question can't be answered unless we know the weights of O_2 and N_2 in the flasks
   (e) None of the above

Let us consider 1 gm of N_2 and 1 gm of O_2 and V = 1 L and T = 298.15 K.

Number of N_2 molecules = $6.022 \times 10^{23}/28.01 = 2.15 \times 10^{22}$

Number of O_2 molecules = $6.022 \times 10^{23}/32.00 = 1.88 \times 10^{22}$

Number of mole of N_2 = 1/28.01

Number of mole of O_2 = 1/32.00

$PV = nRT$

Pressure in the flask that contains N_2 = $(1/28.01) \times 0.08206 \times 298.15 = 0.87$ atm

Pressure in the flask that contains O_2 = $(1/32.00) \times 0.08206 \times 298.15 = 0.76$ atm   Answer(b)

7-14. Two identical flasks are at the same temperature. One is filled with 1 mol of hydrogen

gas and the other with 1 mol of nitrogen gas. Which property would be different for the two

samples?

(MW: N_2 = 28.01 g/mol, H_2 = 2.02 g/mol)

   (a) Pressure   (b) Average kinetic energy   (c) Density

   (d) The number of molecules in each container   (e) The weight of the containers

There is one mole of hydrogen in one flask.

There is one mole of nitrogen in another flask.

Number of N_2 molecules = 6.022×10^{23}

Number of H_2 molecules = 6.022×10^{23}

Pressure in the flask that contains N_2 = $0.08206 \times T$ atm

Pressure in the flask that contains H_2 = 0.08206×T atm

Since T is same, pressure will also be same.

Two gases at the same temperature will have the same average kinetic energy.

Density in the flask that contains N_2 = 28.01/V g/L

Density in the flask that contains H_2 = 2.02/V g/L

Since V is same, density will be different for the two samples.   Answer(c)

7-15. Which would always lead to an increase in the average kinetic energy of a gas?

  (a) Increasing the volume by decreasing the pressure
  (b) Increasing the pressure by decreasing the volume
  (c) Increasing the pressure by increasing the number of molecules of gas
  (d) Increasing the volume by increasing the temperature of the gas
  (e) All of the above are equally effective ways of increasing the average kinetic energy of a gas

Average kinetic energy of a gas can be increased only by increasing the temperature. Answer(d)

7-16. Which isn't one of the postulates of the kinetic molecular theory?

  (a) At a constant temperature, all of the particles have the same speed
  (b) Gas particles are in constant motion
  (c) Gas particles move in straight line between collisions
  (d) The volume of the particles is negligibly small compared to the volume of the container
  (e) There are no forces of attraction between gas particles

The kinetic molecular theory assumes that the average kinetic energy of a collection of

gas particles depends on the temperature of the gas and nothing else. It does not assume that

at a constant temperature, all of the gas particles have the same speed.   Answer(a)

7-17. Which isn't the postulate of the kinetic molecular theory?

  (a) The mass of a molecule is negligibly small
  (b) The volume of a molecule is a negligibly small fraction of the volume of the container
  (c) The number of molecules in a gas is very large
  (d) The molecules are in a state of constant, random motion
  (e) The attractive forces between molecules is negligibly small

The kinetic molecular theory assumes that the volume of a molecule is a negligibly

small fraction of the volume of the container; it does not assume that the mass of a molecule

is negligibly small.   Answer(a)

7-18. Uranium reacts with fluorine to produce a compound that is a gas at 57 °C. The density of

this gas is 13.0 g/L at 57 °C and 1 atm pressure. What is the molecular formula of this compound? (AW: F= 19.0 amu, U = 238 amu)

(a) UF_2 (b) UF_3 (c) UF_4 (d) UF_5 (e) UF_6

$PV = nRT$

Let us consider 1 L of gas.

$n = 1/(0.08206 \times 330.15) = 0.0369$ mole

Molar mass $= 13/0.0369 = 352.3$

Number of F atoms $= (352.3 - 238)/19 = 6$

Molecular formula is UF_6.   Answer(e)

7-19. Which of the following gases would have the largest density at 25 ^{o}C and 1.00 atm

pressure.

(a) Methane CH_4 (b) Acetylene C_2H_2 (c) Ethylene C_2H_4 (d) Ethane C_2H_6

(e) Propane C_3H_8 (AW: H = 1.01 amu, C = 12.01 amu)

$PV = nRT$

Let us consider 1 L of gas.

$n = PV/RT = 1/ (0.08206 \times 298.15) = 0.041$ mole

Density g/L

Methane CH_4 : $\{12.01 + (1.01 \times 4)\} \times 0.041 = 0.658$

Acetylene C_2H_2 : $\{(12.01 \times 2) + (1.01 \times 2)\} \times 0.041 = 1.068$

Ethylene C_2H_4 : $\{(12.01 \times 2) + (1.01 \times 4)\} \times 0.041 = 1.150$

Ethane C_2H_6 : $\{(12.01 \times 2) + (1.01 \times 6)\} \times 0.041 = 1.233$

Propane C_3H_8 : $\{(12.01 \times 3) + (1.01 \times 8)\} \times 0.041 = 1.809$ Answer(e)

7-20. Which of the following gases has a density of 1.72 g/L at 10 ^{o}C and 1 atm pressure?

(AW: H = 1.01 amu, He = 4.00 amu, C =12.01 amu, Ne = 20.18 amu, P = 30.97 amu, Ar =

39.95 amu)

(a) He   (b) Ne   (c) Ar   (d) CH_4 (e) P_4

$PV = nRT$

Let us consider 1 L of gas. $n = PV/RT = 1/(0.08206 \times 283.15) = 0.043$ mole

Density g/L

He: $4.00 \times 0.043 = 0.17$

Ne: $20.18 \times 0.043 = 0.88$

Ar: $39.95 \times 0.043 = 1.72$

CH$_4$: $\{12.01 + (4 \times 1.01)\} \times 0.043 = 0.69$

P$_4$: $30.97 \times 4 \times 0.043 = 5.33$ Answer(c)

7-21. What is the density of NO (g) at 25 $^{\circ}$C and 783 mm Hg? (AW: N = 14.01 amu, O = 16.00 amu)

    (a) 1.26 g/L   (b) 2.68 g/L   (c) 3.12 g/L   (d) 3.76 g/L   (e) 22.4 g/L

$PV = nRT$

Let us consider 1 L of gas.

$n = PV/RT = (783/760)/(0.08206 \times 298.15) = 0.0421$ mole

Density of NO (g) = $(14.01 + 16.00) \times 0.0421 = 1.26$ g/L   Answer(a)

7-22. Which of the gases in Group 18 of the periodic table has a density of 3.74 g/L at 0 $^{\circ}$C and 1 atm?

    (a) He   (b) Ne   (c) Ar   (d) Kr   (e) Xe

$PV = nRT$

Let us consider 1 L of gas. $n = PV/RT = 1/(0.08206 \times 273.15) = 0.0446$ mole

Density g/L

He: $4.00 \times 0.0446 = 0.18$

Ne: $20.18 \times 0.0446 = 0.90$

Ar: $39.95 \times 0.0446 = 1.78$

Kr: $83.80 \times 0.0446 = 3.74$

Xe: $131.29 \times 0.0446 = 5.86$ Answer(d)

7-23. Calculate the molecular formula of diazomethane if this compound is 28.6% C, 4.8% H and 66.6% N by weight and the density of this gas is 1.72 g/L at 25 $^{\circ}$C and at 1 atm.

$PV = nRT$

Let us consider 1 L of gas. $n = PV/RT = 1/(0.08206 \times 298.15) = 0.0409$ mole

Molar mass $= 1.72/0.0409 = 42.05$ g/mole

Number of Carbon atoms $= \{(42.05) \times (28.6/100)\}/12.01 = 1$

Number of Hydrogen atoms $= \{(42.05) \times (4.8/100)\}/1.01 = 2$

Number of Nitrogen atoms $= \{(42.05) \times (66.6/100)\}/14.01 = 2$

Molecular Formula CH_2N_2

7-24. Methane (CH_4) combines with oxygen (O_2) to produce carbon-di-oxide (CO_2) and water (H_2O). When 3.0 L methane is burned in 5.0 L of oxygen how many L of carbon-di-oxide are formed, assuming that the volumes of all gases are measured at 400 $^{\circ}$C and 4.00 atm pressure? (AW: H = 1.01 amu, C = 12.01 amu, O = 16.00 amu)

   (a) 2.5 L   (b) 3.0 L   (c) 5.0 L   (d) 8.0 L   (e) None of the above

At constant temperature and pressure equal volumes of different gases contain the same number of particles.

CH_4 (g) + $2O_2$ (g) $\rightarrow CO_2$(g) + $2H_2O$ (g)

1 volume   2 volumes   1 volume   2 volumes

Oxygen is the limiting reagent. 5.0 L of oxygen will react with 2.5 L of methane to produce 2.5 L of carbon-di-oxide.   Answer(a)

7-25. The first step of the Ostwald process for making nitric acid involves a reaction between ammonia and oxygen to form nitrogen oxide and water.

$4NH_3$ (g) + $5O_2$ (g) $\rightarrow$ $4NO$ (g) + $6H_2O$ (g)

How many liters of O_2 gas at 650 $^{\circ}$C and 1.00 atm pressure are needed to react with 48 L of NH_3 at the same temperature and pressure?

   (a) 32 L   (b) 38 L   (c) 48 L   (d) 60 L   (e) 72 L

At constant temperature and pressure equal volumes of different gases contain the same number of particles.

4 NH_3 (g)   +   5 O_2 (g) $\rightarrow$ 4NO (g)   +   6H_2O (g)

4 volumes   5 volumes   4 volumes   6 volumes

5 volumes of oxygen are required to react with 4 volumes of ammonia.

So 60.00 L of oxygen will be required to react with 48 L of oxygen.   Answer(d)

7-26. An anesthetic is 85.63% carbon and 14.37% hydrogen by weight. Calculate the molecular formula if 0.45 L of this anesthetic reacts with excess oxygen at 120 °C and 0.72 atm to form 1.35 L of CO_2 and 1.35 L of water vapor.

At constant temperature and pressure equal volumes of different gases contain the same number of particles.

The compound contains 85.63 % carbon and 14.37 % hydrogen by weight. So the empirical formula is CH_2.

$2CH_2$ + $3O_2$ → $2CO_2$ + $2H_2O$

0.45 L   0.45 L   0.45 L

C_2H_4 + $3O_2$ → $2CO_2$ + $2H_2O$

0.45 L   0.90 L   0.90L

$2C_3H_6$ + $9O_2$ → $6CO_2$ + $6H_2O$

0.45 L   1.35 L   1.35 L

The correct molecular formula is C_3H_6.

7-27. A gas containing only carbon and hydrogen is an anesthetic. If 0.45 L of this gas at 120 °C and 0.72 atm reacts with excess O_2 to give 1.35 L of CO_2 (g) and 1.35 L of H_2O (g) at the same temperature and pressure, what is the percent by weight of carbon in this gas?

    (a) 33.3 %   (b) 46.8 %   (c) 45.0 %   (d) 85.6 %   (e) 90.0 %

aC_xH_y + bO_2 → cCO_2 + dH_2O

0.45 L   1.35 L   1.35 L

At constant temperature and pressure equal volumes of different gases contain the same number of particles. Number of moles of CO_2 produced is equal to the number moles of H_2O produced.

Number of moles of CO_2 produced = PV/RT = (0.72×1.35)/(0.08206×393.15) = 0.03

Weight of carbon in the gas = 12.01×0.03 = 0.3604

Weight of hydrogen in the gas = 2.01×0.03 = 0.0606

% of carbon in in the gas = {0.3604/(0.3604 + 0.0606)}×100 = 85.60   Answer(d)

7-28. Nitrogen gas reacts with hydrogen gas to produce gaseous ammonia. How many liters of product at STP can be formed if 6 L of hydrogen gas at STP reacts with a stochiometric amount of nitrogen gas?

(a) 1 L   (b) 2 L   (c) 3 L   (d) 4 L   (e) 6 L

At STP (constant temperature and pressure) equal volumes of different gases contain the same number of particles.

$N_2 (g)$ + $3H_2 (g)$ $\rightarrow$ $2NH_3 (g)$

1 volume   3 volumes   2 volumes

6 L of hydrogen will react with 2 L of nitrogen to produce 4L of ammonia.   Answer(d)

7-29. What is the empirical formula of the product of the following reaction

___$N_2O_3 (g)$ + ___$O_3 (g)$ $\rightarrow$

if 3 L of N_2O_3 react with 1 L of O_3 to give 6 L of the product, all gases measured at STP?

(a) NO   (b) NO_2 (c) NO_3 (d) N_2O_5 (e) None of the above

___$N_2O_3 (g)$ + ___$O_3 (g)$ $\rightarrow$

3 L   1 L

At STP (constant temperature and pressure) equal volumes of different gases contain the same number of particles. 3 moles of N_2O_3 will react with 1 mole of O_3.

$3N_2O_3 (g) + O_3 (g) \rightarrow 6NO_x$

x = 2

Product: NO_2 Answer(b)

7-30. 15 L of acetylene (C_2H_2) were burned in 15 L of O_2 to form CO_2 and H_2O.

$2C_2H_2 (g) + 5O_2 (g) \rightarrow 4CO_2 (g) + 2H_2O (g)$

What would be the total volume of products after the reaction mixture cools back to room temperature?

(a) 18 L   (b) 25 L   (c) 30 L   (d) 40 L   (e) 45 L

$2C_2H_2 (g)$ + $5O_2 (g)$ $\rightarrow$ $4CO_2 (g)$ + $2H_2O (g)$

At constant temperature and pressure equal volumes of different gases contain the same number of particles. O_2 is the limiting reagent. 15 L of O_2 (g) will react with 6 L of

C_2H_2 (g) to produce 12 L of CO_2 (g) and 6 L of H_2O (g).

Total volume of the products = 12 L + 6 L = 18 L   Answer(a)

7-31. What volume of NH_3 (g) at STP is required to prepare 49.8 g of $(NH_4)_2SO_4$ by the reaction of ammonia with H_2SO_4? (AW: H = 1.01 amu , N = 14.01 amu, O = 16.0 amu, S = 32.1 amu)

   (a) 0.377 L   (b) 6.41 L   (c) 8.4 L   (d) 12.84 L   (e) 16.9 L

$2NH_3 + H_2SO_4 \rightarrow (NH_4)_2SO_4$

Molecular weight of $(NH_4)_2SO_4$ = 132.2 g/mole

Number of mole of $(NH_4)_2SO_4$ to be prepared = 49.8/132.2

For ammonia PV = nRT

n = (2×49.8)/132.2

Volume of NH_3 required = (2×49.8×0.08206×273.15)/132.2 = 16.9 L   Answer(e)

7-32. Which sample would have the largest volume at 25 ^{o}C and 750 mm Hg?

   (a) 100 g CH_4 (b) 100 g CO_2 (c) 100 g NO   (d) 100 g SO_2

PV = nRT   V = nRT/P

100g CH_4:   V = (760/750)×(100/16)×0.08206×298.15 = 154.5 L

100g CO_2:   V = (760/750)×(100/44.01)×0.08206×298.15 = 56.3 L

100g NO: V = (760/750)×(100/30.01)×0.08206×298.15 = 82.6 L

100g SO_2:   V = (760/750)×(100/64.10)×0.08206×298.15 = 38.7 L   Answer(a)

7-33. What is the volume of 1 mole of an ideal gas at 25 ^{o}C and atm pressure?

   (a) 0.0409 L   (b) 2.05 L   (c) 22.414 L   (d) 24.466 L   (e) None of the above

PV = nRT

V = nRT/P = 0.08206×298.15/1 = 24.466 L   Answer(d)

7-34. A small gas cylinder of He used in chemistry lecture demonstrations has a volume of 275 mL at a pressure of 1823 kPa at 24.7 ^{o}C. The volume of He measured at 772 torr and 24.7 ^{o}C needed to fill the cylinder is:

   (a) 45.9 mL   (b) 303 mL   (c) 4870 mL   (d) 4.95×10^4 mL   (e) 4.86×10^5 mL

$P_1V_1 = P_2V_2$

$V_2 = P_1V_1/P_2 = (1823/101.325)\times(275/1000)\times(760/772) = 4.87$ L

$V_2 = 4870$ mL   Answer(c)

7-35. A small gas cylinder of He used in chemistry lecture demonstrations has a volume of 275 mL at a pressure of 1823 kPa at 24.7 $^\circ$C. At what temperature would the He gas cylinder exhibit a pressure of 25.0 atm?

   (a) 4.96 $^\circ$C   (b) 34.3 $^\circ$C   (c) 58.3 $^\circ$C   (d) 140.7 $^\circ$C   (e) 307 $^\circ$C

$T_1/T_2 = P_1/P_2$

$T_2 = T_1\times(P_2/P_1) = (297.85\times25)/(1823/101.325) = 413.87$ K

$T_2 = 140.7$ $^\circ$C   Answer(d)

7-36. A small gas cylinder used in chemistry lecture demonstrations has a volume of 275 mL at a pressure of 1823 kPa at 24.7 $^\circ$C. The number of grams of He in the cylinder is:

   (a) 0.202 g   (b) 0.81 g   (c) 1.24 g   (d) 4.96 g

$PV = nRT$ $n = PV/RT$

$n = (1823/101.325)\times(275/1000)/(0.08206\times297.85) = 0.2024$ mole

Amount of He in the cylinder = $4\times0.2024 = 0.81$ g   Answer(b)

7-37. Polypropylene is a plastic formed by polymerising propylene which has the empirical formula CH_2. If a sample of propylene with a mass of 21.0 g occupies a volume of 11.2 L at STP (0 $^\circ$C and 1 atm), what is the molecular formula of this gas?

   (a) CH_2 (b) C_2H_4 (c) C_3H_6 (d) C_4H_8 (e) None of these

$PV = nRT$

$n = (1\times11.2)/(0.08206\times273.15) = 0.5$ mole

Molecular weight = $21.0/0.5 = 42.0$ g/mol

Empirical formula weight = 14.03   Molecular formula: C_3H_6 Answer(c)

7-38. 2.91 g sample of a gaseous compound that contains only boron and hydrogen has a volume of 1.22 L at 25 $^\circ$C and 1.09 atm. What is the formula of this compound? (AW: H = 1.01 amu, B = 10.8 amu)

   (a) B_2H_6 (b) B_4H_{10} (c) B_5H_9 (d) B_6H_{10} (e) B_6H_{12}

$PV = nRT$ $n = PV/RT$

$n = (1.09 \times 1.22)/(0.08206 \times 298.15) = 0.0543$ Molecular weight $= 2.91/0.0543 = 53.6$ g/mol

Compound	Molecular weight g/mole	
B_2H_6	27.66	
B_4H_{10}	53.30	Answer(b)

7-39. For which of the following compounds would a 1.00 g sample occupy a volume of 390 cm^3 at 25 °C and 0.993 atm?

(a) B_2H_6 (b) B_4H_{10} (c) B_5H_9 (d) B_6H_{10} (e) B_6H_{12}

$PV = nRT$

$n = PV/RT = 0.993 \times (390/1000)/(0.08206 \times 298.15) = 0.01583$

Molecular weight $= 1/0.01583 = 63.2$ g/mol

Compound	Molecular weight g/mol	
B_2H_6	27.66	
B_4H_{10}	53.30	
B_5H_9	63.10	Answer(c)

7-40. Equal volumes of oxygen and an unknown gas weigh 3.00 grams and 7.5 g respectively. Which of the following is the unknown gas?

(a) CO_2 (b) NO   (c) NO_2 (d) SO_2 (e) SO_3

At constant temperature and pressure equal volumes of different gases contain the same number of particles.

O_2: Number of mole of oxygen $= 3/32 = 0.09375$ mole

Molecular weight of unknown gas $= 7.5/0.09375 = 80.0$ g/mol

Compound	Molecular weight g/mol	
CO_2	44.01	
NO	30.01	
NO_2	46.01	
SO_2	64.01	
SO_3	80.01	Answer(e)

7-41. What is the identity of an unknown metal if 1.00 g of this metal reacts with excess acid

164

according to the equation $M (s) + 2H^+ (aq) \rightarrow M^{2+} + H_2 (g)$ to produce 374 mL of H_2 gas at 25 °C and 1 atm pressure?

(a) Mg, 24.3 g/mol   (b) Ca, 40.1 g/mol   (c) Mn, 54.9 g/mol

(d) Zn, 65.3 g/mol   (e) Sr, 87.6 g/mol

$PV = nRT$ $n = PV/RT = 1\times(374/1000)/(0.08206\times298.15) = 0.0153$

Weight of 1 mole of M (s) = $1/0.0153 = 65.36$ g/mol   Answer(d)

7-42. Calculate the weight of a flask if the flask filled with oxygen weighs 125.00 g while the flask filled with argon weighs 125.384 g.

At constant temperature and pressure equal volumes of different gases contain the same number of particles.

Let 'x' be the weight of the flask.

$(125 - x)/32 = (125.384 - x)/39.948$

$x = 123.452$ g

7-43. Assume that several mL of ether are placed in a bulb with a volume of 293 mL, and the bulb is immersed in water at 36 °C until the last drop of the liquid disappears leaving the bulb filled with ether vapor. The bulb is then removed from the water bath, and the weight of ether that condenses in the bulb is measured. If 0.841 g of ether collect in this experiment at a pressure of 746 mm Hg, what is the molecular weight of ether?

$PV = nRT$ $n = PV/RT = (746/760)\times(293/1000)/(0.08206\times309.15) = 0.011337$

Molecular weight of ether = $0.841/0.011337 = 74.18$ g/mol

7-44. How many cm^3 of liquid SO_2 (d = 1.46 g/cm^3) can be obtained by compressing 1.00 L of the gas collected at 25 °C and 1 atm?

$PV = nRT$ $n = PV/RT = 1/(0.08206\times298.15)$

Molecular weight of $SO_2 = 64.1$ g/mol

Amount of $SO_2 = 64.1/(0.08206\times298.15)$ g

Volume of liquid $SO_2 = \{64.1/(0.08206\times298.15)\}/1.46 = 1.79$ cm^3

7-45. Calculate the volume of HBr that can be obtained by reacting 10.0 g of PBr_3 with excess water at 21 °C and 753 mm Hg.

PBr_3 (l) + $3H_2O$ (l) → $3HBr$ (g) + H_3PO_3 (aq)

Molecular weight of PBr_3 = 30.974 + (79.904)×3 = 270.686

Number of mole of PBr_3 = 10/270.686 = 0.03694

Number of mole of HBr produced = 3×0.03694

$PV = nRT$ $V = nRT/P = \{(3×0.03694)×0.08206×294.15\}/(753/760) = 2.65$ L

7-46. Calculate the volume of oxygen at 0 ^{o}C and 1 atm pressure that can be produced by decomposing 100 mL of H_2O_2 if this solution is 27.6% H_2O_2 by weight and the density is 1.09 g/cm^3.

$2H_2O_2$ (l) → $2H_2O$ (l) + O_2 (g)

Molecular weight of H_2O_2 = 34.02 g/mol

Number of mole of H_2O_2 = $\{1.09×100×(27.6/100)\}/34.02$

Number of mole of O_2 = $\{1.09×100×(27.6/100)\}/(34.02×2)$

$PV = nRT$ $V = nRT/P = \{1.09×100×(27.6/100)\}×0.08206×273.15/(34.02×2) = 9.91$ L

7-47. Calculate the volume of H_2S gas collected at 25 ^{o}C and 1 atm needed to precipitate all of the Cu^{2+} from a 250 mL of a 0.10 M Cu^{2+} solution.

Cu^{2+} (aq) + H_2S (g) → CuS (s) + $2H^+$ (aq)

$PV = nRT$ n = (0.10/1000)×250 = 0.025 mole

$V = nRT/P = (0.025×0.08206×298.15)/1 = 0.61$ L

7-48. 0.500 mole of Ar gas occupies a volume of 4.07 L at 25 ^{o}C and 3 atm pressure. What volume will it occupy at STP (0 ^{o}C and 1 atm pressure)?

  (a) 1.24 L   (b) 1.48 L   (c) 4.07 L   (d) 11.2 L   (e) 13.3 L

$PV = nRT$ $V = nRT/P = (0.5×0.08206×273.15)/1 = 11.2$ L   Answer(d)

7-49. What is the final temperature if a sample of ammonia gas, initially at a pressure of 3.00 atm, a temperature of 500 K, and a volume of 275 L is changed to a volume of 200 L and a presssure of 2.5 atm?

  (a) 303 K   (b) 436 K   (c) 573 K   (d) 825 K   (e) None of the above

$(P_1V_1)/(P_2V_2) = T_1/T_2$

$T_2 = T_1/\{(P_1V_1)/(P_2V_2)\} = (2.5×200×500)/(3×275) = 303$ K   Answer(a)

7-50. At 25 °C and one atm pressure, 4/5ths of the pressure of the atm is due to N_2 and 1/5th is due to O_2. What fraction of the pressure at 10 atm and 100 °C would be due to N_2?

(a)

$P = 1$ atm and $T_1 = 298.15$

$P_{Total} = P_{Nitrogen} + P_{Oxygen}$

$P_{Nitrogen} = n_{Nitrogen}RT_1/V_1 = (4/5) \times 1 = 0.8$

$P_{Oxygen} = n_{Oxygen}RT_1/V_1 = (1/5) \times 1 = 0.2$

$P_{Nitrogen} / P_{Oxygen} = n_{Nitrogen}/ n_{Oxygen} = 4$

(b)

$P = 10$ atm and $T_2 = 373.15$

$P_{Nitrogen} / P_{Oxygen} = n_{Nitrogen}/ n_{Oxygen} = 4$

$4P_{Oxygen} + P_{Oxygen} = P_{Total} = 5 P_{Oxygen}$

$P_{Oxygen} = (1/5)P_T$

$P_{Nitrogen} = (4/5) P_T$

7-51. A sealed container holds 150 g of ammonia, 150 g of CO_2 and 150 g of nitrogen gases. What is the mole fraction of CO_2? (AW: H = 1.01 amu, C = 12.01 amu, N = 14.01 amu, O = 16 amu)

   (a) 0.00758   (b) 0.0227   (c) 0.194   (d) 0.333   (e) 3.41

Number of moles of NH_3 = 150/{14.01 + (3×1.01)} = 8.80

Number of moles of CO_2 = 150/(12.01 + 32) = 3.41

Number of moles of N_2 = 150/28.02 = 5.35

Mole fraction of CO_2 = 3.41/(8.8+3.41+5.35) = 0.194   Answer(c)

7-52. What is the total pressure when 0.400 g of H_2, 2.00 g of N_2, and 10.5 g of CO_2 are injected into a 10.0 liter flask at 273 K?

   (a) 0.571 atm   (b) 0.877 atm   (c) 1.14 atm   (d) 1.75 atm

$P_{Total} = P_{Hydrogen} + P_{Nitrogen} + P_{Carbon-di-oxide}$

$P_{Hydrogen}$ = (0.400/2.02)×(0.08206×273)/10 = 0.4436

$P_{Nitrogen}$ = (2.0/28.02)×(0.08206×273)/10 = 0.1600

$P_{\text{Carbon-di-oxide}} = (10.5/44.01) \times (0.08206 \times 273)/10 = 0.5345$

$P_{\text{Total}} = P_{\text{Hydrogen}} + P_{\text{Nitrogen}} + P_{\text{Carbon-di-oxide}} = 0.4436 + 0.1600 + 0.5345 = 1.14$ atm   Answer(c)

7-53. Calculate the total pressure in a 10 liter flask at 27 $^{\circ}$C of a sample of gas that contains 6.0 g of H_2, 15.2 g of N_2, and 16.8 g of He.

$P_{\text{Total}} = P_{\text{Hydrogen}} + P_{\text{Nitrogen}} + P_{\text{Helium}}$

$P_{\text{Hydrogen}} = (6/2.02) \times (0.08206 \times 300.15)/10 = 7.32$

$P_{\text{Nitrogen}} = (15.2/28.02) \times (0.08206 \times 300.15)/10 = 1.34$

$P_{\text{Helium}} = (16.8/4.00) \times (0.08206 \times 300.15)/10 = 10.34$

$P_{\text{Total}} = P_{\text{Hydrogen}} + P_{\text{Nitrogen}} + P_{\text{Helium}} = 7.32 + 1.34 + 10.34 = 19.00$ atm

7-54. Calculate the volume of the flask that would contain 0.40 mole of O_2, 0.60 mole of N_2, and 1.5 moles of H_2 at 25 $^{\circ}$C and a total pressure of 1.80 atm. Calculate the partial pressure of each gas.

$P_{\text{Oxygen}} = 0.4 \times 0.08206 \times 298.15/V = 9.786/V$

$P_{\text{Nitrogen}} = 0.6 \times 0.08206 \times 298.15/V = 14.68/V$

$P_{\text{Hydrogen}} = 1.5 \times 0.08206 \times 298.15/V = 36.7/V$

$P_{\text{Total}} = (9.786 + 14.68 + 36.7)/V = 1.8$

$V = 61.17/1.8 = 34.0$ L

$P_{\text{Oxygen}} = 9.786/34 = 0.288$ atm

$P_{\text{Nitrogen}} = 14.68/34 = 0.432$ atm

$P_{\text{Hydrogen}} = 36.7/34 = 1.079$ atm

7-55. 0.300 L of H_2 gas was collected over water at 27 $^{\circ}$C on a day when the atmospheric pressure was 745 torr. What would be the volume of H_2 gas at 760 torr after the water vapor was removed?

($P_{\text{Water}} = 27$ torr at 27 $^{\circ}$C)

   (a) 0.283 L   (b) 0.294 L   (c) 0.300 L   (d) 0.306 L   (e) 0.318 L

$P_{\text{Total}} = P_{\text{Hydrogen}} +$ Vapor pressure of water at 27 $^{\circ}$C

$P_{\text{Hydrogen}} = P_{\text{Total}} -$ Vapor pressure of water at 27 $^{\circ}$C $= 745 - 27 = 718$ mm Hg

$P_1 V_1 = P_2 V_2$

$V_2 = (718 \times 0.3)/760 = 0.283$ L $\hspace{5cm}$ Answer(a)

7-56. O_2 is bubbled through water before it is given to a patient in a hospital. What volume of pure O_2 gas at 21 °C and 750 torr pressure would a patient receive if the patient breathed 250 mL of O_2 bubbled through water at this temperature and pressure? ($P_{Water} = 19$ torr at 21 °C)

$\hspace{1cm}$ (a) 0.00633 L $\hspace{0.5cm}$ (b) 0.244 L $\hspace{0.5cm}$ (c) 0.250 L $\hspace{0.5cm}$ (d) 0.256 L

$P_{Total} = P_{Oxygen} +$ Vapor pressure of water at 21 °C

$P_{Oxygen} = P_{Total} -$ Vapor pressure of water at 21 °C $= 750 - 19 = 731$ mm Hg

$P_1V_1 = P_2V_2$

Volume of pure oxygen gas $= (250/1000) \times (731/750) = 0.244$ L

Volume of pure oxygen gas $= 0.244$ L $\hspace{5cm}$ Answer(b)

7-57. A metal reacts with acid to form hydrogen gas as shown by the following equation:

M (s) $+ 2H^+$ (aq) $\rightarrow M^{2+}$ (aq) $+ H_2$ (g)

What is the weight of 1 mole of this metal if 3.49 g of this metal generates enough hydrogen collected over water to fill a bottle with a volume of 2.20 L at 25 °C and 1.00 atm pressure under conditions where the vapor pressure of the solution is 23.8 torr?

$\hspace{1cm}$ (a) 19.4 g/mol $\hspace{0.5cm}$ (b) 37.6 g/mol $\hspace{0.5cm}$ (c) 38.8 g/mol $\hspace{0.5cm}$ (d) 40.0 g/mol

$P_{Total} = P_{Hydrogen} +$ Vapor pressure of water at 25 °C

$P_{Hydrogen} = P_{Total} -$ Vapor pressure of water at 25 °C $= 760 - 23.8 = 736.2$ mm Hg

$PV = nRT$ $\hspace{0.3cm}$ $n = PV/RT = (736.2/760) \times 2.2/(0.08206 \times 298.15) = 0.0871$

Weight of 1 mole of the metal $= 3.49/0.0871 = 40$ g/mol $\hspace{3cm}$ Answer(d)

7-58. If the average speed of O_2 molecules at STP is 4.3×10^2 m/s, what is the average speed of H_2 molecules at the same temperature and pressure?

$\hspace{1cm}$ (a) 1.1×10^2 $\hspace{0.5cm}$ (b) 2.2×10^2 $\hspace{0.5cm}$ (c) 8.6×10^2 $\hspace{0.5cm}$ (d) 1.7×10^3

$v_{Hydrogen}/v_{Oxygen} = \sqrt{m_{Oxygen}/m_{Hydrogen}} = \sqrt{32/2.02} = 3.98$

$v_{Hydrogen} = 3.98 \times 4.3 \times 10^2 = 1.7 \times 10^3$ m/s $\hspace{4cm}$ Answer(d)

7-59. The root-mean-square speed of CH_4 molecules at 25 °C is about 0.56 km/s. What is the root-mean-square speed of a hydrogen at 25 °C?

(a) 0.070 km/s $\hspace{0.5cm}$ (b) 0.20 km/s $\hspace{0.5cm}$ (c) 1.1 km/s $\hspace{0.5cm}$ (d) 1.6 km/s $\hspace{0.5cm}$ (e) 4.5 km/s

Root-mean-square speed of hydrogen/Root mean square speed of methane

$$= \sqrt{m_{Methane}/m_{Hydrogen}} = \sqrt{16.05/2.02} = 2.8187$$

Root-mean-square speed of hydrogen = 0.56×2.8187 = 1.6 km/s   Answer(d)

7-60. Bromine vapor at a given temperature is roughly 5 times denser than oxygen gas.

Calculate the relative rates at which Br_2 (g) and O_2 (g) diffuse.

$$Rate_{Oxygen\ diffusion}/Rate_{Bromine\ diffusion} = \sqrt{d_{Bromine}/d_{Oxygen}} = \sqrt{5} = 2.24$$

Oxygen should diffuse 2.24 times faster.

7-61. N_2O and NO are often known by the trivial names, nitrous oxide and nitric oxide.

Associate correct formula with the appropriate trivial name if nitric oxide diffuses through a pin

hole 1.21 times as fast as nitrous oxide.

$$Rate_{Nitric\ oxide\ diffusion}/Rate_{Nitrous\ oxide\ diffusion} = \sqrt{m_{Nitrous\ oxide}/m_{Nitric\ oxide}} = \sqrt{44.02/30.01} = 1.21$$

Molecular weight of NO = 30.01   Molecular weight of N_2O = 44.02

Correct formula for Nitric oxide: NO   Correct formula for Nitrous oxide: N_2O

7-62. Which of the following would diffuse fastest at room temperature?

   (a) NH_3 (b) CO   (c) H_2S (d) F_2 (e) CO_2

Rate of diffusion α $1/\sqrt{Density}$

Rate of diffusion α $1/\sqrt{Molecular\ weight}$

Compound	Molecular weight g/mole
NH_3	17.04
CO	28.01
H_2S	34.12
F_2	38.00
CO_2	44.01

NH_3 has the lowest molecular weight. So it will have the fastest diffusion rate.   Answer(a)

7-63. Nitrogen and oxygen are allowed to effuse through a porous barrier at 295 K. If nitrogen

effuses at a rate of 0.0355 mol/min, what is the rate of effusion of oxygen?

   (a) 0.0311 mol/min   (b) 0.0322 mol/min   (c) 0.0380 mol/min

   (d)0.045 mol/min   (e) 31.1 mol/min

$$\text{Rate}_{\text{Oxygen effusion}}/\text{Rate}_{\text{Nitrogen effusion}} = \sqrt{m_{\text{Nitrogen}}/m_{\text{Oxygen}}} = \sqrt{28.02/32}$$

$$\text{Rate}_{\text{Oxygen effusion}} = \text{Rate}_{\text{Nitrogen effusion}} \times \sqrt{28.02/32} = 0.0355 \times \sqrt{28.02/32} = 0.0332 \text{ mol/min}$$

Answer(b)

7-64. Assume that a container is filled with a mixture of SO_3 and Ne. The molecular weight of SO_3 is 80 g/mol and the weight of 1 mole of Ne is 20 g/mol. The average velocity of SO_3 molecule is:

(a) One-fourth that of a Ne atom   (b) One-half that of a Ne atom

(c) The same as a Ne atom   (d) Two times that of a Ne atom

(e) 4 times that of a Ne atom

$$v_{SO_3}/v_{Ne} = \sqrt{m_{Ne}/m_{SO_3}} = \sqrt{20/80} = 0.5$$

$$v_{SO_3} = v_{Ne} \times 0.5 \qquad\qquad \text{Answer(b)}$$

7-65. A lecture room contains 50 rows of seats numbered 1 through from front to back. Ammonia is released from the front of the room at the same instant that hydrogen chloride gas is released from the back. Assuming Graham's law of diffusion, over which row (counting from the front) of students will a white cloud of ammonium chloride form?
(MW: NH_3 = 17.04 g/mol, HCl = 36.46 g/mol)

(a) 10   (b) 20   (c) 30   (d) 40   (e) 50

$$v_{NH_3}/v_{HCl} = \sqrt{36.46/17.04} = 1.46$$

$$v_{NH_3} = 1.46 \times v_{HCl}$$

Let the row number be 'Y' $\qquad\qquad Y/(v_{NH_3}) = (50 - Y)/v_{HCl}$

$Y/(1.46 \times v_{HCl}) = (50 - Y)/v_{HCl} \qquad Y = 30 \qquad\qquad$ Answer(c)

7-66. A lecture hall has 45 rows of seats numbered 1 through 45 from front to back. If laughing gas, N_2O, is released from the front of the room at the same time HCN is released from the back of the room, in roughly which row (counting from the front) will students first begin to die laughing?
(MW: HCN = 27.0 g/mol, N_2O = 44.0 g/mol)

(a) 10   (b) 20   (c) 25   (d) 30   (e) 40

$$v_{N_2O}/v_{HCN} = \sqrt{27/44} = 0.7833$$

v_{N_2O} = 0.7833×v_{HCN}

Let the row number be 'Y'   Y/ v_{N_2O} = (45 – Y)/ v_{HCN}

Y/(0.7833× v_{HCN}) = (45 – Y)/ v_{HCN} Y = 20   Answer(b)

7-67. ^{235}U and ^{238}U are separated on the basis of the relative rates of diffusion of $^{235}UF_6$ and $^{238}UF_6$. Calculate the relative rates of diffusion of these two compounds.

Rate of diffusion of $^{235}UF_6$/Rate of diffusion of $^{238}UF_6$

$= \sqrt{\text{Molecular weight of } ^{238}UF_6/\text{Molecular weight of } ^{235}UF_6}$

$= \sqrt{352.06/349.03}$

$= 1.0043$

Rate of diffusion of $^{235}UF_6$ = 1.0043 × Rate of diffusion of $^{238}UF_6$

7-68. The atomic weight of radon was first determined by measuring its rate of diffusion compared with mercury vapor. If mecury vapor diffuses 1.052 times as fast as radon, what is the atomic weight of radon?

Rate of diffusion of Hg/Rate of diffusion of radon

$= \sqrt{m_{Rn}/m_{Hg}}$ = 1.052

$m_{Rn}/m_{Hg} = (1.052)^2$

Molar mass of radon = $m_{Hg}×(1.052)^2 = 200.59×(1.052)^2 = 221.99$ g/mol

7-69. 2.11 moles of C_3H_8 are allowed to burn in the presence of excess O_2 to form CO_2 and H_2O.

C_3H_8 (g) + $5O_2$ (g)   → $3CO_2$ (g) + $4H_2O$ (g)

The CO_2 is isolated from all other gases and stored in a 1.25 L container at 0 °C. What is the pressure in this container if the gas obeys the ideal gas equation?

   (a) 18.8 atm   (b) 38.5 atm   (c) 113.5 atm   (d) 1130.3 atm

PV = nRT

P = nRT/V = (3×2.11)×0.08206×273.15/1.25 = 113.5 atm   Answer(c)

7-70. According to van der Waals' equation, what is the actual pressure in the container described in the previous question? (For CO_2 : a = 3.592 L^2-atm/mol^2 , b = 0.04267 L/mol)

   (a) 52.66 atm   (b) 92.1 atm   (c) 114 atm   (d) 145 atm

$\{P + a\,(n^2/V^2)\}\,(V - nb) = nRT$

$\{P + 3.592(40.07/1.5625)\} = (6.33\times0.08206\times273.15)/\{1.25 - (6.33\times0.04267)\}$

$P + 92.12 = 144.78$

$P = 52.66$ atm   Answer(a)

CHAPTER 8

THERMOCHEMISTRY

8-1. Which of the following is most likely to be exothermic?

(a) H_2 (g) $\rightarrow$ 2H (g) (b) 2F (g) $\rightarrow$ F_2 (g) (c) C (s) $\rightarrow$ C (g)

(d) CCl_4 (g) $\rightarrow$ C (g) + 4Cl (g) (e) HCl (g) $\rightarrow$ H (g) + Cl (g)

a: Energy is required to break a bond.

b: Energy is released when a bond is formed.

c: Energy is required to convert solid to gas.

d: Energy is required to break bonds.

e: Energy is required to break a bond. Answer(b)

8-2. Which of the following is most likely to be exothermic?

(a) $CaCO_3$ (s) $\rightarrow$ CaO (s) + CO_2 (g)
(b) Ca (g) $\rightarrow$ Ca^{2+} (g) + $2e^-$
(c) Cl_2 (g) $\rightarrow$ 2Cl (g)
(d) Cl (g) + e^- $\rightarrow$ Cl^- (g)
(e) $CaCl_2$ (g) $\rightarrow$ Ca^{2+} (g) + $2Cl^-$ (g)

a: Energy is required to form CO_2 from $CaCO_3$.

b: Energy is required to ionize (ionization energy).

c: Energy is required to break a bond.

d: Energy is released when electron is added to chlorine atom (electron affinity) in

gas phase.

e: Energy is required to break forces of attraction. Answer(d)

8-3. Which of the following is most likely to be endothermic?

(a) Mg (s) $\rightarrow$ Mg (g)
(b) Na^+ (g) + Cl^- (g) $\rightarrow$ NaCl (s)
(c) Na^+ (g) + e^- $\rightarrow$ Na (g)
(d) H^+ (aq) + OH^- (aq) $\rightarrow$ H_2O (l)
(e) H_2 (g) + 1/2 O_2 (g) $\rightarrow$ H_2O (l)

a: Energy is required to convert solid to gas.

b: Energy is given off (Lattice energy of NaCl).

c: Energy is released when electron is added to sodium ion.

d: Energy is released when a bond is formed.

e: Energy is released when gas is condensed to liquid and when a bond is formed.   Answer(a)

8-4. At what temperature are standard-state enthalpy of reaction measurements most often made?

(a) 0 K   (b) 273.15 K   (c) 0 °C   (d) 25 °C   (e) More than one of the above

Standard-state enthalpy of reaction measurements are most often made at 25 °C.   Answer(d)

8-5. How much energy is required to heat 10.0 grams of gold from 20 °C to 100 °C if the specific heat of gold is 0.129 J/g-°C?

Energy required = 10.0×(100 − 20)×0.129 J = 103.2 J

8-6. A piece of copper metal weighing 145 grams was heated to 100 °C and then dropped into 250 grams of water at 25 °C. The copper metal cooled down and the water became warmer until both were at a temperature of 28.8 °C. Calculate the amount of heat absorbed by the water. Assuming that heat lost by the copper was absorbed by the water, what is the molar heat capacity of copper metal? (C_{water} = 75.376 J/mol-K)

(145/63.546)×C_{Cu}×(373 − 301.8) = (250/18.02)×75.376×(301.8 − 298) = 3973.76 J

C_{Cu} = (13.873×75.376×3.8)/(2.282×71.2) = 24.46 J/mol-K

8-7. For which reaction is ΔH roughly equal to ΔE?

(a) $2H_2$ (g) + O_2 (g) → $2H_2O$ (g)
(b) $Pb(NO_3)_2$ (s) + 2KI (s) → PbI_2 (s) + $2KNO_3$ (s)
(c) 2Na (s) + $2H_2O$ (l) → $2Na^+$ (aq) + $2OH^-$ (aq) + H_2 (g)
(d) NaOH (s) + CO_2 (g) → $NaHCO_3$ (s)

$\Delta H = \Delta E + \Delta(PV)$

ΔH is roughly equal to ΔE for reactions that involve only liquids and solids because there is little, if any, change in volume of the system during the reaction.   Answer(b)

8-8. How much heat is produced by mixing 50 mL of 1.0 M HBr at 25.6 °C with 50.0 mL of 1.0 M KOH at 25.6 °C if this reaction produces 100 mL of a solution with a temperature of 32.3 °C? (assume the heat capacity of water is 4.18 J/g-°C and the density of these solutions is 1.00 g/cm³).

Heat produced = $100 \times 4.18 \times (32.3 - 25.6) = 2800.6$ J

8-9. Using the data in the previous question, calculate ΔH for the following reaction:

HBr (aq) + KOH (aq)$\rightarrow$ KBr (aq) + H_2O (l)

(1/20) HBr (aq) + (1/20) KOH $\rightarrow$ (1/20) KBr (aq) + (1/20) H_2O (l) $\Delta H = -2.8$ kJ

HBr (aq) + KOH (aq) $\rightarrow$ KBr (aq) + H_2O (l) $\Delta H = -(20 \times 2.8)$ kJ $= -56$ kJ

8-10. If mixing 50.0 mL of 1.0 M HBr and 50.0 mL of 1.0 M KOH produces a temperature increase of 6.70 °C, then mixing 100 mL of 1.0 M HBr and 100 mL of 1.0 M KOH will produce a temperature increase of

   (a) 1.68 °C   (b) 3.35 °C   (c) 6.70 °C   (d) 13.4 °C   (e) impossible to predict

(1/20) HBr (aq) + (1/20) KOH $\rightarrow$ (1/20) KBr (aq) + (1/20) H_2O (l) $\Delta H = -2.8$ kJ

(1/10) HBr (aq) + (1/10) KOH $\rightarrow$ (1/10) KBr (aq) + (1/10) H_2O (l) $\Delta H = -5.6$ kJ

$200 \times 4.18 \times \Delta t = 5.6$ kJ $\Delta t = 6.7$ °C   Answer(c)

8-11. How much heat is produced when 0.200 mole of H_2 (g) reacts with 0.30 mole of Cl_2 (g) if the enthalpy of reaction for the production of one mole of HCl is -92.3 kJ/mol?

½ H_2 (g) + ½ Cl_2 (g) $\rightarrow$ HCl (g) $\Delta H = -92.3$ kJ/mol HCl

H_2 (g) is the limiting agent. Only 0.200 mole of Cl_2 will be used.

0.2 M H_2 (g) + 0.2 M Cl_2 (g) $\rightarrow$ 0.4 M HCl (g)

Heat produced = $92.3 \times 0.4 = 36.92$ kJ

8-12. How much energy is given off when 4.8 g of carbon are burned to produce carbon-di-oxide, if ΔH^0_{rxn} for the combustion of carbon to form CO_2 is -394 kJ/mol?

C (s) + O_2 (g) $\rightarrow$ CO_2 (g) $\Delta H^0 = -394$ kJ/mol CO_2

(4.8/12) C (s) + (4.8/12) O_2 (g) $\rightarrow$ 0.4 CO_2 (g)

Energy given off = $394 \times 0.4 = 157.6$ kJ

8-13. Calculate ΔH^0_{rxn} in kJ/mol for the following reaction if 1.00 gram of magnesium reacts with excess fluorine to give off 46.22 kJ of heat :

Mg (s) + F_2 (g) $\rightarrow$ MgF_2 (s)

Mg is the limiting agent.

Number of mole of Mg = 1/24.305 = 0.0411

Only 0.0411 mole of fluorine will be used.

0.0411M Mg (s) + 0.0411 M F_2 (g) $\rightarrow$ 0.0411M MgF_2 (g)   Heat given off = 46.22 kJ

Mg (s) + F_2 (s) $\rightarrow$ MgF_2 (s) ΔH^0_{rxn} = $-$ (46.22/0.0411) = $-$ 1124.57 kJ

8-14. Calculate ΔH^0 for the reaction: H_2O (l) $\rightarrow$ H_2O (g) from the following data.

$2H_2$ (g) + O_2 (g) $\rightarrow$ $2H_2O$ (g)	ΔH^0_{rxn} = $-$ 483.6 kJ
$2H_2$ (g) + O_2 (g) $\rightarrow$ $2H_2O$ (l)	ΔH^0_{rxn} = $-$ 571.6 kJ
H_2O (l) $\rightarrow$ H_2 (g) + $\frac{1}{2}$ O_2 (g)	ΔH = 285.8 kJ
H_2 (g) + $\frac{1}{2}$ O_2 (g) $\rightarrow$ H_2O (g)	ΔH = $-$ 241.8 kJ

H_2O (l) $\rightarrow$ H_2O (g) ΔH^0_{rxn} = 285.8 -241.8 = 44 kJ

8-15. What is ΔH^0 for the reaction: 2CO (g) + O_2 (g) $\rightarrow$ $2CO_2$ (g) if

C (s) + $\frac{1}{2}$ O_2 (g) $\rightarrow$ CO (g)	ΔH^0_{rxn} = $-$ 111 kJ
C (s) + O_2 (g) $\rightarrow$ CO_2 (g)	ΔH^0_{rxn} = $-$ 393 kJ
2CO (g) $\rightarrow$ 2C (s) + O_2 (g)	ΔH = 222 kJ
2C (s) + $2O_2$ (g) $\rightarrow$ $2CO_2$ (g)	ΔH = $-$ 786 kJ

2CO (g) + O_2 (g) $\rightarrow$ $2CO_2$ (g) ΔH^0_{rxn} = 222 $-$ 786 = $-$ 564 kJ

8-16. Calculate ΔH^0 for the reaction: C(s) + $2H_2$ (g) $\rightarrow$ CH_4 (g) from the following data.

C (s) + O_2 (g) $\rightarrow$ CO_2 (g)	ΔH^0_{rxn} = $-$ 393.9 kJ
H_2 (g) + $\frac{1}{2}$ O_2 (g) $\rightarrow$ H_2O (l)	ΔH^0_{rxn} = $-$ 285.8 kJ
CH_4 (g) + $2O_2$ (g) $\rightarrow$ CO_2 (g) + $2H_2O$ (l)	ΔH^0_{rxn} = $-$ 890.4 kJ
CO_2 (g) + $2H_2O$ (l) $\rightarrow$ CH_4 (g) + $2O_2$ (g)	ΔH = 890.4 kJ
$2H_2$ (g) + O_2 (g) $\rightarrow$ $2H_2O$ (l)	ΔH = ($-$285.8×2) = $-$ 571.6 kJ
C (s) + O_2 (g) $\rightarrow$ CO_2 (g)	ΔH = $-$ 393.9 kJ

C (s) + $2H_2$ (g) $\rightarrow$ CH_4 (g) ΔH^0_{rxn} = 890.4 $-$ 571.6 $-$ 393.9 = $-$ 75.1 kJ

8-17. Calculate ΔH^0 for the reaction:

$4NH_3$ (g) + $3O_2$ (g) $\rightarrow$ $2N_2$ (g) + $6H_2O$ (l) from the following data.

$3H_2$ (g) + N_2 (g) $\rightarrow$ $2NH_3$ (g)	ΔH^0_{rxn} = $-$ 92.4 kJ
$2H_2$ (g) + O_2 (g) $\rightarrow$ $2H_2O$ (l)	ΔH^0_{rxn} = $-$ 571.7 kJ

$4NH_3 (g) \rightarrow 6H_2 (g) + 2N_2 (g)$ $\qquad$ $\Delta H = 92.4 \times 2 = 184.8$ kJ

$6H_2 (g) + 3O_2 (g) \rightarrow 6H_2O (l)$ $\qquad$ $\Delta H = -(571.7 \times 3) = -1715.1$ kJ
_____
$4NH_3 (g) + 3O_2 (g) \rightarrow 2N_2 (g) + 6H_2O (l)$ $\quad$ $\Delta H^0_{rxn} = -1530.3$ kJ

8-18. Calculate ΔH^0 for the reaction: $2H_2O_2 (aq) \rightarrow 2H_2O (l) + O_2 (g)$ from the following data.

$2H_2 (g) + O_2 (g) \rightarrow 2H_2O (l)$ $\qquad$ $\Delta H^0_{rxn} = -571.6$ kJ

$H_2 (g) + O_2 (g) \rightarrow H_2O_2 (aq)$ $\qquad$ $\Delta H^0_{rxn} = -187.8$ kJ

$2H_2O_2 (aq) \rightarrow 2H_2 (g) + 2O_2 (g)$ $\qquad$ $\Delta H = 187.8 \times 2 = 375.6$ kJ

$2H_2 (g) + O_2 (g) \rightarrow 2H_2O (l)$ $\qquad$ $\Delta H = -571.6$ kJ
_____
$2H_2O_2 (aq) \rightarrow 2H_2O (l) + O_2 (g)$ $\quad$ $\Delta H^0_{rxn} = 375.6 - 571.6 = -196.0$ kJ

8-19. Calculate ΔH^0_{rxn} for the reaction: $N_2 (g) + O_2 (g) \rightarrow 2NO (g)$ from the following data.

$N_2 (g) + 2O_2 (g) \rightarrow 2NO_2 (g)$ $\qquad$ $\Delta H^0_{rxn} = 66.4$ kJ

$2NO (g) + O_2 (g) \rightarrow 2NO_2 (g)$ $\qquad$ $\Delta H^0_{rxn} = -114.2$ kJ

$N_2 (g) + 2O_2 (g) \rightarrow 2NO_2 (g)$ $\qquad$ $\Delta H = 66.4$ kJ

$2NO_2 (g) \rightarrow 2NO (g) + O_2 (g)$ $\qquad$ $\Delta H = 114.2$ kJ
_____
$N_2 (g) + O_2 (g) \rightarrow 2NO (g)$ $\qquad$ $\Delta H^0_{rxn} = 66.4 + 114.2 = 180.6$ kJ

8-20. Calculate the heat of combustion of propane, C_3H_8,

$C_3H_8 (g) + 5O_2 (g) \rightarrow 3CO_2 (g) + 4H_2O (g)$ from the following data.

$3C (s) + 4H_2 (g) \rightarrow C_3H_8 (g)$ $\qquad$ $\Delta H^0_{rxn} = -103.85$ kJ

$C (g) + O_2 (g) \rightarrow CO_2 (g)$ $\qquad$ $\Delta H^0_{rxn} = -393.51$ kJ

$H_2 (g) + \frac{1}{2} O_2 (g) \rightarrow H_2O (g)$ $\qquad$ $\Delta H^0_{rxn} = -241.83$ kJ

$C_3H_8 (g) \rightarrow 3C (s) + 4H_2 (g)$ $\qquad$ $\Delta H = 103.85$ kJ

$3C (g) + 3O_2 (g) \rightarrow 3CO_2 (g)$ $\qquad$ $\Delta H = (-393.51 \times 3) = -1180.53$ kJ

$4H_2 (g) + 2O_2 (g) \rightarrow 4H_2O (g)$ $\qquad$ $\Delta H = (-241.83 \times 4) = -967.32$ kJ
_____
$C_3H_8 (g) + 5O_2 (g) \rightarrow 3CO_2 (g) + 4H_2O (g)$ $\quad$ $\Delta H^0_{rxn} = 103.85 - 1180.53 - 967.32 = -2044.0$ kJ

8-21. Calculate ΔH^0 for the reaction: $2N_2 (g) + 5O_2 (g) \rightarrow 2N_2O_5 (g)$ from the following data.

$N_2 (g) + 3O_2 (g) + H_2 (g) \rightarrow 2HNO_3 (aq)$ $\qquad$ $\Delta H^0_{rxn} = -207.4$ kJ

$N_2O_5 (g) + H_2O (g) \rightarrow 2HNO_3 (aq)$ $\qquad$ $\Delta H^0_{rxn} = 218.4$ kJ

$2H_2 (g) + O_2 (g) \rightarrow 2H_2O (g)$ $\qquad$ $\Delta H^0_{rxn} = -571.6$ kJ

$2N_2$ (g) + $6O_2$ (g) + $2H_2$ (g) → $4HNO_3$ (aq)   ΔH = (− 207.4×2) = − 414.8 kJ

$4HNO_3$ (aq) → $2N_2O_5$ (g) + $2H_2O$ (g)   ΔH = (−218.4×2) = − 436.8 kJ

$2H_2O$ (g) → $2H_2$ (g) + O_2 (g)   ΔH = 571.6 kJ
_____

$2N_2$ (g) + $5O_2$ (g) → $2N_2O_5$ (g) ΔH^0_{rxn} = 571.6 – 414.8 – 436.8 = − 280 kJ

8-22. What is the sign of the enthalpy of reaction for the following reaction?

P_4O_6 (s) + $2O_2$ (g) + $6H_2O$ (g) → $4H_3PO_4$ (s) (assume that all compounds are present in their

most stable state at 25 °C and at 1 atm pressure).

Compound	ΔH^0_{ac} (kJ/mol)
P_4O_6 (s)	− 4393.7
O_2 (g)	− 498.34
H_2O (g)	− 926.29
H_3PO_4 (s)	− 3243.3

ΔH^0_{rxn} = Σ ΔH^0_{ac} products − Σ ΔH^0_{ac} reactants

= − (3243.3×4) − { (−4393.7) + (−498.34×2) + (−926.29×6) = − 2025 kJ

8-23. Calculate the energy required to transform 18 g of ice at 0 °C to water vapor at 100 °C

from the following data.

H_2O (s) → H_2O (l) ΔH^0_{rxn} = 6.03 kJ

H_2O (l, 0° C) → H_2O (l, 100 °C) ΔH^0_{rxn} = 7.53 kJ

H_2O (l) → H_2O (g) ΔH^0_{rxn} = 40.67 kJ

Energy required = 6.03 + 7.53 + 40.67 = 54.23 kJ

8-24.

ΔH^0_{ac} H_2 (g) = − 435.3 kJ/mol ΔH^0_{ac} F_2 (g) = − 157.98 kJ/mol

ΔH^0_{ac} N_2 (g) = − 945.41 kJ/mol ΔH^0_{ac} C (s) = − 716.68 kJ/mol

ΔH^0_{ac} O_2 (g) = − 498.34 kJ/mol ΔH^0_{ac} NH_3 (g) = − 1171.8 kJ/mol

ΔH^0_{ac} H_2O (g) = − 926.3 kJ/mol ΔH^0_{ac} HF (g) = − 567.7 kJ/mol

(i)   What is ΔH^0 for the following reaction?

3N(g) → (3/2) N_2 (g)

ΔH^0_{rxn} = 1.5× (− 945.41) = − 1418 kJ

(ii)   What is the bond energy of $N \equiv N$?

Bond energy of $N \equiv N$ is equal to 945.41 kJ/mol

(iii)   What is the ΔH^0 for the following reaction as written?

$$4NH_3 \text{ (g)} + 3O_2 \text{ (g)} \rightarrow 2N_2 \text{ (g)} + 6H_2O \text{ (g)}$$

$$\Delta H^0_{rxn} = \sum \Delta H^0_{ac} \text{ products} - \sum \Delta H^0_{ac} \text{ reactants}$$

$$= \{(-945.41 \times 2) + (-926.3 \times 6)\} - \{(-1171.8 \times 4) + (-498.34 \times 3)\}$$

$$= -1266.4 \text{ kJ}$$

(iv)   What is the bond energy, H_2N-H, for one of the N–H bonds in NH_3 (g)?

One N–H bond energy = $(1171.8/3) = 390.6$ kJ/mol

(v)   How much heat is absorbed or evolved when just enough H_2 (g) reacts with just enough F_2 (g) to produce 0.200 mole of HF (g)?

$$H_2 \text{ (g)} + F_2 \text{ (g)} \rightarrow 2HF \text{ (g)} \qquad (1/10) H_2\text{(g)} + (1/10)F_2 \text{ (g)} \rightarrow (2/10)HF \text{ (g)}$$

$$\Delta H = \{(-567.7 \times 0.2)\} - \{(-435.3 \times 0.1) \times (-157.98 \times 0.1)\} = -54.2 \text{ kJ}$$

54.2 kJ of heat is evolved.

(vi)   What is ΔH^0 for the following reaction?

$$N_2 \text{ (g)} + 3H_2 \text{ (g)} \rightarrow 2NH_3$$

$$\Delta H^0_{rxn} = \{(-1171.8 \times 2)\} - \{(-945.41) + (-435.3 \times 3)\} = -92.3 \text{ kJ}$$

8-25. Calculate the ΔH^0 for the following reaction:

$4NH_3 \text{ (g)} + 5O_2 \text{ (g)} \rightarrow 4NO \text{ (g)} + 6H_2O \text{ (g)}$ from the following data.

Compound	ΔH^0_{ac} (kJ/mol)
NH_3 (g)	-1171.76
O_2 (g)	-498.34
NO (g)	-631.62
H_2O (g)	-926.29

$$\Delta H^0_{rxn} = \sum \Delta H^0_{ac} \text{ products} - \sum \Delta H^0_{ac} \text{ reactants}$$

$$= \{(-926.29 \times 6) + (-631.62 \times 4)\} - \{(-1171.76 \times 4) + (-498.34 \times 5)\}$$

$$= -905.5 \text{kJ}$$

8-26. Calculate the ΔH^0 for the reaction:

$3NO_2$ (g) + H_2O (l) $\rightarrow$ $2HNO_3$ (aq) + NO (g) from the following data.

Compound	ΔH^0_{ac} (kJ/mol)
NO_2 (g)	-937.86
H_2O (l)	-970.30
HNO_3 (aq)	-1645.22
NO (g)	-631.22

$\Delta H^0_{rxn} = \sum \Delta H^0_{ac}$ products $- \sum \Delta H^0_{ac}$ reactants

$= \{(-1645.22 \times 2) + (-631.22)\} - \{(-937.86 \times 3) + (-970.30)\}$

$= -137.8$ kJ

8-27. Both ethanol (CH_3CH_2OH) and methanol have been considered as fuels for automobiles.

Which is a better fuel, on a per gram basis, when burned with O_2?

Compound	ΔH^0_{ac} (kJ/mol)
CH_3CH_2OH (g)	-3223.53
CH_3OH (g)	-2037.11
O_2 (g)	-498.34
H_2O (g)	-926.29
CO_2 (g)	-1608.53
H_2O (l)	-970.30

    (i) CH_3OH

$2CH_3OH$ (g) + $3O_2$ (g) $\rightarrow$ $2CO_2$ (g) + $4H_2O$ (g)

$\Delta H^0_{rxn} = \sum \Delta H^0_{ac}$ products $- \sum \Delta H^0_{ac}$ reactants

$= \{(-1608.53 \times 2) + (-926.29 \times 4)\} - \{(-498.34 \times 3) + (-2037.11 \times 2)\}$

$= -1353$ kJ

Molecular weight of CH_3OH = 32.04

Heat given off per gram of CH_3OH = 1353/(32.04×2) = 21 kJ/g

    (ii) CH_3CH_2OH

CH_3CH_2OH (g) + $3O_2$ (g) $\rightarrow$ 2 CO_2 (g) + $3H_2O$ (g)

$\Delta H^0_{rxn} = \sum \Delta H^0_{ac}$ products $- \sum \Delta H^0_{ac}$ reactants

$= \{(-1608.53\times2) + (-926.29\times3)\} - \{(-3223.53) + (-498.34\times3)\}$ $= -1277$ kJ

Molecular weight of $CH_3CH_2OH = 46.06$

Heat given off per gram of $CH_3CH_2OH = 1277/(46.06) = 28$ kJ/g

8-28. Calculate the value of ΔH^0_{ac} for sucrose if $\Delta H^0_{rxn} = -5645$ kJ for the following reaction:

$C_{12}H_{22}O_{11}$ (s) $+ 12O_2$ (g) $\rightarrow 12CO_2$ (g) $+ 11H_2O$ (l)

Compound	ΔH^0_{ac} (kJ/mol)
O_2 (g)	-498.34
CO_2 (g)	-1608.53
H_2O (l)	-970.30

$\Delta H^0_{rxn} = \sum \Delta H^0_{ac}$ products $- \sum \Delta H^0_{ac}$ reactants

$\{(-970.30\times11) + (-1608.53\times12)\} - \{(\Delta H^0_{ac} \text{ sucrose}) + (-498.34\times12)\} = -5645$

ΔH^0_{ac} sucrose $= -18350$ kJ/mol

8-29. Calculate the ΔH^0 for the following reactions:

(i) $C(s) + H_2O$ (g) $\rightarrow CO$ (g) $+ H_2$ (g)
(ii) $C(s) + H_2O$ (l) $\rightarrow CO$ (g) $+ H_2$ (g) from the following data.

Compound	ΔH^0_{ac} (kJ/mol)
C (s)	-717
H_2O (g)	-926
H_2O (l)	-970
CO (g)	-1076
H_2 (g)	-435

(i) $\Delta H^0_{rxn} = \sum \Delta H^0_{ac}$ products $- \sum \Delta H^0_{ac}$ reactants

$= \{(-1076 - 435)\} - \{(-717 - 926)\}$

$= 132$ kJ

(ii) $\Delta H^0_{rxn} = \sum \Delta H^0_{ac}$ products $- \sum \Delta H^0_{ac}$ reactants

$= \{(-1076 - 435)\} - \{(-717 - 970)\}$

$= 176$ kJ

CHAPTER 9

LIQUIDS AND SOLUTIONS

9-1. Hemoglobin is a large molecule that carries oxygen in human blood. A water solution that contains 0.263 g of hemoglobin in 10.0 mL solution has an osmotic pressure of 7.51 torr at 25 $^{\circ}$C. What is the molar mass of hemoglobin?

Osmotic Pressure $\prod = nRT/V$

$$n = \prod V/RT$$

$$\prod = 7.51/760 = 9.88 \times 10^{-3} \text{ atm}$$

$$V = 10/1000 = 0.01 \text{ L}$$

Number of mole $n = (9.88 \times 10^{-3} \times 0.01)/(0.08206 \times 298) = 4.04 \times 10^{-6}$

Molar mass $= (0.263)/(4.04 \times 10^{-6}) = 6.51 \times 10^{4}$ g/mole

9-2. Calculate the freezing point of an aqueous solution containing 30.2% ethylene glycol by mass. 30.2% ethylene glycol by mass means that 302 g of ethylene are in 698 g of water.

Molecular weight of ethylene glycol (HOH_2CCH_2OH) = 62.07 g/mole

Number of moles of ethylene glycol = 302/62.07 = 4.87

Molality = Number of moles/1000 g of solvent = $(4.87/698) \times 1000 = 6.98$

$\Delta T_{FP} = - k_f \, m = -1.853 \times 6.98 = - 13 \,^{\circ}C$

Freezing point of pure water = 0.0 $^{\circ}$C

Freezing point of 30.2% solution of ethylene glycol in water = $- 13 \,^{\circ}C$

9-3. One of the byproducts formed during the synthesis of C_{60} is a deep red solid containing only carbon. A solution of 205 mg of this compound in 10.0 g of CCl_4 has a freezing point of $-23.38 \,^{\circ}C$. What are the molar mass and the most probable formula of this substance?

(CCl_4 : Freezing point = $-22.62 \,^{\circ}C$; k_f for CCl_4 = 31.4 $^{\circ}$C/m)

$\Delta T_{FP} = - (23.38 - 22.62) \,^{\circ}C = - 0.76 \,^{\circ}C$

$\Delta T_{FP} = - k_f \, m$

$m = -\Delta T_{FP}/k_f = 0.76/31.4 = 0.0242$

Number of mole of carbon in 10.0 g of $CCl_4 = (0.0242 \times 10)/1000 = 2.42 \times 10^{-4}$

Molar mass $= 205/(1000 \times 2.42 \times 10^{-4}) = 847$ g/mol

Most probable formula $= C_{70}$ $(C_{847/12})$

9-4. Assume that the fluids inside a sausage are approximately 0.80 M in dissolved particles due to the salt and sodium nitrite used to prepare them. Calculate the osmotic pressure inside the sausage at 100 ^{o}C to learn why experienced cooks pierce the semipermeable skin of sausages before cooking them.

$\prod = MRT = 0.80 \times 0.08206 \times 373 = 24$ atm

9-5. Arrange the following aqueous solutions in order of increasing freezing points: 0.2 m NaCl, 0.3 m aceticacid, 0.1 m $CaCl_2$, and 0.2 m sucrose (0.3 m acetic acid in water is 1.3 % ionized). (k_f for $H_2O = 1.853$ ^{o}C/m)

	Actual concentration of dissolved species
NaCl:	$0.2 \times 2 = 0.4$ m
Aceticacid:	$0.3 \times 1.013 = 0.3039$ m
$CaCl_2$:	$0.1 \times 3 = 0.3$ m
Sucrose:	$0.2 \times 1 = 0.2$ m

NaCl: $\Delta T_{FP} = -1.853 \times 0.4 = -0.741$ ^{o}C   FP $= -0.741$ ^{o}C

Aceticacid: $\Delta T_{FP} = -1.853 \times 0.3039 = -0.563$ ^{o}C   FP $= -0.563$ ^{o}C

$CaCl_2$: $\Delta T_{FP} = -1.853 \times 0.3 = -0.556$ ^{o}C   FP $= -0.556$ ^{o}C

Sucrose: $\Delta T_{FP} = -1.853 \times 0.2 = -0.371$ ^{o}C   FP $= -0.371$ ^{o}C

0.2 m NaCl < 0.3 m aceticacid < 0.1 m $CaCl_2$ < 0.2 m sucrose

9-6. Calculate the van't Hoff factor for a 0.05 m aqueous solution of $MgCl_2$ that has a measured freezing point of -0.25 ^{o}C.

van't Hoff factor ($i_{observed}$) = Number of particles observed/1 formula unit of the solute

(For $MgCl_2$, $i_{ideal} = 3$)

Freezing point of water $= 0.0$ ^{o}C   Measured freezing point $= -0.25$ ^{o}C

$\Delta T_{FP} = -0.25$ ^{o}C

$\Delta T_{FP} = - k_f \times m \times i_{observed} = - 1.853 \times (0.05) \times i_{observed} = - 0.25\ ^oC$

$i_{observed} = 0.25\ /(1.853 \times 0.05) = 2.7$

9-7. A 1.79 M $CuCl_2$ solution has a density of 1.205 g/cm^3. Calculate the molality of this solution.

Molecular weight of $CuCl_2$ = 63.546 + (35.45×2) = 134.45 g/mol

Weight of $CuCl_2$ in 1 liter of solution = 1.79×134.45 = 240.7 g/L

Weight of 1 liter of solution = 1.205×1000 = 1205 g

Weight of water in 1 liter of solution = 1205 − 240.7 = 964.3 g

Molality = Number of moles of solute/Number of kilograms of solvent

$\qquad$ = (240.7/134.45)×(1000/964.3) = 1.86 m

9-8. A 1.86 molal $CuCl_2$ solution has a density of 1.205 g/cm^3. Calculate the molarity and the mass percent of this solution.

Molecular weight of $CuCl_2$ = 63.546 + (35.45×2) = 134.45 g/mol

Weight of $CuCl_2$ in 1000 g of solvent = 1.86×134.45 = 250 g

Weight of ($CuCl_2$ + 1000 g of solvent) = 1250 g

Density of the solution = 1.205 g/cm^3

Volume of 1250 g of solution = 1250/1.205 = 1037 mL

Molarity = (250/134.45)×(1000/1037) = 1.79 M

Mass percent = (250/1250)×100 = 20 %

9-9. A 1.79 M $CuCl_2$ solution has a density of 1.205 g/cm^3. Calculate the mass percent of this solution.

Molecular weight of $CuCl_2$ = 63.546 + (35.45×2) = 134.45 g/mol

Weight of $CuCl_2$ in 1000 ml of solution = 1.79×134.45 = 240.7 g

Density of the solution = 1.205 g/cm^3

Weight of 1000 ml of solution = 1.205×1000 = 1205 g

Mass percent = (240.7/1205)×100 = 20 %

9-10. A solution was prepared by dissolving 27.5 g of NaCl, 2.4 g of $MgCl_2$, 3.4 g of $MgSO_4$, 0.75 g of KCl and 1.1 g of $CaCl_2$ in 1000 g of pure water. What is the vapor pressure of this solution at 25 $^{\circ}$C if the vapor pressure of pure water at 25 $^{\circ}$C is 23.8 mm Hg ?(MW: NaCl =

58.43 g/mol, $MgCl_2$ = 95.21 g/mol, $MgSO_4$ = 120.37 g/mol, KCl = 74.548 g/mol,

$CaCl_2$ = 110.98 g/mol, and H_2O = 18.02 g/mol)

Number of moles of H_2O = 1000/18.02 = 55.5

	Actual concentration of dissolved species
NaCl:	(27.5/58.43)×2 = 0.94 mole
$MgCl_2$:	(2.4/95.21)×3 = 0.075 mole
$MgSO_4$:	(3.4/120.37)×2 = 0.056 mole
KCl:	(0.75/74.548)×2 = 0.02 mole
$CaCl_2$:	(1.1/110.98)×3 = 0.03 mole

Mole fraction of solvent: 55.5/ (55.5 + 0.94 + 0.075 + 0.056 + 0.02 + 0.03) = 0.98

Raoult's Law: P = $X_{solvent}$ × 23.8 = 0.98×23.8 = 23.324 mm Hg

9-11. Which of the following correctly describes the behaviour of the vapor pressure of liquids?

   (a) It remains constant as the temperature of the liquid increases
   (b) It remains constant when a solute is dissolved in the liquid
   (c) It increases when a solute is dissolved in the liquid
   (d) If the vapor pressure of CH_3OCH_3 is much larger than CH_3CH_2OH at room temperature, CH_3OCH_3 should boil at a lower temperature than CH_3CH_2OH.
   (e) All of the above are correct.

(a): Vapor pressure will increase if the temperature increases.

(b): Vapor pressure will decrease if a solute is dissolved in the liquid

(c): Vapor pressure will decrease if a solute is dissolved in the liquid.

(d): Liquid boils when its vapor pressure is equal to the pressure exerted on the liquid by

   its surroundings. So a liquid with a much larger vapor pressure at room temperature

   will have a lower boiling point.   Answer(d)

9-12. Which compound would you expect to have the largest vapor pressure at 25 $^{\circ}$C?

   (a) C_7H_{16} (BP = 98.4 $^{\circ}$C)   (b) C_8H_{18} (BP = 125.7 $^{\circ}$C)   (c) C_9H_{20} (BP =150.8 $^{\circ}$C)

   (d) $C_{10}H_{22}$ (BP = 174.3 $^{\circ}$C)

   (e) It is impossible to tell which compound would have the largest vapor pressure because there is no relationship between vapor pressure and the boiling point of a compound

Liquid boils when its vapor pressure is equal to the pressure exerted on the liquid by its surroundings. So the liquid with the lowest boiling point will have the largest vapor pressure at 25 °C. Answer(a)

9-13. Which of the following would you expect to have the largest boiling point?

(a) Methane CH_4 (b) Chloromethane CH_3Cl (c) Dichloromethane CH_2Cl_2

(d) Chlorofrm $CHCl_3$ (e) Carbon tetrachloride CCl_4

CH_4 and CCl_4 are both non-polar. But Cl is larger than H. So stronger intermolecular attractions in CCl_4 and the larger mass of CCl_4 will overpower the polarity differences and its boiling point will be the largest. Answer(e)

9-14. In which of the following substances do you expect to find the strongest hydrogen bonds?

(a) NH_3 (b) H_2S (c) H_2O (d) CF_4 (e) NaH

CF_4 is a symmetrical non-polar molecule without a H atom.

Electronegativity

H 2.1

Na 0.9

N 3.04

O 3.44

S 2.58

Hydrogen bonding is a special type of dipole-dipole attraction between a hydrogen atom (bonded to a stongly electronegative atom) and a strongly electronegative element like O,N, and F (the more the electronegativity, stronger the hydrogen bond).

Hydrogen bonding is not possible in NaH. H_2O will have the strongest hydrogen bond.

Answer(c)

9-15. What is the molality of a bromine solution made by dissolving 39.95 g of Br_2 in 500 g of CCl_4?

Molecular weight of $Br_2 = 79.90 + 79.90 = 159.8$ g/mole

Number of mole of Br_2 in 500 g of $CCl_4 = 39.95/159.8 = 0.25$

Molality $= (0.25/500) \times 1000 = 0.50$

9-16. Which of the following is true?

(a) Molarity (M) is calculated by dividing the number of moles of solute by the number of liters of solvent.

(b) Molality (m) is calculated by dividing the number of moles of solute by the number of liters of solvent

(c) Mole fraction is calculated by dividing the number of moles of solvent by the number of moles of solute

(d) Mole fraction is calculated by dividing the number of moles of solute by the number of moles of the solvent

(e) Statements (a) through (d) are false

Molarity = Number of moles of solute/ Number of liters of solution

Molality = Number of moles of solute/ Number of kilograms of solvent

Mole fraction of solute = Number of moles of solute/(Number of moles of solutes + Number of moles of solvent)

Mole fraction of solvent = Number of moles of solvent/(Number of moles of solutes + Number of moles of solvent)   Answer(e)

9-17. Calculate the molality of a 10% by weight solution of NaCl in water.

(NaCl: MW = 78.5 g/mol)

Molality = Number of moles of NaCl in 1000 g of water = $(10/78.5) \times (1000/90) = 1.42$

9-18. Calculate the molarity of a 20.0% by weight solution of $CuCl_2$ in water if the density of this solution is 1.205 g/cm^3.

Molecular weight of $CuCl_2$ = $63.546 + (35.45 \times 2) = 134.45$

20% by weight solution of $CuCl_2$ contains 20 g of $CuCl_2$ in 100 g of solution.

Number of mole of $CuCl_2$ in 100 g of solution = 20/134.45

The density of the solution = 1.205 g/mL

So the volume of 100 g of solution = 100/1.205 = 82.99 mL

Molarity = Number of moles of solute/ number of liters of solution

$= (20/134.45) \times (1000/82.99) = 1.79$ M

9-19. What is the mole fraction of CCl_4 (MW = 154 g/mol) in a solution prepared by dissolving 32 g of CCl_4 in 75 g of C_6H_6?

Molecular weight of CCl_4 = 154 g/mol

Number of moles of CCl_4 = 32/154 = 0.2078

Molecular weight of C_6H_6 = $(12.01 \times 6) + (1.01 \times 6) = 78.12$

Number of moles of C_6H_6 = 75/78.12 = 0.96

Mole fraction of CCl_4 = Number of moles of CCl_4/(Number of moles of CCl_4 + Number of moles of C_6H_6)

$$= 0.2078/(0.2078 + 0.96) = 0.18$$

9-20. What volume of a solution of sulfuric acid that is 60.0 % by weight H_2SO_4 (d =1.50 g/cm^3) would be required to prepare 500 ml of a 1.5M solution? (H_2SO_4 : MW = 98 g/mol).

Molecular weight of H_2SO_4 = 98 g/mol

Weight of H_2SO_4 in 1.5 M solution = (98×1.5) g/1000 mL

Weight of H_2SO_4 in 500 ml of 1.5 M solution = (98×1.5)/2 = 73.5 g

60% by weight solution of H_2SO_4 contains 60 g of H_2SO_4 in 100 g of solution.

Volume of 100 g of solution = 100/1.5 = 66.67 mL

Volume of H_2SO_4 solution required = (66.67/60)×73.5 = 81.7 mL

9-21. Calculate the mole ratio of solute to solvent for a solution of 2 g of solute (MW = 200 g/mol) dissolved in 200 g of solvent (MW = 20 g/mol).

Number of moles of solute = 2/200 = 0.01

Number of moles of solvent = 200/20 = 10

Mole ratio of solute to solvent = 0.01/10 = $1.0×10^{-3}$

9-22. Which of the following concentration units would change when the temperature of the solution changes?

   (I)   Molarity   (II) Molality   (III) Mole fraction of solute   (IV) Mole fraction of solvent

   (a) I   (b) I and II   (c) I,II, and III   (d) All of the above   (e) None of the above

Only molarity will change with temperature because density of a solvent is sensitive to temperature. For example let us consider water. The density of water at 20 °C is 0.99821 g/cm^3 and the density of water at 50 °C is 0.98803 g/cm^3. So the volume of a given amount of water will increase when the temperature increases. This means that the molarity of an aqueous solution will decrease when the temperature increases.   Answer(a)

9-23. Which statement is not consistent with the information contained in the phase diagram of a compound?

(a) Liquids can be made to boil by raising the pressure at constant temperature
(b) Solids can be made to melt by raising the temperature at constant pressure
(c) Gases are most likely to be found at high temperatures and low pressures
(d) The boiling point of a compound is the temperature at which the vapor pressure is equal to atmospheric pressure
(e) All of the statements are consistent with the information in the phase diagram of a compound

As per phase diagram a liquid can be made to boil at constant temperature by reducing the pressure. For example water will not boil at 90 ^{o}C at 1 atm (760 mm Hg). But at 90 ^{o}C water will boil if the pressure is reduced to 526 mm Hg.   Answer(a)

9-24. Gases are most likely to be found in a phase diagram under conditions of:

(a) High temperature and high pressure   (b) High temperature and low pressure

(c) Low temperature and high pressure   (d) Low temperature and low pressure

(e) Moderate temperature and pressure

As per phase diagram gases are most likely to be found at high temperature and low pressure.

Answer(b)

9-25. Which of the following will be most soluble in a non-polar solvent such as CCl_4?

(a) KI   (b) H_2O (c) NH_3 (d) CBr_4 (e) HF

KI is a polar solute. H_2O, NH_3 and HF are polar molecules (due to difference in the electronegativities of the components) with hydrogen bonds. CBr_4 is a symmetrical non-polar molecule in which the four Br atoms point towards the corners of a tetrahedron. CCl_4 is a symmetrical non-polar solvent. Like dissolves like. So CBr_4 will be most soluble in CCl_4.

Answer(d)

9-26. Which alcohol should be most soluble in a non-polar solvent such as benzene?

(a) CH_3OH (b) CH_3CH_2OH (c) CH_3CH_2OH (d) $CH_3CH_2CH_2OH$

(e)$CH_3CH_2CH_2CH_2OH$

As the hydrocarbon chain becomes longer the non-polar character increases. Like dissolves like. So the alcohol with the longest hydrocarbon chain should be most soluble in a non-polar solvent such as benzene.   Answer(e)

9-27. Carboxylic acids with the general formula $CH_3(CH_2)_nCO_2H$ have a non-polar CH_3-

CH_2..... tail and a polar CO_2H head. What effect would increasing the value of "n" have on the solubility of carboxylic acids?

   (a) The solubility would increase in both water and in non-polar solvents such as CCl_4
   (b) The solubility would decrease in both water and in non-polar solvents such as CCl_4
   (c) The solubility would increase in water but would decrease in CCl_4
   (d) The solubility would decrease in water but would increase in CCl_4
   (e) The solubility of carboxylic acids in both water and CCl_4 would remain the same.

As the value of "n" increases, the hydrocarbon chain becomes longer and the non-polar

character of the molecule increases. So the solubility would decrease in water (polar solvent)

but would increase in CCl_4 (a non-polar solvent).   Answer(d)

9-28. An alcohol is a compound with the formula $CH_3(CH_2)_nOH$, in which the $CH_3(CH_2)_n$

group is non-polar and the OH group is polar. As "n" increases for different alcohols, we would

expect the solubilities of the alcohol to:

   (a) Increase in both polar and non-polar solvents
   (b) Increase in polar solvents and decrease in non-polar solvents
   (c) Decrease in both polar and non-polar solvents
   (d) Decrease in polar solvents and increase in non-polar solvents

As the value of "n" increases, the hydrocarbon chain becomes longer and the non-polar

character of the molecule increases. So the solubility would decrease in polar solvents and

increase in non-polar solvents.   Answer(d)

9-29. The best explanation for the fact that water and ethyl alcohol are completely miscible is

that

   (a) Both compounds are non-polar
   (b) Water is polar and ethyl alcohol is non-polar
   (c) Water has an unusually large dielectric constant
   (d) Hydrogen bonding takes place between the two solvents

The "OH" group in ethyl alcohol is "hydrophilic" and can form hydrogen bonds to

neighboring water molecules. So ethyl alcohol is infinitely soluble in water.   Answer(d)

9-30. Potassium Iodide reacts with iodine in aqueous solution to form the triiodide ion:

KI (aq) + I_2 (aq) $\rightarrow$ KI_3 (aq)

What would happen if we added CCl_4 to this reaction?

   (a) The KI would tend to dissolve in the CCl_4 layer
   (b) The I_2 would tend to dissolve in the CCl_4 layer
   (c) Both KI and I_2 would dissolve in the CCl_4 layer
   (d) Neither KI nor I_2 would dissolve in the CCl_4 layer

(e) Because CCl_4 and water are miscible, no distinct CCl_4 layer would form

CCl_4 is not miscible with water. So when CCl_4 is added the two liquid phases will be clearly visible. CCl_4 is a symmetrical non-polar solvent in which the "Cl" atoms point toward the corners of a tetrahedron. Water is a polar solvent. I_2 is a non-polar solute. Like dissolves like. So I_2 will readily leave the aqueous layer and dissolve in the CCl_4 layer.   Answer(b)

9-31. The equilibrium constant for the following reaction can be measured by extracting this solution with a non-polar solvent and then titrating both the non-polar solvent and the aqueous solution with sodium thiosulfate.

$$I_2 \text{ (aq)} + I^- \text{ (aq)} \rightarrow I_3^- \text{ (aq)}$$

Which of the following dissolves in the non-polar solvent?

    (a) I_3^- (b) I^- (c) I_3^- and I^- (d) I_2 (e) I_2, I^-, and I_3^-

Among the three only I_2 is non-polar. Like dissolves like. So only I_2 will dissolve in a non-polar solvent.   Answer(d)

9-32. Raoult's law predicts that the vapor pressure of solvent over a solution of a non-volatile solute

    (a) increases with increasing mole fraction of solvent
    (b) increases with increasing mole fraction of solute
    (c) is independent of the mole fraction of solute
    (d) decreases with increasing temperature

Raoult's Law: $P = X_{solvent} \times P^0$

P increases with increasing mole fraction of solvent.   Answer(a)

9-33. When a non-volatile solute is dissolved in a volatile solvent, the vapor pressure of the solvent over the solution

    (a) increases as the mole fraction of solute increases
    (b) decreases as mole fraction of solute increases
    (c) is independent of the mole fraction of solute
    (d) changes in an unpredictable way with changes in the mole fraction of solute

$P = X_{solvent} \times P^0$

$\quad = (1 - X_{solute}) \times P^0$

$P^0 - P = \Delta P = X_{solute} \times P^0$

ΔP increases as the mole fraction of solute increases. So the vapor pressure of the solvent over the solution will decrease as the mole fraction of solute increases. Answer(b)

9-34. For a non-volatile solute dissolved in a volatile solvent, the vapor pressure, freezing point and boiling point

    (a) are all higher for the solution than for the pure solvent
    (b) are all lower for the solution than for the pure solvent
    (c) change differently, with vapor pressure increasing and boiling and freezing points decreasing
    (d) change differently, with boiling point increasing and vapor pressure and freezing point decreasing

When a solute is added to a solvent, the vapor pressure of the solvent decreases, the boiling

point increases and the freezing point decreases. Answer(d)

9-35. For a solution of two volatile compounds, the ratio of the vapor pressure of the two

compounds in the gas phase over the liquid is

    (a) proportional to the ratio of vapor pressures of the pure compounds
    (b) proportional to the sum of the vapor pressures of the pure compounds
    (c) independent of the vapor pressures of the pure compounds
    (d) proportional to the sum of the mole fractions of the two compounds
    (e) proportional to the ratio of the mole fractions of the two compounds

$P_A = X_A P^0_A$

$P_B = X_B P^0_B$

$P_A/P_B = X_A P^0_A / X_B P^0_B = K (X_A/X_B)$

P_A/P_B is proportional to X_A/X_B Answer(e)

9-36. Given dilute, equimolal solutions of NaCl and $ZnCl_2$ in water, we would expect the

freezing points for the solutions relative to pure water to

    (a) change by the same amount
    (b) change by a ratio of 2:3, with the change for NaCl being 2/3rds that for $ZnCl_2$
    (c) change by a ratio of 1:3, with the change for NaCl being 1/3rd that for $ZnCl_2$
    (d) change by a ratio of 2:3, with the change for $ZnCl_2$ being 2/3rds that for NaCl
    (e) change by a ratio of 1:3, with the change for $ZnCl_2$ being 1/3rd that for NaCl

Let us say that the concentration is 0.1 m. NaCl and $ZnCl_2$ are strong electrolytes.

ΔT_{FP} (NaCl) $= -1.853 \times 0.1 \times 2 = -0.3706$ °C

ΔT_{FP} ($ZnCl_2$) $= -1.853 \times 0.1 \times 3 = -0.5559$ °C

ΔT_{FP} (NaCl)/ΔT_{FP} ($ZnCl_2$) $= 2/3$ Answer(b)

9-37. If freezing point depression was used to determine the molecular weight of an acid dissolved in water, and the dissociation of the acid was not taken into account, the determined value of the molecular weight of the acid would be

(a) accurate   (b) high   (c) low

(d) inaccurate, but in a direction that can not be predicted.

Molality = (Number of moles of the solute)/(Number of kilograms of the solvent)

For w_2 g of acid having molecular weight M_2 present in w_1 g of solvent

molality = $(w_2/M_2)/(w_1/1000)$

Let us say that 18 g of acid (molecular weight M_2) was dissolved in 1000 g of water and the freezing point was -0.563 $^\circ$C and the dissociation was 1.3%.

Molecular weight (when dissociation was not taken into account) = $(18/0.563) \times 1.853$

$= 59.24$ g/mol

Molecular weight (when dissociation was taken into account) $= (18/0.563) \times 1.853 \times 1.013$

$= 60.01$ g/mol

So if the dissociation was not taken into account the determined molecular weight would be lower.   Answer(c)

9-38. Which of the following equations would be appropriate for calculating the change in the freezing point of an aqueous $NaHSO_4$ solution? $NaHSO_4$ dissociates in solution in a two step process.

$NaHSO_4$ (s) $\rightarrow$ Na^+ (aq) + HSO_4^- (aq) (100 % dissociation)

HSO_4^- (aq) $\rightarrow$ H^+(aq) + SO_4^{2-} (aq) (10 % dissociation)

(a) $\Delta T_f = -(1 + 0.1)$ k$_f$m   (b) $\Delta T_f = -(0.1)$ k$_f$m   (c) $\Delta T_f = -\{1 + 2(0.1)\}$ k$_f$m

(d) $\Delta T_f = -(0.2)$ k$_f$m   (e) $\Delta T_f = -(2 + 0.1)$ k$_f$m

$\Delta T_f = -$ k$_f$mi

i = Number of particles observed/1 formula unit of the solute

$NaHSO_4$ (s) $\rightarrow$ Na^+ (aq) + HSO_4^- (aq) (100 % dissociation)

1   1

$HSO_4^- (aq) \rightarrow H^+(aq) + SO_4^{2-} (aq)$ (10 % dissociation)

1 – 0.1   0.1   0.1

$i = 1 + (1 – 0.1) + 0.1 + 0.1 = 2.1$

$\Delta T_f = – k_f m$ (2.1)   Answer(e)

9-39. A 0.010 molal solution of NH_3 in water is 4.1 % ionized.

$NH_3 (aq) + H_2O (l) \rightarrow NH_4^+ (aq) + OH^- (aq)$

Given that pure water freezes at 0 °C and k_f for water is 1.853 °C/m, which of the following

expressions gives the freezing point depression of the aqueous ammonia solution?

     (a) $\Delta T_{FP} = – 1.853 (0.01)$ (b) $\Delta T_{FP} = – 1.853 (0.01)(1.041)$

     (c) $\Delta T_{FP} = – 1.853 (0.01)(2.082)$ (d) $\Delta T_{FP} = – 1.853 (0.01)(4.1)$

     (e) $\Delta T_{FP} = – 1.853 (0.01)(8.2)$

$\Delta T_{FP} = – k_f m i$

i = Number of particles observed/1 formula unit of the solute

$NH_3 (aq) + H_2O (l) \rightarrow NH_4^+ (aq) + OH^- (aq)$ (4.1 % dissociation)

1 – 0.041   0.041   0.041

$i = (1 – 0.041) + 0.041 + 0.041 = 1.041$

$\Delta T_{FP} = – k_f m i = – 1.853 (0.01)(1.041)$ Answer(b)

9-40. Which of the following correctly describes a graph of the freezing point

of a solution versus the molality of a solution?

     (a) The freezing point increases linearly with the molality of the solution
     (b) The freezing point decreases linearly with the molality of the solution
     (c) The freezing point remains constant as the molality of the solution increases
     (d) The freezing point increases exponentially with the molality of the solution
     (e) The freezing point decreases exponentially with the molality of the solution

Freezing point of a solution = Freezing point of solvent – $k_f m$

                         = – $k_f m$ + Freezing point of solvent

          y = – $k_f m$ + b

A plot of y versus m (molality) will have –ve slope. So the freezing point will decrease linearly

with the molality of the solution.   Answer(b)

9-41. A 0.10 molal solution of sulfuric acid in water freezes at – 0.371 °C. Which of the following statements agrees with this experimental observation? (k_f for H_2O = 1.853 $^oC/m$)

(a) H_2SO_4 does not dissociate in water
(b) H_2SO_4 dissociates into H_3O^+ and HSO_4^- ions in water
(c) H_2SO_4 dissociates in water to form two H_3O^+ ions and one SO_4^{2-} ion
(d) H_2SO_4 dissociates in water to form $(H_2SO_4)_2$ dimers
(e) None of the above are consistent with experimental observation

a: $\Delta T_{FP} = -1.853 \times 0.1 = -0.1853\ ^oC$

b: $\Delta T_{FP} = -1.853 \times 0.1 \times 2 = -0.371\ ^oC$ 　　　　　　　　　　Answer(b)

9-42. KCl is a strong electrolyte while sucrose is not. Which of the following lists 0.10 m KCl,

0.10 m sucrose, and pure water in order of increasing boiling point?

(a) KCl<Sucrose<Water　　　(b) Water<KCl=Sucrose　　　(c) KCl = Sucrose<Water

(d) Water<Sucrose<KCl　　　(e) Sucrose<KCl<Water

(k_b for water = 0.515 $^oC/m$)

Boiling point of water = 100 ^oC

Boiling point of solution = (100 + ΔT_{BP}) ^oC

ΔT_{BP} for 0.1 m sucrose = $k_b \times 0.1 = 0.0515\ ^oC$

ΔT_{BP} for 0.1 m KCl = $k_b \times 0.1 \times 2 = 0.103\ ^oC$ (KCl is a strong electrolyte and i_{ideal} for KCl =2)

Boiling point of water = 100 ^oC Boiling point of 0.1 m sucrose = 100.0515 ^oC

Boiling point of 0.1 m KCl = 100.103 ^oC 　　　　　　　　　　Answer(d)

9-43. 0.400 mole of CCl_4 and 0.600 mole of $CHCl_3$ are mixed at 43 ^oC. The vapor pressure of

pure CCl_4 at this temperature is 0.354 atm and the vapor pressure of pure $CHCl_3$ at this

temperature is 0.526 atm. What is the vapor pressure of the solution?

Vapor pressure of the solution = $X_A P^0_A + X_B P^0_B$ = (0.4×0.354) + (0.6×0.526) = 0.457 atm

9-44. What would be the freezing point of a saturated solution of caffeine ($C_8H_{10}O_2N_4$) in water

if it takes 45.6 g of water to dissolve 1 g of caffeine (Caffeine: MW = 194.19 ;

H_2O: k_f = 1.853 $^oC/m$).

Molality (m) = (1/194.19)×(1000/45.6) = 0.1129 m

ΔT_{FP} = $-1.853 \times 0.1129 = -0.209\ ^oC$　　　　Freezing point of water = 0 ^oC

Freezing point of a saturated solution of caffeine in water = $-0.209\ ^oC$

9-45. The 'Tip of the week' in a recent newspaper suggested using a fertilizer such as ammonium nitrate or ammonium sulfate instead of salt to melt snow and ice on sidewalks because salt can damage lawns. Which of the following compounds would give the largest freezing point depression when 100 grams of the compound are dissolved in 1 kg of water?

    (a) NaCl (MW = 58.44 g/mol)   (b) NH_4NO_3 (MW = 80.05 g/mol)

    (c) $(NH_4)_2SO_4$ (MW = 132.14 g/mol)

    (d) These solutions all would have the same freezing point depression

Actual concentration of dissolved species

NaCl: $(100/58.44) \times 2$ = 3.422 m

NH_4NO_3: $(100/80.05) \times 2$ = 2.498 m

$(NH_4)_2SO_4$: $(100/132.14) \times 3$ = 2.271 m

ΔT_{FP}

NaCl: $-1.853 \times 3.422 = -6.341\ ^{\circ}C$

NH_4NO_3: $-1.853 \times 2.498 = -4.629\ ^{\circ}C$

$(NH_4)_2SO_4$: $-1.853 \times 2.271 = -4.208\ ^{\circ}C$ Answer(a)

9-46. The melting point of pure benzene is 278.7 K and the molal freezing point depression constant is 5.12 K/m. When 4.20 g of an unknown non-electrolyte is added to 100 g of benzene, the freezing point of the solution is 277.60 K. What is the molecular weight of unknown?

$\Delta T_{FP} = -k_f m = -5.12m = -(278.7 - 277.6) = -1.1$

m = 1.1/5.12 = 0.2148 m

Number of mole in 1000 g of solvent = 0.2148

Number of mole in 100 g of solvent = 0.02148

Molecular weight = 4.2/0.02148 = 195.5 g/mol

9-47. If 5.17 g of an unknown substance X that is a non-electrolyte lowers the freezing point of 200 g of benzene (k_f = 5.12 K/m) by 0.84 K, what is the molecular weight of X?

$\Delta T_{FP} = -k_f m = -0.84$ K   m = 0.84/5.12 = 0.1641 m

Number of mole of non-electrolyte in 1000 g of solvent = 0.1641

Number of mole of non-electrolyte in 200 g of solvent = (0.1641/1000)×200 = 0.0328

Molecular weight = 5.17/ 0.0328 = 157.62 g/mol

9-48. When a 2.15 g sample of a non-electrolyte is dissolved in 105 g of water, the freezing point of the solution is −0.62 °C. If the molal freezing point depression constant for water is 1.853 °C/m, what is the molecular weight of the non-electrolyte?

$\Delta T_{FP} = - k_f m = - 0.62$ m = 0.62/1.853 = 0.3346

Number of mole of non-electrolyte in 1000 g of solvent = 0.3346

Number of mole in 105 g of solvent = (0.3346/1000)×105 = 0.0351

Molecular weight of the non-electrolyte = 2.15/0.0351 = 61.25 g/mol

9-49. The molal freezing point depression constant for water is 1.853 °C/m and the feezing point of pure water is 0 °C. When 1.50 g of a non-electrolyte is added to 47.8 of water, the freezing point of the solution is found to change by 0.174 °C. Calculate the molecular weight of the solute.

$\Delta T_{FP} = - k_f m = - 0.174$ m = 0.174/1.853 = 0.0939

Number of mole in 1000 g of solvent = 0.0939

Number of mole in 47.8 g of solvent = (0.0939/1000)×47.8 = 4.488×10^{-3}

Molecular weight of the solute = $1.5/(4.488 \times 10^{-3})$ = 334.2 g/mol

9-50. Pure benzene melts at 5.5 °C. A solution of 1.25 g of CCl4 in 100 g of benzene would have what freezing point? (Benzene: k_f = 5.12 °C/m)

$\Delta T_{FP} = - k_f m$

Molecular weight of CCl4 = 12.01 + (35.45×4) = 153.81 g/mol

Number of mole of CCl4 in 100 g of benzene = 1.25/153.81

Molality = (1.25/153.81)×(1000/100) = 0.08127 m

$\Delta T_{FP} = - 5.12 \times 0.08127 = - 0.416$ °C

Freezing point of solution = 5.5 – 0.416 = 5.084 °C

9-51. Pure ethanol has a boiling point of 78.0 °C. What is the boiling point of a solution that contains 12.7 g of KI (MW= 166 g/mol) dissolved in 125 g of ethanol. Assume KI is a strong electrolyte in ethanol. (Ethanol: k_b = 1.22 °C/m)

$\Delta T_{BP} = k_b m$

Number of mole of solute = 12.7/166 = 0.0765

KI will dissociate to provide 2 ions.

Actual concentration of KI = (0.0765/125)×1000×2 m

ΔT_{BP} = 1.22×(0.0765/125)×1000×2 = 1.5 ^{o}C

Boiling point of solution = 78 + 1.5 = 79.5 ^{o}C

CHAPTER 10

THE STRUCTURE OF SOLIDS

INVALUABLE INFORMATION

$1 \text{ cm} = 10^7 \text{ nm}$

$1 \text{ Angstrom} = 10^{-10} \text{ m}$

$1 \text{ nm} = 10^{-9} \text{ m}$

$1 \text{ pm} = 10^{-12} \text{ m}$

10-1. Which of the following descriptions of hydrogen bonding is correct?

  (a) It is stronger than a van der Waals interaction
  (b) It is stronger than a covalent bond
  (c) It is stronger than an ionic bond

The H-bond (5 to 30 kJ/mol) is stronger than a van der Waals interaction but weaker than

covalent or ionic bonds. Answer(a)

10-2. The bonding in diamond involves sp^3 hybridized carbon. Diamond is an example of a:

  (a) Molecular solid   (b) A covalent solid   (c) An ionic solid   (d) A solid solution

  (e)A metallic solid

Diamond is like a giant molecule with endless number of covalent bonds. Each carbon atom

(sp^3 hybridized) is covalently bonded to four other carbon atoms oriented towards the

corners of a tetrahedron. Answer(b)

10-3. Rare gases, such as xenon, can be frozen to solids. The intermolecular forces responsible

for maintaining the solid are best described as:

  (a) Hydrogen bonding forces
  (b) Dipole-dipole forces
  (c) Dipole-induced dipole forces
  (d) Induced dipole-induced dipole forces

Rare gases, such as xenon, can be frozen to solids. A rare gas atom is perfectly symmetrical.

So the intermolecular forces responsible for maintaining the solid are neither dipole-dipole nor

dipole-induced dipole forces. Hydrogen bonding force also is not possible. Movement of

200

electrons around the nuclei of a pair of neighboring rare gas atoms leads to induced dipole moment in these atoms. So the inermolecular forces responsible for maintaing thc solid are induced dipole-induced dipole forces.   Answer(d)

10-4. SiO_2 is a covalent solid. Which of the following is most likely to be true of SiO_2?

  (a) SiO_2 would have low melting point and would conduct electricity when melted
  (b) SiO_2 would conduct electricity as a solid and its melting point cannot be predicted
  (c) SiO_2 would have high melting point and would conduct electricity when melted
  (d) SiO_2 would have high melting point and would not conduct electricity when melted
  (e) SiO_2 would have a low melting point and would not conduct electricity when melted

Covalent bonds are strong; so covalent solids will have high melting points. SiO_2 is a covalent solid and it will have high melting point. Electrons in molten covalent solids are localized and are not free to migrate; so molten covalent solids such as, SiO_2, will not conduct electricity.

                                                                                  Answer(d)

10-5. Which of the following types of solids is most likely to be soft, have low melting point, and not conduct electricity when melted?

   (a) Ionic   (b) Molecular   (c) Covalent (d) Metallic   (e) None of the above

Electrons in molten metallic solids are delocalized and are free to migrate ; so molten metallic solids are good conductors of electricity.

Covalent and Ionic bonds are strong; so covalent and ionic solids have high melting points. Molecular solids (example: sugar) have strong intramolecular bonds, but weak intermolecular bonds; the melting points will be low due to weak intermolecular bonds. Electrons in molten molecular solids are localized and are not free to migrate ; so molten molecular solids will not conduct electricity.   Answer(b)

10-6. Which of the following is held together by an extended network of covalent bonds?

   (a) Sodium chloride   (b) Gold   (c) Calcium carbonate   (d) Diamond

   (e) Dry ice (solid CO_2)

Diamond is like a giant molecule with endless number of covalent bonds. Each carbon atom (sp^3 hybridized) is covalently bonded to four other carbon atoms oriented towards the corners of a tetrahedron.   Answer(d)

10-7. Which force makes the most important contribution to the lattice energy of solid CO_2?

(a) Metallic bonding   (b) Ionic bonding   (c) Hydrogen bonding   (d) van der Waals forces

(e) All of the above

Solid CO_2 (dry ice) is a molecular solid. van der Waals forces hold the CO_2 molecules

together; but they are so weak that it passes from solid to gas phase at $-78\ ^oC$.   Answer(d)

10-8. A compound that is a poor conductor of electricity when solid, but a very good conductor

of electricity when molten is most likely to fit into which category?

(a) Molecular solid   (b) Covalent solid   (c) Ionic solid   (d) Metallic solid

(e) Any of the above

Electrons in metallic solids, are delocalized and are free to migrate; so metallic solids

are good conductors of electricity.

Electrons in molten molecular solids and molten covalent solids are localized and are not free

to migrate; so molten molecular solids and molten covalent solids are poor conductors

of electricity. Electrons in ionic solids are localized and are not free to migrate; so ionic solids

are poor conductors of electricity. But electrons in molten ionic solids are delocalised and

are free to migrate ; so molten ionic solids are good conductors of electricity.   Answer(c)

10-9. Sodium crystallizes in a structure in which the coordination number is 8. Which structure

best describes this crystal?

(a) Simple cubic   (b) Body-centered cubic   (c) Cubic closest-packing

(d) Hexagonal closest-packing   (e) None of the above.

Structure	Coordination number	
Simple cubic	6	
Body-centered cubic	8	
Cubic closest-packing	12	
Hexagonal closest-packing	12	Answer(b)

10-10. Cubic holes can be found in which structure?

(a) Simple cubic   (b) Body-centered cubic   (c) Cubic closest-packed

(d) Hexagonal closest-packed   (e) None of the above

Tetrahedral holes and octahedral holes are found in closest-packed structures. Cubic holes are

202

found in simple cubic structures.   Answer(a)

10-11. In which of the following structures would Xe atoms be able to form the largest number of induced dipole-induced dipole interactions?

(a) Simple cubic   (b) Body-centered cubic   (c) Either cubic or hexagonal closest-packing

Rare gases, such as xenon, can be frozen to solids. Xe atoms are perfectly symmetrical. But movement of electrons around the nuclei of a pair of neighboring Xe atoms leads to induced dipole moment in these atoms. So the intermolecular forces responsible for maintaing the solid are induced dipole-induced dipole forces. Xe atoms will be able to form the largest number of induced dipole-induced dipole interactions when the coordination number is maximum.

Structure	Coordination number
Simple cubic	6
Body-centered cubic	8
Cubic closest-packing	12
Hexagonal closest-packing	12

The coordination number is maximum in both cubic closest-packing and hexagonal closest-packing. So in either cubic closest-packing or hexagonal closest-packing Xe atoms will be able to form the largest number of induced dipole-induced dipole interactions.   Answer(c)

10-12. In KF, the K^+ and F^- ions are almost exactly the same size: 0.134 nm. Which would be larger: a neutral potassium atom or a neutral fluorine atom?

When a neutral atom loses a electron to form a cation, other electrons will be more strongly attracted to the nucleus and the radius will get smaller.So the size of a neutral potassium will be > 0.134 nm. When an electron is added to a neutral atom to form an anion, the attraction of the nucleus will decrease and the radius will get larger. So the size of a neutral fluorine atom will be < 0.134 nm. So a neutral potassium atom will be larger than a neutral fluorine atom.

10-13. The crystal structure of CoO would be expected to be most like which of the following crystals? (Radii: Co^{2+} = 0.072 nm ; O^{2-} = 0.140 nm)

(a) NaCl (b) ZnS (c) CsCl (d) CaF_2 (e) TiO_2

Radius ratio rules:

Radius ratio = (Radius of the cation)/(Radius of the anion) = $(r_+)/(r_-)$

Radius ratio	Coordination number	Holes in which +ve ions pack
0.225 – 0.414	4	Tetrahedral holes
0.414 – 0.732	6	Octahedral holes
0.732 – 1	8	Cubic holes
1	12	Closest-packed structure

CoO: 1:1 salt. $(r_+)/(r_-) = (0.072)/(0.140) = 0.514$

So filling of octahedral holes are expected.

Number of octahedral holes = Number of spheres that form the closest-packed structure

So in the case of CoO (1:1 salt) all the octahedral holes are filled by the Co^{2+} ions.

NaCl: 1:1 salt. Na^+ ions occupy all the octahedral holes in a closest-packed array of Cl^- ions.

ZnS: 1:1 salt. Zn^{2+} ions occupy half of the tetrahedral holes in a closest-packed array of S^{2-} ions.

CsCl: 1:1 salt. Crystallizes in a simple cubic unit cell of Cl^- ions with a Cs^+ ion at the center of the body of the cell.

CaF_2: 1:2 salt.

TiO_2: 1:2 salt.

So the crystal structure of CoO would be expected to be most like the crystal structure of NaCl.

Answer(a)

10-14. The crystal structure of CaTe ($Ca^{2+} = 0.099$ nm ; $Te^{2-} = 0.221$ nm) is most like which of the following?

   (a) CaF_2 (b) CsCl (c) Cu_2O (d) NaCl (e) ZnS

Radius ratio rules:

Radius ratio = (Radius of the cation)/(Radius of the anion) = $(r_+)/(r_-)$

Radius ratio	Coordination number	Holes in which +ve ions pack
0.225 – 0.414	4	Tetrahedral holes
0.414 – 0.732	6	Octahedral holes
0.732 – 1	8	Cubic holes

CaTe: 1:1 salt. $(r_+)/(r_-) = (0.099)/(0.221) = 0.448$

So filling of octahedral holes are expected.

Number of octahedral holes = Number of spheres that form the closest-packed structure

So in the case of CaTe (1:1 salt) all the octahedral are filled by the Ca^{2+} ions.

CaF_2: 1:2 salt.

CsCl: 1:1 salt.   Crystallises in a simple cubic unit cell of Cl^- ions with a Cs^+ ion at the center of the body of the cell.

Cu_2O: 2:1 salt.

NaCl: 1:1 salt. Na^+ ions occupy all the octahedral holes in a closest-packed array of Cl^- ions.

ZnS: 1:1 salt. Zn^{2+} ions occupy half of the tetrahedral holes in a closest-packed array of S^{2-} ions.

So the crystal structure of CaTe is most like the crystal structure of NaCl.   Answer(d)

10-15. What is the coordination number of a cation packed in an octahedral hole?

(a) 2   (b) 4   (c) 6   (d) 8   (e) 12

Each cation in an octahedral hole will touch six negative ions oriented towards the corners of an octahedron. So the coordination number will be 6.   Answer(c)

10-16. Which of the following compounds could be described as a closest-packed array of anions with Li cations in all of the tetrahedral holes?

(a) Li_4C (b) Li_3N (c) Li_2S (d) LiF

Number of tetrahedral holes = 2 × Number of spheres that form the closest-packed structure

So if there are 'n' anions in a closest-packed array of anions, there will be '2n' tetrahedral holes. To fill all these tetrahedral holes there must be '2n' Li cations. So the salt must be a 2:1 salt. So Li_2S could be described as a closest-packed array of anions with Li^+ ions in all of the tetrahedral holes.   Answer(c)

10-17.

(i)   What is the formula of a compound of niobium and nitrogen that crystallizes in a hexagonal closest-packed array of nitride anions with niobium cations in half of the tetrahedral holes?

(a) Nb_4N (b) Nb_2N (c) NbN (d) NbN_2 (e) NbN_4

Number of tetrahedral holes = 2 × Number of spheres that form the closest-packed structure

So if there are 'n' anions in a hexagonal closest-packed array of anions, there will be '2n' tetrahedral holes. To fill half of these tetrahedral holes there must be 'n' Nb cations. So the salt must be a 1:1 salt. So the formula must be NbN.   Answer(c)

(ii)   What is the coordination number of the niobium in the compound in the previous question?

(a) 4   (b) 6   (c) 8   (d) 12   (e) 14

The niobium cation that packs in a tetrahedral hole will touch 4 nitride ions oriented towards the corners of a tetrahedron. So the coordination number will be 4.   Answer(a)

10-18. Scandium oxide, Sc_2O_3, crystallizes with the oxide ions in a closest-packed array with the scandium ions in octahedral holes. What fraction of the octahedral holes are filled?

(a) All   (b) 2/3   (c) 1/2   (d) 1/3   (e) 1/4

Number of octahedral holes = Number of spheres that form the closest-packed structure

So in a closest-packed array of oxide ions there will be 'n' octahedral holes if there are 'n' oxide ions. In a 1:1 salt all the octahedral holes will be filled by cations. In a 1:2 salt only half of the octahedral holes will be filled by cations. In a 2:3 salt only 2/3 of the octahedral holes will be filled by cations. Sc_2O_3 is a 2:3 salt and only 2/3 of the octahedral holes will be filled by scandium ions.   Answer(b)

10-19. Titanium carbide is a high melting, chemically inert compound that is almost as hard as diamond. The crystal structure can be described as a closest-packed array of carbide ions with titanium ions in all of the octahedral holes. What is the empirical formula of this compound?

(a) Ti_2C (b) TiC (c) TiC_2 (d) Ti_2C_3 (e) TiC_4

Number of octahedral holes = Number of spheres that form the closest-packed structure

So in a closest-packed array of carbide ions there will be 'n' octahedral holes if there are 'n' carbide ions. In a 1:1 salt all the octahedral holes will be filled by cations.

So the empirical formula is TiC.   Answer(b)

10-20. What is the formula of an oxide of Titanium that crystallizes as a closest- packed array of oxide ions with titanium ions in two thirds of the octahedral holes?

(a) TiO (b) TiO_2 (c) Ti_2O_3 (d) Ti_3O_2 (e) None of the above

Number of octahedral holes = Number of spheres that form the closest-packed structure

So in a closest-packed array of oxide ions there will be 'n' octahedral holes if there are 'n' oxide ions. In a 1:1 salt all the octahedral holes will be filled by cations. In a 1:2 salt only half of the octahedral holes will be filled by cations. In a 2:3 salt only 2/3 of the octahedral holes will be filled by cations. So the formula is Ti_2O_3.   Answer(c)

10-21. Cadmium iodide crystallizes in a structure that could be described as a closest-packed array of iodide ions with cadmium ions in half of the octahedral holes. What is the oxidation state of cadmium in this compound?

(a) 0 (b) +1 (c) +2 (d) +3 (e) +4

Number of octahedral holes = Number of spheres that form the closest-packed structure

So in a closest-packed array of iodide ions there will be 'n' octahedral holes if there are 'n' iodide ions. In a 1:1 salt all the octahedral holes will be filled by cations. Half of the octahedral holes will be filled by cations in a 1:2 salt. So the formula of the compound is CdI_2 and the oxidation state of cadmium in this compound is +2.   Answer(c)

10-22. What is the oxidation state of cobalt in a compound in a closest-packed array of oxide ions with cobalt ions in 1/8 th of the tetrahedral holes and 1/2 of the octahedral holes?

(a) 5/8 (b) 3/2 (c) 2 (d) 8/3 (e) 3

Number of octahedral holes = Number of spheres that form the closest-packed structure

Number of tetrahedral holes = 2 × Number of spheres that form the closest-packed structure

So for each oxide ion there is 2/8 cobalt ion in tetrahedral hole and 1/2 cobalt ion in octahedral hole. So for each oxide ion there is $\{(1/4) + (1/2)\}$ cobalt ion.

$\{(1/4) + (1/2)\} = 3/4$

If 'Y' is the oxidation state of cobalt , $(3/4) \times Y = 2$

$Y = 8/3$ The oxidation state of cobalt in this compound = 8/3   Answer(d)

10-23. Thallium cyanide, $Tl(CN)_x$, crystallizes in a simple cubic array of CN^- ions with thallium ions in all of the cubic holes. What is the oxidation state of thallium in this compound?

(a) – 1 (b) 0 (c) +1 (d) +2 (e) +3

In a simple cubic array of CN^- ions, if there are 'n' CN^- ions there will be 'n' cubic holes. In a 1:1 salt all the cubic holes will be filled by thallium ions. So the formula for the compound is TlCN and the oxidation state of thallium in this compound is +1.   Answer(c)

10-24. Thallium iodide (TlI) crystallizes with the same structure as CsCl. How many nearest neighbor iodide ions does each Tl^+ have?

(a) 1 (b) 4 (c) 6 (d) 8 (e) 12

CsCl crystallizes in a simple cubic unit cell of Cl^- ions with a Cs^+ ion in the center of the body of the unit cell. If TlI crystallizes in this structure then each Tl^+ ion will have 8 nearest neighbor iodide ions.   Answer(d)

10-25. Which of the following correctly describes a body-centered cubic structure?

(a) CN = 6, 1 atom per unit cell
(b) CN = 6, 2 atoms per unit cell
(c) CN = 6, 4 atoms per unit cell
(d) CN = 8, 1 atom per unit cell
(e) CN = 8, 2 atoms per unit cell

Body-centered cubic unit cell is the simplest repeating unit in a body-centered cubic structure. There are identical particles at the 8 eight corners of the cubic unit cell. There is also a nineth identical particle at the center of the body of the unit cell. So the CN = 8.

Atoms per unit cell = $\{(8 \times 1/8) + 1\} = 2$.   Answer(e)

10-26. Which of the following correctly describes the structure of CaF_2?

(a) CN of Ca^{2+} = 8, CN of F^- = 4, $4Ca^{2+}$ and $8F^-$ per unit cell.
(b) CN of Ca^{2+} = 4, CN of F^- = 8, $8Ca^{2+}$ and $4F^-$ per unit cell.
(c) CN of Ca^{2+} = 4, CN of F^- = 8, $4Ca^{2+}$ and $8F^-$ per unit cell.
(d) CN of Ca^{2+} = 8, CN of F^- = 4, $8Ca^{2+}$ and $4F^-$ per unit cell.
(e) CN of Ca^{2+} = 4, CN of F^- = 4, $4Ca^{2+}$ and $8F^-$ per unit cell.

CaF_2 has 'fluorite' structure. It has FCC array of Ca^{2+} ions and all the tetrahedral holes are occupied by fluoride ions. There are $\{(8 \times 1/8)+(6 \times 1/2)\} = 4$ Ca^{2+} ions and 8 F^- ions per unit cell. Each F^- anion touches 4 Ca^{2+} cations and the CN of F^- = 4. Each Ca^{2+} touches 8 F^- anions which fill the tetrahedral holes and the CN of Ca^{2+} = 8.   Answer(a)

10-27. Diamond crystallizes in a cubic unit cell with carbon atoms in the following coordinates:

0,0,0 ; 1/2 ,1/2,0 ; 1/2 ,0,1/2 ; 0,1/2,1/2 ;

1/4,1/4,1/4 ; 1/4,3/4,3/4 ; 3/4,1/4,3/4 ; 3/4,3/4,1/4

What is the coordination number of the carbon atom in diamond?

(a) 2   (b) 4   (c) 6   (d) 8   (e) 10

In diamond each carbon atom is covalently bonded to 4 carbon atoms oriented towards the

corners of a tetrahedron. The CN of carbon in diamond = 4.   Answer(b)

10-28. What is the ratio of the length of the body diagonal of a cube to the face diagonal of a

cube?

(a) 0.817   (b) 1.00   (c) 1.22   (d) 1.41   (e) 1.73

Cell edge length = a

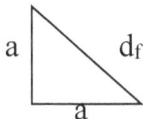   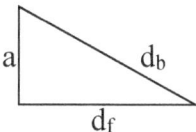

Length of the face diagonal $= d_f = a\sqrt{2}$

Length of the body diagonal $= d_b = a\sqrt{3}$

$(d_b) / (d_f) = (a\sqrt{3})/(a\sqrt{2}) = 1.22$ Answer(c)

10-29. What is the relationship between the unit cell edge length (a) and the radius of a metal

atom when the metal crystallizes in a face-centered cubic unit cell?

(a) $r = a\sqrt{2}$ (b) $(a\sqrt{2})/4$ (c) $(a\sqrt{2})/2$ (d) $(a\sqrt{3})/4$ (e) None of the above

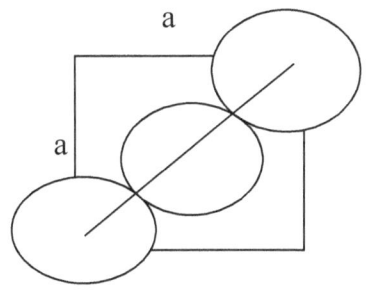

$4r = d_f = \sqrt{a^2 + a^2} = a\sqrt{2}$

$r = (a\sqrt{2})/4$ Answer(b)

10-30. Ca titanate crystallizes in a cubic unit cell with titanium ions at the corners of the cell,

oxide ions in the middle of the edges of the cell, and calcium ion in the center of the cell. What is the formula of calcium titanate?

(a) CaTiO (b) CaTiO$_3$ (c) caTiO$_4$ (d) Ca$_2$TiO$_4$ (e) CaTi$_8$O$_2$

Titanium ions are at the 8 corners of a cubic unit cell. So there is $8 \times 1/8 = 1$ titanium ion per unit cell. There is a calcium ion at the center of the body of the cubic unit cell. So there is 1 calcium ion per unit cell. Oxide ions are at the centers of the 12 edges of the cubic unit cell. So there are $12 \times 1/4 = 3$ oxide ions per cubic unit cell. So the formula is CaTiO$_3$.   Answer(b)

10-31. At very low temperatures, argon crystallizes in a structure in which argon atoms are located at the following positions:

0,0,0 ; 0,1/2,1/2 ; 1/2,0,1/2 ; 1/2,1/2,0

What is the unit cell?

(a) Simple cubic   (b) Body-centered cubic   (c) Face-centered cubic

(d) It is impossible to determine the unit cell of argon from this information

Argon atom is at 0,0,0. This implies that there are argon atoms at the 8 corners of the cubic unit cell. Argon atoms are also at 0,1/2,1/2 ; 1/2,0,1/2 ; 1/2,1/2,0. This implies that there are argon atoms at the centers of the 6 faces of the cubic unit cell. So this is a face-centered cubic unit cell.   Answer(c)

10-32.The mineral cuprite crystallizes in a structure with oxide ions at 0,0,0 and 1/2,1/2,1/2 and copper ions at 1/4,1/4,1/4 ; 1/4,3/4,3/4 3/4,1/4,3/4 and 3/4,3/4,1/4. What is the unit cell of cuprite?

(a) Simple cubic   (b) Body-centered cubic   (c) Face-centered cubic

(d) It is impossible to determine the unit cell of cuprite from this information

Oxide ion is at 0,0,0 . This implies that there are oxide ions at the 8 corners of a cubic unit cell. Oxide ion is also at 1/2,1/2,1/2. This means that there is an oxide ion at the center of the body of the unit cell. So this is a body-centered cubic unit cell.

Copper ions are at 1/4,1/4,1/4 ; 1/4,3/4,3/4 ; 3/4,1/4,3/4 and 3/4,3/4,1/4. This means that 4 terahedral holes are filled by copper ions. So this is a body-centered cubic unit cell of oxide ions and 4 tetrahedral holes are filled by copper ions.   Answer(b)

10-33. What is the empirical formula for the mineral cuprite, whose structure was described in the previous question?

(a) Cu_2O (b) CuO (c) CuO_2 (d) Cu_2O_3 (e) Cu_3O_2

It is a body-centered cubic (BCC) unit cell of oxide ions; so there are $\{(8\times1/8)+1\} = 2$ oxide ions per cubic unit cell. All the 4 tetrahedral holes are filled by copper ions; so there are 4 copper ions per unit cell. So the empirical formula is Cu_2O.   Answer(a)

10-34. Gallium arsenide is a semiconductor with several advantages over silicon. It crystallizes in a structure with Gallium ions at 0,0,0 ; 1/2,1/2,0 ; 1/2,0,1/2 ; 0,1/2,1/2 and arsenide ions at 1/4,1/4,1/4 ; 1/4,3/4,3/4 ; 3/4,1/4,3/4 ; 3/4,3/4,1/4. Describe the unit cell of this ompound, the kind of holes in which the arsenide ions are found, the fraction of the holes occupied , and the empirical formula of this compound.

Gallium ion is at 0,0,0. This implies that there are gallium ions at the 8 corners of the cubic unit cell. Gallium ions are also at 1/2,1/2,0 ; 1/2,0,1/2 ; 0,1/2,1/2. This implies that there are gallium ions at the centers of the 6 faces of the cubic unit cell. So this is a face-centered (FCC) cubic unit cell. There are $\{8\times1/8) + (6\times1/2)\} = 4$ gallium ions per cubic unit cell. There are $4\times2 = 8$ tetrahedral holes per unit cell.

Arsenide ions are at 1/4,1/4,1/4 ; 1/4,3/4,3/4 ; 3/4,1/4,3/4 ; 3/4,3/4,1/4. This means that 4 tetrahedral holes are occupied by arsenide ions. So only half of the total 8 tetrahedral holes are occupied by arsenide ions. There are 4 arsenide ions per unit cell.

There are 4 gallium ions per unit cell and 4 arsenide ions per unit cell. So the empirical formula is GaAs.

10-35. The mineral perovskite crystallizes in a cubic unit cell with a titanium ion at 0,0,0 and a calcium ion at 1/2,1/2,1/2 , and oxide ions at 1/2,0,0 ; 0,1/2,0 ; 0,0,1/2. Describe the unit cell and calculate the empirical formula of this compound.

Titanium ion is at 0,0,0. This implies that there are titanium ions at the 8 corners of the cubic unit cell. There is $8\times1/8 =1$ titanium ion per unit cell. There are no titanium ions at the centers of the 6 faces of the cubic unit cell. There is no titanium ion at the center of the body of the unit cell. So this is a simple cubic unit cell.

Calcium ion is at 1/2,1/2,1/2. This means that there is one calcium ion at the center of the body of the unit cell. There is one calcium ion per unit cell.

Oxide ions are at 1/2,0,0 ; 0,1/2,0 ; 0,0,1/2. This implies that there are oxide ions at the centers of the 12 edges of the cubic unit cell. There are $12 \times 1/4 = 3$ oxide ions per unit cell.

There is one titanium ion and one calcium ion and 3 oxide ions per unit cell. So the empirical formula of this compound is $CaTiO_3$.

10-36. Calculate the number of chloride and ammonium ions per unit cell if NH_4Cl is a simple cubic unit cell of NH_4^+ ions with a Cl^- ion in the center of the body of the unit cell.

(a) $1NH_4^+$ ion $1Cl^-$ ion (b) $2NH_4^+$ ions and $2Cl^-$ ions (c) $4NH_4^+$ ions and $4Cl^-$ ions

(d) $8NH_4^+$ ions and $8Cl^-$ ions (e) None of the above

The unit cell is a simple cubic unit cell of NH_4^+ ions. This implies that there are NH_4^+ ions at the 8 corners of the cubic unit cell. So there is $8 \times 1/8 = 1$ NH_4^+ ion per unit cell. There is a Cl^- ion at the center of the body of the cubic unit cell. So there is 1 Cl^- ion per unit cell.

Answer(a)

10-37. Which of the following correctly describes the structure of NaOH if this compound crystallizes in the same structure as CsCl?

(a) Na^+ CN =6, 1 Na^+ and 1 OH^- per unit cell
(b) Na^+ CN =6, 2 Na^+ and 2 OH^- per unit cell
(c) Na^+ CN =8, 1 Na^+ and 1 OH^- per unit cell
(d) Na^+ CN =8, 2 Na^+ and 2 OH^- per unit cell
(e) Na^+ CN =8, 4 Na^+ and 4 OH^- per unit cell

CsCl crystallizes in a simple cubic unit cell of Cl^- ions with a Cs^+ ion at the center of the body of the unit cell. If NaOH crystallizes in this structure then there is $8 \times 1/8 = 1$ OH^- ion and 1 Na^+ ion per unit cell. The Na^+ ion will touch 8 OH^- ions at the corners of the cubic unit cell. So Na^+ CN = 8.

Answer(c)

10-38. Potassium metal has a density of 0.862 g/cm^3 and a cell edge length of 0.532 nm. What is the lattice type of potassium metal? (AW: K = 39.1 amu)

(a) Simple cubic (b) Body-centered cubic (c) Face-centered cubic

(d) The NaCl structure (e) The ZnS structure

Cell edge length = 0.532 nm = 0.532×10^{-7} cm = a

Volume of the cubic unit cell = $a^3 = (0.532 \times 10^{-7})^3 = 0.15 \times 10^{-21}$ cm^3

Mass of the potassium atom = $(39.1)/(6.022 \times 10^{23})$ g

Let us say that the unit cell contains 'Y' potassium atoms.

Density = $(Y \times 39.1)/\{(6.022 \times 10^{23}) \times (0.15 \times 10^{-21})\}$ g/cm^3 = 0.862 g/cm^3

Y = 2

In a simple cubic unit cell there will be $(8 \times 1/8) = 1$ potassium atom.

In a body-centered cubic unit cell there will be $\{(8 \times 1/8) + 1\} = 2$ potassium atoms.

In a face-centered cubic unit cell there will be $\{(8 \times 1/8) + (6 \times 1/2)\} = 4$ potassium atoms.

There are 2 potassium atoms per unit cell. So the lattice type is body-centered cubic. Answer(b)

10-39. CdO crystallizes in a cubic unit cell with a cell edge length of 0.47 nm. The density of CdO is 8.2 g/cm^3 . How many cadmium and oxide ions are present in a single unit cell? (AW: O = 16.0, Cd = 112 amu)

(a) 1 Cadmium ion and 1 oxide ion   (b) 2 Cadmium ions and 2 oxide ions

(c) 3 Cadmium ions and 3 oxide ions   (d) 4 Cadmium ions and 4 oxide ions

(e) 8 Cadmium ions and 8 oxide ions

Cell edge length = a = 0.47×10^{-7} cm

Volume of the cubic unit cell = a^3 = $(0.47 \times 10^{-7})^3$ = 1.0×10^{-22} cm^3

This is a 1:1 salt. So the number of cadmium ions and the number of oxide ions should be same in a unit cell. Let 'Y' be the number of cadmium and oxide ions per unit cell.

Density = $\{Y \times (16 + 112)\}/\{(6.022 \times 10^{23}) \times (1.0 \times 10^{-22})\}$ = 8.2 g/cm^3

Y = 4

There are 4 cadmium ions and 4 oxide ions in a single unit cell.   Answer(d)

10-40. Pyrite FeS$_2$ crystallizes in a cubic unit cell with a cell edge length of 0.541 nm. The density of pyrite is 5.02 g/cm^3. How many Fe^{2+} and S_2^{2-} ions are present in a sigle unit cell? (AW: S =32.1, Fe = 55.8 amu)

(a) $1Fe^{2+}$ and $1S_2^{2-}$ (b) $2Fe^{2+}$ and $2S_2^{2-}$ (c) $4Fe^{2+}$ and $4S_2^{2-}$ (d) $6Fe^{2+}$ and $6S_2^{2-}$

(e) $8Fe^{2+}$ and $8S_2^{2-}$

Cell edge length = a = 0.541 nm = 0.541×10^{-7} cm

Volume of the cubic unit cell = a^3 = $(0.541 \times 10^{-7})^3$ = 0.16×10^{-21} cm^3

This is a 1:1 salt. For each Fe^{2+} ion there will be one S_2^{2-} ion in the unit cell.

Let 'Y' be the number of Fe^{2+} ions and S_2^{2-} ions per unit cell.

Density = {Y×(55.8+32.1 +32.1)}/ {(6.022×10²³)×(0.16×10⁻²¹)} = 5.02 g/cm³

$$Y = 4$$

There are 4 Fe^{2+} and 4 S_2^{2-} ions per unit cell. $\qquad$ Answer(c)

10-41. What is the density of sodium if this metal crystallizes in a body-centered cubic unit cell with a cell edge length of 0.429 nm? (AW: Na = 23.0 amu)

$\quad$ (a) 0.484 g/cm³ $\quad$ (b) 0.582 g/cm³ $\quad$ (c) 0.968 g/cm³ $\quad$ (d) 1.94 g/cm³ $\quad$ (e) 9.67×10⁻²⁵ g/cm³

Cell edge length = a = 0.429 nm = 0.429×10⁻⁷ cm

Volume of the cubic unit cell = a³ = (0.429×10⁻⁷)³ = 7.895×10⁻²³ cm³

Na crystallizes in body-centered cubic unit cell. So there are 2 sodium atoms per unit cell.

Density $\,$ = (2×23)/{(6.022×10²³)×(7.895×10⁻²³)} = 0.968 g/cm³ $\qquad$ Answer(c)

10-42. What is the approximate density of titanium metal if the atomic radius of a titanium atom is 0.145 nm and the atoms form a body-centered cubic unit cell? (AW: Ti = 47.9 amu)

$\quad$ (a) 0.1 g/cm³ $\quad$ (b) 1 g/cm³ $\quad$ (c) 4 g/cm³ $\quad$ (d) 25 g/cm³ $\quad$ (e) 30 g/cm³

Titanium atoms form a body-centered cubic unit cell. So there are two titanium atoms per unit cell.

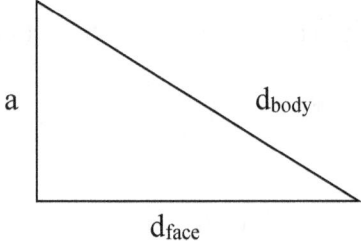

Cell edge length = a

Atomic radius of titanium = r = 0.145 nm = 0.145×10⁻⁷ cm

In a body-centered cubic unit cell $d_{body} = 4r = \sqrt{a^2 + (d_{face})^2} = \sqrt{a^2 + (a^2 + a^2)} = a\sqrt{3}$

a = 4r/ ($\sqrt{3}$) = {4× (0.145×10⁻⁷)}/ ($\sqrt{3}$) = 3.349×10⁻⁸ cm

Volume of the cubic unit cell = a³ = (3.349×10⁻⁸)³ = 3.755×10⁻²³ cm³

Density = (2×47.9)/{(6.022×10²³)×(a³)} = (2×47.9)/{(6.022×10²³)×(3.755×10⁻²³)}

$\qquad$ = 4.24 g/cm³ ≈ 4 g/cm³ $\qquad\qquad$ Answer(c)

10-43. Determine whether molybdenum crystallizes in a simple cubic, body-centered cubic or face-centered cubic unit cell if the unit cell edge length is 0.3147 nm and the density of this metal is 10.2 g/cm^3. (AW: Mo = 95.94 amu)

(a) Simple cubic   (b) Body-centered cubic   (c) Face-centered cubic

(d) Impossible to determine from the information given

Cell edge length = a = 0.3147 nm = 0.3147×10^{-7} cm

Volume of the cubic unit cell = $a^3 = (0.3147 \times 10^{-7})^3 = 0.031 \times 10^{-21}$ cm^3

Let 'Y' be the number of molybdenum atoms per unit cell.

Density = $(Y \times 95.94)/\{(6.022 \times 10^{23}) \times (0.031 \times 10^{-21})\} = 10.2$ g/cm^3

Y = 2

In a simple cubic unit cell there will be $(8 \times 1/8) = 1$ molybdenum atom.

In a body-centered cubic unit cell there will be $\{(8 \times 1/8)+1\} = 2$ molybdenum atoms.

In a face-centered cubic unit cell there will be $\{(8 \times 1/8)+(6 \times 1/2)\} = 4$ molybdenum atoms.

There are 2 molybdenum atoms in a cubic unit cell. So molybdenum crystallizes in a body-centered cubic unit cell.   Answer(b)

10-44. Which of the following metals crystallizes in a face-centered cubic unit cell with a cell edge length of 0.3608 nm if the density of the metal is 8.95 g/cm^3? (AW: Na= 23.0 , Ca = 40.1, Ti = 47.9, Cu = 63.5 and Au = 197.0 amu)

(a) Na   (b) Ca   (c) Ti   (d) Cu   (e) Au

In a face-centered cubic unit cell there are $\{(8 \times 1/8)+(6 \times 1/2)\} = 4$ metal atoms.

 Let 'Y' be the number of atoms per unit cell.

(a) Cell edge length = a = 0.3608 nm = 0.3608×10^{-7} cm

Volume of the cubic unit cell = $a^3 = (0.3608 \times 10^{-7})^3 = 4.697 \times 10^{-23}$ cm^3

Density = $(Y \times 23)/\{(6.022 \times 10^{23}) \times (4.697 \times 10^{-23})\} = 8.95$ g/cm^3

Y = 11.0

(b) Cell edge length = a = 0.3608 nm = 0.3608×10^{-7} cm

Volume of the cubic unit cell = $a^3 = (0.3608 \times 10^{-7})^3 = 4.697 \times 10^{-23}$ cm^3

Density = $(Y \times 40.1)/\{(6.022 \times 10^{23}) \times (4.697 \times 10^{-23})\} = 8.95$ g/cm^3

Y = 6.3

(c) Cell edge length = a = 0.3608 nm = 0.3608×10^{-7} cm

Volume of the cubic unit cell = $a^3 = (0.3608 \times 10^{-7})^3 = 4.697 \times 10^{-23}$ cm^3

Density = $(Y \times 47.9)/\{(6.022 \times 10^{23}) \times (4.697 \times 10^{-23})\} = 8.95$ g/cm^3

Y = 5.3

(d) Cell edge length = a = 0.3608 nm = 0.3608×10^{-7} cm

Volume of the cubic unit cell = $a^3 = (0.3608 \times 10^{-7})^3 = 4.697 \times 10^{-23}$ cm^3

Density = $(Y \times 63.5)/\{(6.022 \times 10^{23}) \times (4.697 \times 10^{-23})\} = 8.95$ g/cm^3

Y = 4.0

(e) Cell edge length = a = 0.3608 nm = 0.3608×10^{-7} cm

Volume of the cubic unit cell = $a^3 = (0.3608 \times 10^{-7})^3 = 4.697 \times 10^{-23}$ cm^3

Density = $(Y \times 197)/\{(6.022 \times 10^{23}) \times (4.697 \times 10^{-23})\} = 8.95$ g/cm^3

Y = 1.3

So Cu crystallizes in a face-centered cubic unit cell.   Answer(d)

10-45. Tungsten crystallizes in a cubic unit cell with a cell edge length of 0.3981 nm. The density of tungsten is 19.35 g/cm^3. Calculate the number of atoms per unit cell. (AW: W= 183.35 amu)

   (a) 1   (b) 2   (c) 3   (d) 4   (e) 8

Cell edge length = a = 0.3981 nm = 0.3981×10^{-7} cm

 Volume of the cubic unit cell = $a^3 = (0.3981 \times 10^{-7})^3 = 6.309 \times 10^{-23}$ cm^3

Let 'Y' be the number of tungsten atoms per unit cell.

Density = $(Y \times 183.35)/\{(6.022 \times 10^{23}) \times (6.309 \times 10^{-23})\} = 19.35$ g/cm^3

 Y = 4.

There are 4 tungsten atoms per cubic unit cell.   Answer(d)

10-46. Iron crystallizes in a cubic unit cell with a cell edge length of 0.2866 nm. The density of iron is 7.875 g/cm^3. Calculate the number of atoms per unit cell. (AW: Fe= 55.847 amu)

   (a) 1   (b) 2   (c) 3   (d) 4   (e) 8

Cell edge length = a = 0.2866 nm = 0.2866×10^{-7} cm

Volume of the cubic unit cell = a^3 = $(0.2866 \times 10^{-7})^3$ = 2.354×10^{-23} cm^3

Let 'Y' be the number of iron atoms per unit cell.

Density = $(Y \times 55.847)/\{(6.022 \times 10^{23}) \times (2.354 \times 10^{-23})\}$ = 7.875 g/cm^3

Y = 2.

There are 2 iron atoms per cubic unit cell.   Answer(b)

10-47. Diamond crystallizes in a cubic unit cell with carbon atoms at the following positions:

0,0,0 ; 1/2,1/2,0 ; 1/2,0,1/2 ; 0,1/2,1/2 ; 1/4,1/4,1/4 ; 1/4,3/4,3/4 ; 3/4,1/4,3/4 ;

3/4, 3/4,1/4. If the density of the diamond is 3.515 g/cm^3, calculate the unit cell edge length.

(AW: C = 12 amu)

Cell edge length = a

Volume of the cubic cell = a^3

Carbon atom is at 0,0,0. This implies that there are carbon atoms at the 8 corners of the cubic unit cell contributing $8 \times 1/8 = 1$ carbon to the unit cell.

Carbon atoms are also at 1/2,1/2,0 ; 1/2,0,1/2 ; 0,1/2,1/2. This implies that there are carbon atoms at the centers of the 6 faces of the cubic unit cell contributing $6 \times 1/2 = 3$ atoms to the cubic unit cell.

Carbon atoms are also at 1/4,1/4,1/4 ; 1/4,3/4,3/4 ; 3/4,1/4,3/4 ; 3/4, 3/4,1/4. This means that 4 tetrahedral holes are filled by carbon atoms contributing 4 atoms to the cubic unit cell. So there are 8 atoms in the cubic unit cell.

Density = $(8 \times 12)/\{(6.022 \times 10^{23}) \times (a^3)\}$ = 3.515 g/cm^3

Volume of the cubic unit cell = a^3 = $(8 \times 12)/\{(6.022 \times 10^{23}) \times (3.515)\}$ = 4.54×10^{-23} cm^3

Cell edge length = a = 0.3567×10^{-7} cm = 0.3567 nm

10-48. The metal chromium has an atomic weight of 52.0 amu and crystallizes in a body-centered cubic structure with a density of 7.12 g/cm^3, what is the volume of the unit cell?

   (a) 1.21×10^{-23} cm^3 (b) 2.43×10^{-23} cm^3 (c) 4.86×10^{-23} cm^3 (d) 14.6 cm^3 (e) 7.32 cm^3

Cell edge length = a   Volume of the cubic unit cell = a^3

Chromium crystallizes in a body-centered cubic structure. So there are 2 chromium atoms per unit cell.

Density = $(2 \times 52)/\{(6.022 \times 10^{23}) \times (a^3)\} = 7.12$ g/cm^3

Volume of the cubic unit cell = $a^3 = (2 \times 52)/\{(6.022 \times 10^{23}) \times (7.12)\} = 2.43 \times 10^{-23}$ cm^3.

Answer(b)

10-49. The metal chromium has an atomic weight of 52.0 amu and crystallizes in a body centered cubic structure with a density of 7.12 g/cm^3. What is the metallic radius of a chromium atom.

(a) 2.90×10^{-8} cm   (b) 1.45×10^{-8} cm   (c) 1.26×10^{-8} cm   (d) 1.79×10^{-8} cm

(e) 2.05×10^{-8} cm

Cell edge length = a   Volume of the cubic unit cell = a^3

Chromium crystallizes in a body-centered cubic structure. So there are

$\{(8 \times 1/8) + 1\} = 2$ chromium atoms per unit cell.

Density = $(2 \times 52)/\{(6.022 \times 10^{23}) \times (a^3)\} = 7.12$ g/cm^3

Volume of the cubic unit cell = $a^3 = (2 \times 52)/\{(6.022 \times 10^{23}) \times (7.12)\} = 2.43 \times 10^{-23}$ cm^3.

a = 0.29×10^{-7} cm

Metallic radius = r

In BCC r = $(a \times \sqrt{3})/4 = (0.29 \times 10^{-7} \times \sqrt{3})/4 = 1.26 \times 10^{-8}$ cm   Answer(c)

10-50. Calculate the atomic radius of an argon atom if argon crystallizes at low temperature in a face-centered cubic unit cell with a density of 1.623 g/cm^3. (AW : Ar = 39.948 amu)

Cell edge length = a   Volume of the cubic unit cell = a^3

Argon crystallizes in a face-centered cubic unit cell. So there are

$\{(8 \times 1/8) + (6 \times 1/2)\} = 4$ argon atoms per unit cell.

Density = $(4 \times 39.948)/\{(6.022 \times 10^{23}) \times (a^3)\} = 1.623$ g/cm^3

Volume of the cubic unit cell = $a^3 = (4 \times 39.948)/\{(6.022 \times 10^{23}) \times (1.623)\} = 1.635 \times 10^{-22}$ cm^3.

a = 0.5468×10^{-7} cm

Atomic radius of argon = r

In FCC r = $(a \times \sqrt{2})/4 = (0.5468 \times 10^{-7} \times \sqrt{2})/4 = 0.1933 \times 10^{-7}$ cm

The atomic radius of argon atom = 0.1933 nm

10-51. Calculate the atomic radius of the Ag atom if silver crystallizes in a face-centered cubic unit cell with a cell edge length of 0.4086 nm.

Cell edge length = a   Atomic radius of silver = r

In FCC the atomic radius = $(a\sqrt{2})/4$ = $(0.4086 \times \sqrt{2})/4$ = 0.1445 nm

The atomic radius of silver = 0.1445 nm

10-52. Cesium iodide consists of a simple cubic lattice of I^- ions with Cs^+ ions in the cubic holes. If the cell edge length is 0.445 nm, what is the Cesium ion – Iodide ion interionic distance?

   (a) 0.193 nm   (b) 0.314 nm   (c) 0.385 nm   (d) 0.629 nm   (e) 0.770 nm

In a simple cubic unit cell,

d_{body} = 2× radius of cesium ion + 2× radius of iodide ion

d_{body} = $a\sqrt{3}$ = $0.445 \times \sqrt{3}$ = 0.771 nm

2× radius of cesium ion + 2× radius of iodide ion = 0.771

 radius of cesium ion + radius of iodide ion = 0.771/2 = 0.385 nm   Answer(c)

10-53. NaH crystallizes with a structure similar to NaCl. If the unit cell edge length in this crystal is 0.4880 nm, calculate the average Na-H bond length in this crystal.

NaH crystallizes in a structure similar to that of NaCl. So H^- ions are at the 8 corners of a cubic unit cell and H^- ions are also at the centers of the 6 faces of the cubic unit cell; there is one Na^+ ion at the center of the body of the cubic unit cell and there are Na^+ ions at the centers of the 12 edges of the cubic unit cell.

(2× radius of sodium ion + 2× radius of hydride ion) = Cell edge length = a = 0.4880 nm

(Radius of sodium ion + radius of hydride ion) = 0.4880/2 = 0.2440 nm

The average Na-H bond length = 0.244 nm.

10-54. Barium sulfide crystallizes in a cubic unit cell with an edge length of 0.638 nm with ions at the following positions: S^{2-} ions are at 0,0,0 ; 1/2,1/2,0 ; 1/2,0,1/2 ; 0,1/2,1/2 and Ba^{2+} ions are at 1/2,1/2,1/2 ; 1/2,0,0 ; 0,1/2,0 ; 0,0,1/2

   (i)   The unit cell would be best described as:

   (a) Simple cubic   (b) Body-centered cubic   (c) Face-centered cubic

   (d) hexagonal closest-packed   (e) None of the above

S^{2-} ion is at 0,0,0. This implies that there are S^{2-} ions are at the 8 corners of the cubic unit cell.

S^{2-} ions are also at 1/2,1/2,0 ; 1/2,0,1/2 ; 0,1/2,1/2. This implies that there are S^{2-} ions at the centers of the 6 faces of the cubic unit cell. So this is a face-centered cubic unit cell.

Answer(c)

(ii)   Each unit cell contains how many Ba^{2+} ans S^{2-} ions?

(a) $1Ba^{2+}$ and $1S^{2-}$ (b) $2Ba^{2+}$ and $2S^{2-}$ (c) $2Ba^{2+}$ and $4S^{2-}$ (d) $4Ba^{2+}$ and $2S^{2-}$

(e) $4Ba^{2+}$ and $4S^{2-}$

S^{2-} ion is at 0,0,0. This implies that there are S^{2-} ions at the 8 corners of the cubic unit cell.

S^{2-} ions are also at 1/2,1/2,0 ; 1/2,0,1/2 ; 0,1/2,1/2. This implies that there are S^{2-} ions at the centers of the 6 faces of the cubic unit cell. So the unit cell is a face-centered cubic unit cell.

Number of S^{2-} ions per unit cell = $\{(8 \times 1/8) + (6 \times 1/2)\}$ = 4.

Ba^{2+} ion is at 1/2,1/2,1/2. This means that there is one Ba^{2+} at the center of the body of the cubic unit cell. Ba^{2+} ions are also at 1/2,0,0 ; 0,1/2,0 ; 0,0,1/2. This implies that there are Ba^{2+} ions at the centers of the 12 edges of the cubic unit cell.

Number of Ba^{2+} ions per unit cell = $\{(12 \times 1/4) + 1\}$ = 4   Answer(e)

(iii)   What is the radius of the Ba^{2+} ion? Assume that the Ba^{2+} and S^{2-} ions touch and that the radius of the S^{2-} ion is 0.184 nm.

(a) 0.135 nm (b) 0.267 nm (c) 0.270 nm (d) 0.276 nm (e) 0.368 nm

S^{2-} ions are at the 0,0,0 ; 1,0,0 and Ba^{2+} ion is at 1/2,0,0. So

$2 \times$ radius of Ba^{2+} ion + $2 \times$ radius of S^{2-} ion = cell edge length = 0.638 nm

Radius of Ba^{2+} ion + radius of S^{2-} ion = 0.638/2 = 0.319 nm

Radius of Ba^{2+} ion = 0.319 − radius of S^{2-} ion = 0.319 − 0.184 = 0.135 nm   Answer(a)

10-55.

(i)   A compound of unknown formula crystallizes in a body-centered cubic unit cell with M ions at 0,0,0 and 1/2,1/2,1/2 and with X ions at the centers of 6 faces. What is the empirical formula of this compound?

(a) MX (b) M_2X (c) MX_2 (d) M_2X_3 (e) M_3X_2

M ion is at 0,0,0. This implies that there are M ions at the 8 corners of a body-centered cubic unit cell. M ion is also at 1/2,1/2,1/2. This means that there is one M ion at the center of the body of the body-centered cubic unit cell. So the number of M ions per unit cell

$= \{(8 \times 1/8)+1] = 2$

X ions are at the centers of the 6 faces of the body-centered cubic unit cell. So the number of X ions per unit cell $= (6 \times 1/2) = 3$

The empirical formula of the compound is M_2X_3 Answer(d)

    (ii)   What is the coordination number of the M ion in the previous question?

    (a) 2 (b) 4 (c) 6 (d) 8 (e) 12

M ion is at 1/2,1/2,1/2. It touches 6 X ions at the centers of the 6 faces of the body-centered cubic unit cell. The coordination number of the M ion = 6   Answer(c)

    (iii)   What is the shortest distance between two X ions in this structure?

    (a) $a/\sqrt{2}$ (b) a (c) $a\sqrt{2}$ (d) $(2\sqrt{2})a$ (e) None of the above

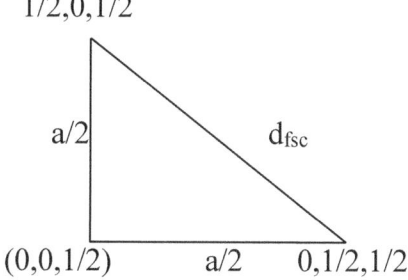

The shortest distance will be the distance between two X ions at the centers of two adjacent faces. X ions are at 1/2,0,1/2 and 0,1/2,1/2. The shortest distance will be the distance between these two points.

Cell edge length = a

Shortest distance between two X ions = $d_{facesubcell}$

$d_{fsc} = d_{facesubcell} = \sqrt{(a/2)^2 + (a/2)^2} = a/\sqrt{2}$ Answer(a)

10-56. X-rays from a molybdenum X-ray tube ($\lambda = 0.7093$ Angstroms) are diffracted at an angle of 7.11° from a sample of metallic iron. Assuming that n =1, what is the distance between the planes that give rise to this deflection? Give your answer in Angstroms and picometers to 3 significant figures.

$2d \sin \theta = n\lambda$

$d = (0.7093)/ (2 \times \sin 7.11°) = 2.87$ Angstroms

$= 287$ picometers

10-57. Cuprite is the name of a mineral that crystallizes in a cubic cell for which the cell edge length is 0.4270nm. The positions of the ions are described by the following coordinates: Oxide ions are at 0,0,0 and 1/2,1/2,1/2 and copper ions are at 1/4,1/4,1/4 ; 1/4,3/4,3/4 ; 3/4,1/4,3/4 ; 3/4,3/4,1/4. How many net ions of each type are contained within the unit cell? What is the empirical formula of this compound? What is the oxidation number of copper in this compound? What are the coordination numbers of copper and oxide ions in this compound?Is this a simple cubic, body-centered cubic or face-centered cubic unit cell? What is the copper ion – oxide ion interionic distance in nm?

Oxide ion is at 0,0,0. This implies that there are oxide ions at the 8 corners of the cubic unit cell. Oxide ion is also at 1/2,1/2,1/2. This means that there is one oxide ion at the center of the body of the cubic unit cell. So this is a body-centered cubic unit cell. Number of oxide ions per cubic unit cell = {(8×1/8)+1} = 2

Copper ions are at 1/4,1/4,1/4 ; 1/4,3/4,3/4 ; 3/4,1/4,3/4 ; 3/4,3/4,3/4. This means that there are copper ions in the 4 tetrahedral holes in the cubic unit cell. Number of Copper ions per cubic unit cell = 4. So the empirical formula of this compound is Cu_2O and the oxidation number of copper is +1.

Copper ion touches two oxide ions and the coordination number of copper ion = 2.

Oxide ion touches 4 copper ions and the coordination number of oxide ion = 4.

Oxide ions are at 0,0,0 and 1/2,1/2,1/2. Copper ion is at 1/4,1/4,1/4.

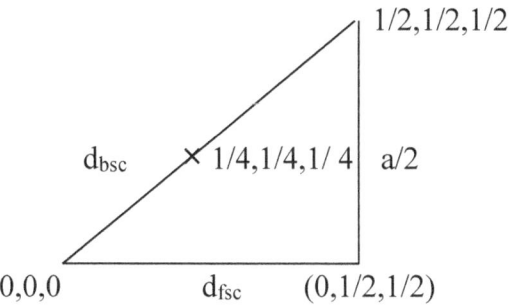

$d_{fsc} = d_{facesubcell} = a/\sqrt{2}$

$d_{bsc} = d_{bodysubcell} = \sqrt{(a/2)^2 + (d_{fsc})^2} = \sqrt{(a/2)^2 + (a^2/2)} = (a/2) \times \sqrt{3} = (0.427/2) \times \sqrt{3}$

$= 0.3698$ nm

2×radius of copper ion + 2× radius of oxide ion = d_{bsc} = 0.3698 nm

Radius of copper ion + radius of oxide ion = $d_{bsc}/2$ = 0.3698/2 = 0.1849 nm

The copper ion – oxide ion interionic distance = 0.1849 nm

10-58. Rhenium oxide crystallizes in a cubic unit cell for which the cell edge length is 0.3751 nm. The positions of the ions are described by the following coordinates: Rhenium ions are at 0,0,0 and oxide ions are at 0,0,1/2 ; 0,1/2,0 ; 1/2,0,0. How many net ions of each type are contained within the unit cell? What is the empirical formula of this compound? What is the oxidation number of Re in this compound? What are the coordination numbers of rhenium and oxide ions in this compound? Is this a simple cubic, body-centered cubic, or face-centered cubic unit cell? What is the rhenium ion – oxide ion interionic distance in nm?

Rhenium ion is at 0,0,0. This implies that there are rhenium ions at the 8 corners of the cubic unit cell. There is no rhenium ion at the center of the body of the cubic unit cell. There are no rhenium ions at the centers of the 6 faces of the cubic unit cell. So this is a simple cubic unit cell. Number of rhenium ion per cubic unit cell = (8×1/8) = 1.

Oxide ions are at 0,0,1/2 ; 0,1/2,0 ; 1/2,0,0. This implies that there are oxide ions at the centers of the 12 edges of the unit cell. Number of oxide ions per unit cell = (12×1/4) = 3.

The empirical formula of this compound is ReO_3 and the oxidation number of Re is +6.

Rhenium ion touches 6 oxide ions and the CN of Rhenium ion = 6.

Oxide ion touches 2 Rhenium ions and the CN of oxide ion = 2.

Oxide ions are at the centers of 12 edges of the cubic unit cell and Re ions are at the 8 corners

of the cubic unit cell.So

2×radius of rhenium ion + 2×radius of oxide ion = cell edge length = 0.3751 nm

Radius of rhenium ion + radius of oxide ion = cell edge length/2 = 0.3751/2 = 0.1876 nm

The rhenium ion - oxide ion interionic distance = 0.1876 nm.

10-59. Potassium magnesium fluoride crystallizes in a cubic unit cell for which the cell edge

length is 0.3973 nm. The positions of the ions are described by the following cordinates:

Potassium ion is at 0,0,0. magnesium ion is at 1/2,1/2,1/2 and fluoride ions are at 0,1/2,1/2 ;

 1/2,0,1/2 ; 1/2,1/2,0. How many net ions of each type are contained within the unit cell? What

is the empirical formula of this compound? What is the oxidation number of magnesium in this

compound? What are the coordination numbers of potassium, magnesium and fluoride ions in

this compound? Is this a simple cubic, body-centered cubic or face-centered cubic unit cell?

What is the magnesium ion – fluoride ion interionic distance in nm?

Potassium ion is at 0,0,0. This implies that there are potassium ions at the 8 corners of the

cubic unit cell. There is no potassium ion at the center of the body and there are no potassium

ions at the centers of the 6 faces of the cubic unit cell. So this is a simple cubic unit cell.

Number of potassium ion per unit cell = (8×1/8) = 1.

Magnesium ion is at 1/2,1/2,1/2. This means that there is one magnesium ion at the center

 of the body of the of the cubic unit cell. Number of magnesium ion per unit cell = 1.

Fluoride ions are at 0,1/2,1/2 ; 1/2,1/2,0 ; 1/2,1/2,0. This implies that there are fluoride ions at

the centers of the 6 faces of the cubic unit cell.

 Number of fluoride ions per unit cell = (6×1/2) = 3.

The empirical formula of this compound is $KMgF_3$ and the oxidation number of

magnesium is +2.

Potassium ion touches 12 fluoride ions and the CN of potassium ion = 12.

Magnesium ion touches 6 fluoride ions and the CN of magnesium ion = 6.

Fluoride ions touch 4 potassium ions and 2 magnesium ions. The coordination number of

fluoride ion = 6.

Magnesium ion is at the center of the body of the cubic unit cell and Fluoride ions are at the centers of the 6 faces of the cubic unit cell. So Radius of the magnesium ion + radius of the fluoride ion

= a/2 = cell edge length/2 = 0.3973/2 = 0.1987 nm.

The magnesium ion – fluoride ion interionic distance = 0.1987 nm.

10-60. Zirconium carbide crystallizes in a cubic unit cell for which the cell edge length is 0.4638 nm. The positions of the ions are described by the following coordinates: Zirconium ions are at 0,0,0 ; 0,1/2,1/2 ; 1/2,0,1/2 ; 1/2,1/2,0 and carbide ions are at 1/2,1/2,1/2 ; 1/2,0,0 ; 0,1/2,0 ; 0,0,1/2. How many net ions of each type are contained within the unit cell? What is the empirical formula of this compound? What are the coordination numbers of zirconium and carbide ions in this compound? Is this a simple cubic, body-centered cubic or face-cetered cubic unit cell? What is the zirconium ion – carbide ion interionic distance?

Zirconium ion is at 0,0,0. This implies that that there are zirconium ions at the 8 corners of the cubic unit cell. Zirconium ions are also at 1,1/2,1/2 ; 1/2,0,1/2 ; 1/2,1/2,0. This implies that there are zirconium ions at the centers of the 6 faces of the cubic unit cell. So this is a face-centered cubic unit cell. Number of Zirconium ions per unit cell = {(8×1/8) + (6×1/2)}

= 4.

Carbide ion is at 1/2,1/2,1/2. This means that there is a carbide ion at the center of the body of the cubic unit cell. Carbide ions are also at 1/2,0,0 ; 0,1/2,0 ; 0,0,1/2. This implies that there are carbide ions at the centers of the 12 edges of the cubic unit cell. Number of carbide ions per unit cell = 1+ (12×1/4) = 4. The empirical formula of this compound is ZrC.

Each zirconium ion touches 6 carbide ions and the CN of zirconium ion = 6.

Each carbide ion touches 6 zirconium ions and the CN of carbide ion = 6.

Zirconium ions are at the 8 corners of the cubic unit cell. Carbide ions are at the centers of the 12 edges of the cubic unit cell. So

2×radius of the zirconium ion + 2×radius of the carbide ion = cell edge length = 0.4638 nm.

Radius of the zirconium ion + radius of the carbide ion = cell edge length/2

= 0.4638/2 = 0.2319 nm.

The zirconium ion – carbide ion interionic distance = 0.2319 nm.

10-61. Gallium arsenide is a light sensitive compound used in photocells, which crystallizes in a cubic unit cell for which the cell edge length is 0.5653 nm. The positions of the ions are described by the following coordinates: Gallium ions are at 0,0,0 ; 0,1/2,1/2 ; 1/2,0,1/2 ; 1/2,1/2,0 and arsenide ions are at 1/4,1/4,1/4 ; 1/4,3/4,3/4 ; 3/4,1/4,3/4 ; 3/4,3/4,1/4. How many net ions of each type are contained within the unit cell? What is the empirical formula of this compound? If gallium is in Group 13 of the periodic table, what is the oxidation number of gallium in this compound? What are the coordination numbers of gallium and arsenide ions in this compound? Is this a simple cubic,body-centered cubic unit cell, or face-centered cubic unit cell? What is the gallium ion – arsenide ion interionic distance in nm?

Gallium ion is at 0,0,0. This implies that there are gallium ions at the 8 corners of the cubic unit cell. Gallium ions are also at 0,1/2,1/2 ; 1/2,0,1/2 ; 1/2,1/2,0. This implies that there are gallium ions at the centers of the 6 faces of the cubic unit cell. So this is a face-centered cubic unit cell. Number of gallium ions per unit cell = {(8×1/8) + (6×1/2)} = 4.

Arsenide ions are at 1/4,1/4,1/4 ; 1/4,3/4,3/4 ; 3/4,1/4,3/4 ; 3/4,3/4,1/4. Half of the tetrahedral holes are filled by the arsenide ions. Number of arsenide ions per unit cell = 4.

The empirical formula of this compound is GaAs and the oxidation number of gallium is +3.

Gallium ion touches 4 arsenide ions and the CN of gallium ion = 4.

Arsenide ion touches 4 gallium ions and the CN of arsenide ion = 4.

Gallium ion is at 0,0,0 and the arsenide ion is at 1/4,1/4,1/4.

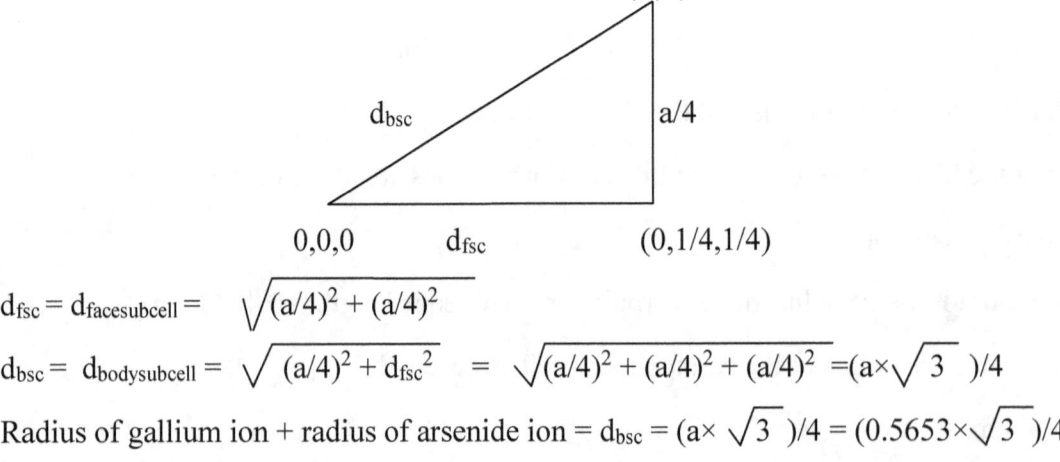

$d_{fsc} = d_{facesubcell} = \sqrt{(a/4)^2 + (a/4)^2}$

$d_{bsc} = d_{bodysubcell} = \sqrt{(a/4)^2 + d_{fsc}^2} = \sqrt{(a/4)^2 + (a/4)^2 + (a/4)^2} = (a \times \sqrt{3})/4$

Radius of gallium ion + radius of arsenide ion = $d_{bsc} = (a \times \sqrt{3})/4 = (0.5653 \times \sqrt{3})/4$

= 0.2448 nm.

The gallium ion – arsenide interionic distance = 0.2448 nm.

10-62. Copper nitride crystallizes in a cubic unit cell for which the cell edge length is 0.3814 nm. The positions of the ions are described by the following coordinates: Nitride ion is at 0,0,0 and copper ions are at 0,0,1/2 ; 0,1/2,0 ; 1/2,0,0. How many net ions of each type are contained within the unit cell? What is the empirical formula of this compound? What is the oxidation number of copper in this compound? What are the coordination numbers of copper and nitride ions in this compound? Is this a simple cubic, body-centered cubic or face-centered cubic unit cell?

What is the copper ion – nitride ion interionic distance in nm?

Nitride ion is at 0,0,0. This implies that there are nitride ions at the 8 corners of the cubic unit cell. There is no nitride ion at the center of body of the cubic unit cell. There are no nitride ions at the centers of the 6 faces of the cubic unit cell. So this is a simple cubic unit cell. Number of nitride ion per unit cell = $(8 \times 1/8)$ = 1.

Copper ions are at 0,0,1/2 ; 0,1/2,0 ; 1/2,0,0. This implies that there are copper ions at the centers of the 12 edges of the cubic unit cell. Number of copper ions per unit cell = $(12 \times 1/2)$ = 3.

The empirical formula of this compound is Cu_3N and the oxidation number of copper is +1.

Each nitride ion touches 6 copper ions and the CN of nitride ion = 6.

Each copper ion touches 2 nitride ions and the CN of copper ion = 2.

Nitride is at 0,0,0 and copper ion is at 1/2,0,0. So

Radius of the nitride ion + Radius of the copper ion = celll edge length/2 = 0.3814/2 = 0.1907 nm.

The copper ion – nitride ion interionic distance = 0.1907.

10-63. A compound that contains only copper and gold ions crystallizes in a cubic unit cell for which the cell edge length is 0.3748 nm. The positions of the ions are described by the following coordinates: Gold ion is at 0,0,0 and copper ions are at 0,1/2,1/2 ; 1/2,0,1/2 ; 1/2,1/2,0. How many net ions of each type are contained within the unit cell? What is the empirical formula of this compound? What is the coordination number of the gold ion? There is no ion at 1/2,1/2,1/2 in this unit cell. If there were, what would be its coordination number? Is this simple cubic, body-centered cubic or face-centered cubic unit cell? What is the shortest distance between the copper ions in this unit cell?

Gold ion is at 0,0,0. This implies that there are gold ions at the 8 corners of the cubic unit cell. There is no gold ion at the center of the body of the cubic unit cell and there are no gold ions at the centers of the 6 faces of the cubic unit cell. So this is a simple cubic unit cell. Number of gold ion per unit cell = (8×1/8) = 1.

Copper ions are at 0,1/2,1/2 ; 1/2,0,1/2 ; 1/2,1/2,0. This implies that there are copper ions at the centers of the 6 faces of the cubic unit cell. Number of copper ions per unit cell = (6×1/2) = 3. The empirical formula of this compound is $AuCu_3$.

Gold ion touches 12 copper ions and the CN of gold ion = 12.

If there were an ion at 1/2,1/2,1/2, it would touch 6 copper ions at the centers of the 6 faces of the cubic unit cell and the coordination number of this ion would be = 6.

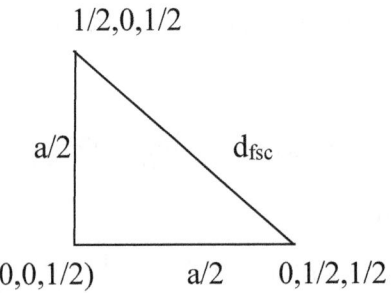

The shortest distance will be the distance between two copper ions at the centers of two adjacent faces. Copper ions are at 1/2,0,1/2 and 0,1/2,1/2. The shortest distance will be the distance between these two points.

Cell edge length = a

Shortest distance between two copper ions = d_{fsc}

$$d_{fsc} = d_{facesubcell} = \sqrt{(a/2)^2 + (a/2)^2} = a/\sqrt{2} = 0.3748/\sqrt{2} = 0.265 \text{ nm}$$

GRAPH-A

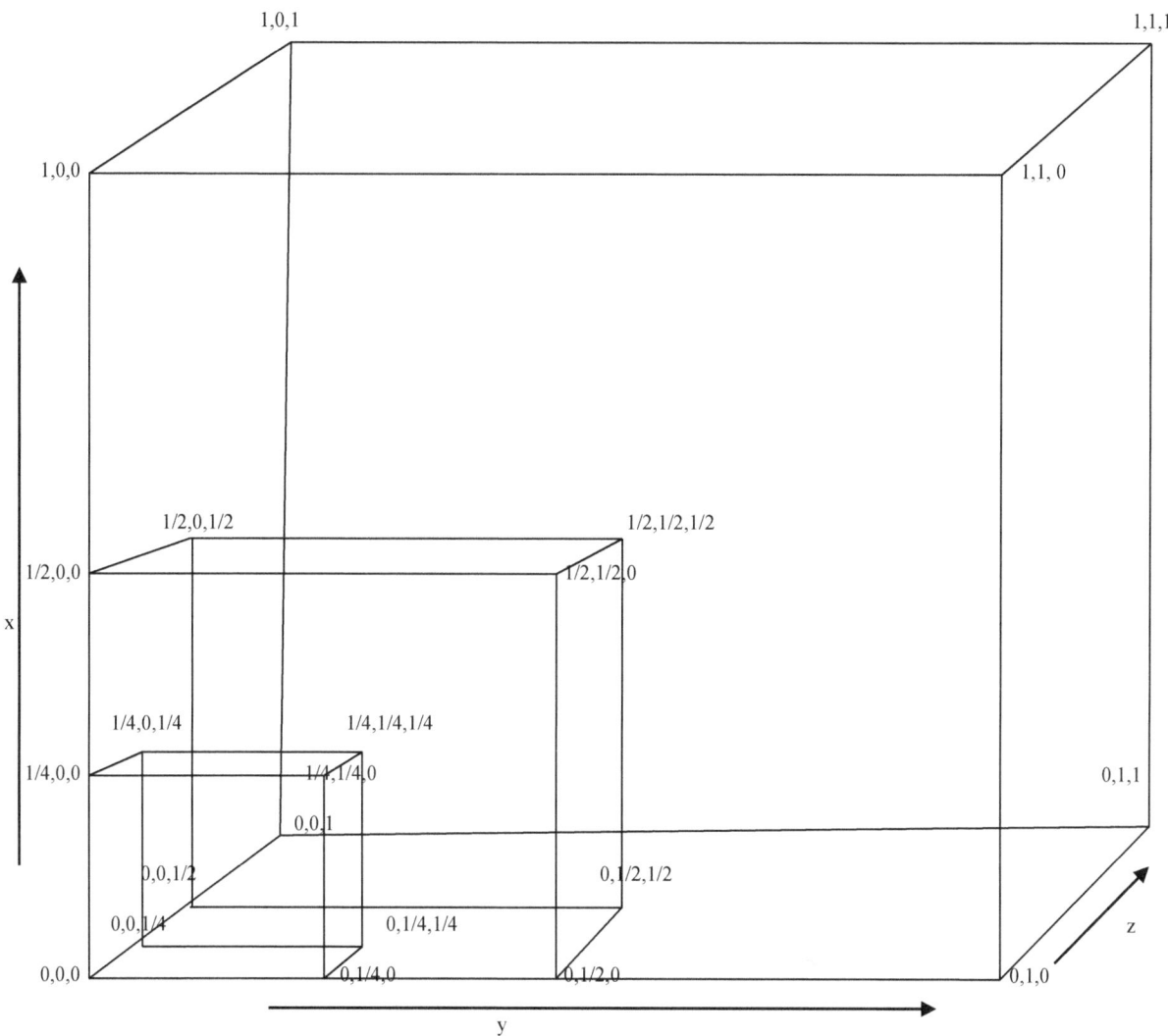

GRAPH-B

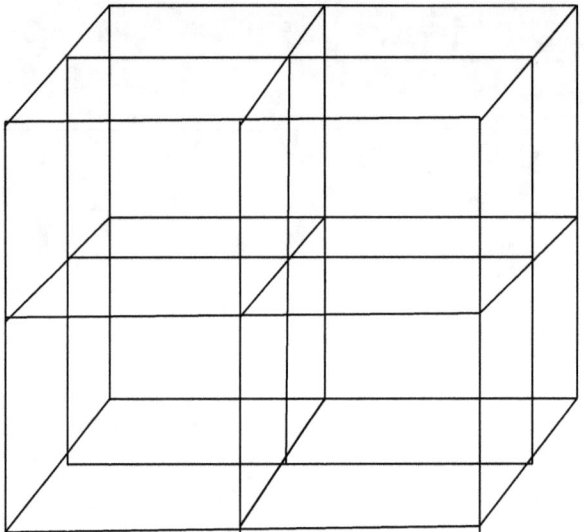

CHAPTER 11

INTRODUCTION TO KINETICS AND EQUILIBRIUM

INVALUABLE INFORMATION

Gas Constant (R) : 0.0820575 (L atm)/(mol K)

11-1. What is the correct equilibrium constant expression for the reaction:

$$Cl_2 (g) + 3F_2 (g) \rightleftharpoons 2ClF_3 (g)$$

$K_c = [ClF_3] [ClF_3] / [Cl_2][F_2][F_2][F_2]$

11-2. What is the correct equilibrium constant expression for the reaction:

$$2NOCl (g) \rightleftharpoons 2NO (g) + Cl_2 (g)$$

$K_c = [Cl_2] [NO][NO] / [NOCl][NOCl]$

11-3. What is the correct equilibrium constant expression for the reaction:

$$2NO_2 (g) \rightleftharpoons 2NO (g) + O_2 (g)$$

$K_c = [NO][NO][O_2] / [NO_2][NO_2]$

11-4. The equilibrium constant for the following reaction is known:

$$2NO_2 (g) \rightleftharpoons 2NO (g) + O_2 (g) \qquad K_1 = 7.4 \times 10^{-16} \qquad (at\ 25\ ^oC)$$

What is the correct value of the equilibrium constant for the opposite reaction?

$$2NO (g) + O_2 (g) \rightleftharpoons 2NO_2 (g) \qquad K_2 = ?$$

(a) $K_2 = K_1 = 7.4 \times 10^{-16}$ (b) $K_2 = 1/K_1 = 1.4 \times 10^{15}$ (c) $K_2 = K_1 (RT) = 1.8 \times 10^{-14}$

(d) $K_2 = K_1 (RT)^{-1} = 3.0 \times 10^{-17}$

$$2NO (g) + O_2 (g) \rightleftharpoons 2NO_2 (g)$$

$K_2 = [NO_2][NO_2] / [NO][NO][O_2] = 1/K_1 = 1/(7.4 \times 10^{-16}) = 1.4 \times 10^{15}$ Answer(b)

11-5. The composition of the following system at equilibrium at 1000 K is 0.202 atm CO_2,

0.594 atm CO, and 5 kg C:

$$2CO (g) \rightleftharpoons CO_2 (g) + C (s)$$

What is the equilibrium constant for this reaction?

(a) 2.94 (b) 14.56 (c) 0.573 (d) 0.337 (e) None of the above

$$2CO \text{ (g)} \rightleftharpoons CO_2 \text{ (g)} + C \text{ (s)}$$

$K_p = P_{CO_2} / P_{CO}^2$

Pure solid does not affect the reactant amount at equilibrium in the reaction; so it is kept at 1 and not included in the equilibrium constant expression.

$K_p = (0.202)/(0.594)^2 = 0.573$ Answer(c)

11-6. For the reaction,

$$CO \text{ (g)} + Cl_2 \text{ (g)} \rightleftharpoons COCl_2 \text{ (g)}$$

k_f is 1×10^7 $M^{-1}s^{-1}$ and $k_r = 2 \times 10^2$ s^{-1}. What is the equilibrium constant for the reaction?

   (a) 2×10^{-5} (b) 4.70 (c) 5×10^4 (d) 2×10^9 (e) None of the above

$K_c = k_f/k_r = (10^7)/(2 \times 10^2) = 5 \times 10^4$ Answer(c)

11-7. The following reaction is used to produce a mixture of CO and H_2 known as synthesis gas:

$$C \text{ (s)} + H_2O \text{ (g)} \rightleftharpoons CO \text{ (g)} + H_2 \text{ (g)}$$

Assume that 1 atm of H_2O (g) and excess solid carbon are sealed in a 1.0 L container at 700 °C. After the reaction reaches equilibrium, the total pressure in the container is 1.65 atm. What is the equilibrium constant for this reaction?

$$C \text{ (g)} + H_2O \text{ (g)} \rightleftharpoons CO \text{ (g)} + H_2 \text{ (g)}$$

Initial:	1 atm	0	0
At equilibrium:	$1 - \Delta P$	ΔP	ΔP

$K_p = (\Delta P)^2 / (1 - \Delta P)$

Pure solid does not affect the reactant amount at equilibrium in the reaction; so it is kept at 1 and not included in the equilibrium constant expression.

$1 - \Delta P + \Delta P + \Delta P = 1.65$

$\Delta P = 0.65$

$K_p = (0.65)^2 / (1-0.65) = 1.21$

11-8. Which statement concerning the following reaction is not true?

$$2HI \text{ (g)} \rightleftharpoons H_2 \text{ (g)} + I_2 \text{ (g)}$$

   (a) At equilibrium, the concentrations of the three gases do not change with time
   (b) $K_p = K_c$

(c) The partial pressures of I_2 and H_2 are equal at equilibrium, regardless of the initial pressures of HI, H_2 and I_2

(d) HI decomposes as fast as it is formed once equilibrium is established

(e) The percent dissociation of HI does not depend on the total pressure

$K_p = K_c \times (RT)^{\Delta n}$

Δn (the difference between the number of moles of products and the number of moles of reactants in the balanced equation) $= (1+1) - (2) = 0$

So $K_p = K_c$ and the equilibrium will not be affected by changes in pressure.

The percent dissociation of HI will not depend on total pressure.

The partial pressures of I_2 and H_2 will be equal at equilibrium only if

the initial pressure of $I_2 = $ the initial pressure of H_2.   Answer(c)

11-9. The reaction ,

$$CO\ (g) + H_2O\ (g) \rightleftarrows CO_2\ (g) + H_2\ (g)$$

has an equilibrium constant $K_c = 5.0$ at room temperature. A 1.00 L vessel contains 0.20 mole of CO, 0.30 mole of H_2O, and 0.90 mole of H_2 in equilibrium at this temperature. What is the the concentration of CO_2 in this equilibrium mixture?

$$CO\ (g) + H_2O\ (g) \rightleftarrows CO_2\ (g) + H_2\ (g)$$

$K_c = [CO_2][H_2]/[CO][H_2O]$

$[CO_2] = K_c [CO][H_2O]/[H_2] = 5\times(0.20\times0.30/0.9) = 0.333$ M (mole per liter)

11-10. What is the relationship between K_p and K_c for the following reaction?

$$Cl_2\ (g) + 3F_2\ (g) \rightleftarrows 2ClF_3\ (g)$$

(a) $K_c = K_p$ (b) $K_c = K_p \times (RT)^{-1}$ (c) $K_c = K_p \times (RT)^{-2}$ (d) $K_c = K_p \times RT$

(e) $K_c = K_p \times (RT)^2$

$K_p = K_c \times (RT)^{\Delta n}$

$\Delta n = 2 - (1+3) = -2$

$K_c = K_p / (RT)^{-2} = K_p \times (RT)^2$ Answer(e)

11-11. The correct value for the equilibrium constant K_p for the reaction,

$$2NO_2\ (g) \rightleftarrows 2NO\ (g) + O_2\ (g)\quad K_c = 7.4 \times 10^{-16} \text{ (at 25 °C)}$$

would be:

(a) $K_p = K_c = 7.4 \times 10^{-16}$ (b) $K_p = 1/K_c = 1.4 \times 10^{15}$ (c) $K_p = K_c (RT) = 1.8 \times 10^{-14}$

(d) $K_p = K_c (RT)^{-1} = 3.0 \times 10^{-17}$ (e) None of the above

$K_p = K_c \times (RT)^{\Delta n}$ $\qquad\qquad\qquad$ $\Delta n = (2+1) - (2) = 1$

$K_p = K_c \times (RT)$

$\qquad = 7.4 \times 10^{-16} \times (0.08206) \times (273.15 + 25) = 1.8 \times 10^{-14}$ $\qquad\qquad$ Answer(c)

11-12. If $K_c = 5.6 \times 10^{-6}$ for the following reaction at 400K

$$2NOCl(g) \rightleftharpoons 2NO (g) + Cl_2 (g),$$

What is the value of K_p for this reaction?

$\qquad$ (a) 1.7×10^{-7} (b) 5.6×10^{-6} (c) 1.8×10^{-4} (d) 6.0×10^{-3} (e) 1.8×10^{-5}

$K_p = K_c \times (RT)^{\Delta n}$

$\Delta n = (2+1) - 2 = 1$

$K_p = K_c \times (RT) = 5.6 \times 10^{-6} \times (0.08206 \times 400) = 1.8 \times 10^{-4}$ $\qquad\qquad$ Answer(c)

11-13. The correct value for the equilibrium constant for the reaction

$$2NO_2 (g) \rightleftharpoons 2NO (g) + O_2 (g) \quad K_c = 7.4 \times 10^{-16} \quad \text{at } 25 \,^oC$$

expressed in terms of pressures of NO_2, NO, and O_2 for this reaction would be:

$\qquad$ (a) $K_p = K_c$ (b) $K_p = 1/K_c$ (c) $K_p = K_c (RT)$ (d) $K_p = K_c (RT)^{-1}$ (e) None of the above

$K_p = K_c \times (RT)^{\Delta n}$

$\Delta n = (2+1) - (2) = 1$ $\qquad$ $K_p = K_c \times (RT)$ $\qquad\qquad\qquad\qquad\qquad$ Answer(c)

11-14. What is the 'Δn' in the following equation, $\qquad K_p = K_c (RT)^{\Delta n}$

for the following reaction?

$$2C_2H_6 (g) + 7O_2 (g) \rightleftharpoons 4CO_2 (g) + 6 H_2O (g)$$

$\qquad$ (a) −1 (b) 0 (c) 1 (d) 2 (e) 3

Δn is the difference between the number of moles of products and the number of moles of reactants in the balanced equation.

$\Delta n = (4+6) - (2+7) = 1$ $\qquad\qquad\qquad\qquad\qquad\qquad\qquad\qquad\qquad$ Answer(c)

11-15. If at any moment in time, we find that Q_c is larger than K_c for the reaction,

$$Cl_2 (g) + 3F_2 (g) \rightleftharpoons 2ClF_3 (g)$$

we can conclude that

(a) The reaction is at equilibrium
(b) The reaction can not reach equilibrium
(c) The reaction must shift to the right to reach equilibrium
(d) The reaction must shift to the left to reach equilibrium
(e) None of the above

If $Q_c = K_c$, the reaction is at equilibrium.

If $Q_c < K_c$, the reaction must shift to the right to reach equilibrium.

If $Q_c > K_c$, the reaction must shift to the left to reach the equilibrium.   Answer(d)

11-16. Assume that the reaction quotient,Q_c, for the following reaction at 25 ^oC is 1.0×10^{-8},

$$2NO_2\ (g) \rightleftharpoons 2NO\ (g) + O_2\ (g)\quad K_c = 7.4\times10^{-16}\ (at\ 25\ ^oC)$$

From this we conclude that

(a) The reaction is at equilibrium
(b) Equilibrium could be reached by adding enough NO or O_2 to the system
(c) The reaction must proceed from left to the right to reach equilibrium
(d) The reaction must proceed from right to left to reach equilibrium
(e) The reaction can never reach equilibrium

$Q_c > K_c$. So the reaction must proceed from right to left to reach equilibrium.   Answer(d)

11-17. If the equilibrium constant for the following reaction is $K_c = 1.0\times10^2$,

$$2C_2H_6\ (g) + 7O_2\ (g) \rightleftharpoons 4CO_2\ (g) + 6H_2O\ (g)$$

and all the concentrations were initially 0.10 M, we can predict that the reaction:

(a) is at equilibrium
(b) must shift from left to right to reach equilibrium
(c) must shift from right to left to reach equilibrium
(d) cannot reach equilibrium

$Q_c = \{(CO_2)^4\times(H_2O)^6\}/\{(C_6H_6)^2\times(O_2)^7\} = \{(0.1)^4\times(0.1)^6\}/\{(0.1)^2\times(0.1)^7)\} = 0.1$

$Q_c < K_c$. So the reaction must shift from left to right to reach equilibrium.   Answer(b)

11-18. What is the correct expression for the reaction quotient for the following reaction?

$$2NO_2\ (g) \rightleftharpoons 2NO\ (g) + O_2\ (g)\quad K_c = 7.4\times10^{-16}\ at\ 25\ ^oC$$

$Q_c = (NO)^2(O_2)/(NO_2)^2$

11-19. For the following reaction,

$$2NOCl\ (g) \rightleftharpoons 2NO\ (g) + Cl_2\ (g)\quad K_c = 5.6\times10^{-6}\ (at\ 400K)$$

If (NOCl) = 0.0222 M , (Cl_2) = 0.0222 M, and (NO) = 0.989 M at some moment in time, we

can conclude that:

(a) The reaction is at equilibrium
(b) Some of the NOCl must be converted to NO and Cl_2 to reach equilibrium
(c) Some of the NO and Cl_2 must be converted to NOCl to reach equilibrium
(d) It is impossible to determine which way the reaction must go to reach equilibrium
(e) The reaction can never reach equilibrium

$Q_c = (NO)^2(Cl_2)/(NOCl)^2 = \{(0.989)^2 \times (0.0222)\}/(0.0222)^2 = 44.060$

$Q_c > K_c$. So the reaction must shift from right to left to reach equilibrium. Some of the NO

and Cl_2 must be converted to NOCl to reach equilibrium. Answer(c)

11-20. Suppose that the reaction quotient, Q_c, for a chemical reaction was found to be close to

unity ($Q_c \approx 1$). Given this information we can accurately predict :

(a) That the reaction will proceed in a direction to generate more product
(b) That the reaction will proceed in a direction to use up the product
(c) That the reaction is close to equilibrium
(d) Nothing about the direction the reaction will proceed with out additional information
(e) None of the above

To predict the reaction direction we need K_c value also. Answer(d)

11-21. The equilibrium constant for the following reaction at a given temperature is 8.24×10^4,

$$2SO_2 (g) + O_2 (g) \rightleftharpoons 2SO_3 (g)$$

If the partial pressure of SO_2 is 1.2×10^{-5} atm, the partial pressure of O_2 is 1 atm, and the partial

pressure of SO_3 is 1 atm, we can accurately state:

(a) That the reaction is at equilibrium
(b) That the reaction must shift from left to right to reach equilibrium
(c) That the reaction must shift from right to left to reach equilibrium
(d) Nothing about the reaction without additional information

$Q_p = (SO_3)^2/\{(SO_2)^2(O_2)\} = (1)^2/\{(1) \times (1.2 \times 10^{-5})^2\} = 6.94 \times 10^9$

$Q_p > K_p$. So the reaction must shift from right to left to reach equilibrium. Answer(c)

11-22. Given the equilibrium constant for the reaction,

$$H_2 (g) + Cl_2 (g) \rightleftharpoons 2HCl (g) \quad K_p = 2.5 \times 10^4$$

If equimolar amounts of H_2 and Cl_2 are allowed to reach equilibrium in a

system in which no HCl was present initially, the ratio of partial pressure of

H_2 to partial pressure of Cl_2 at equilibrium will be:

(a) Larger than 1   (b) Smaller than 1   (c) Equal to 1

(d) Have a value that cannot be generalized by any of the above statements

$P = (n/V)RT$ Since the initial concentrations of H_2 and Cl_2 are equal,

Initial pressure of H_2 (P_1) = Initial pressure of Cl_2 (P_2)

$$H_2 \text{ (g)} + \quad Cl_2 \text{ (g)} \rightleftharpoons \quad 2HCl \text{ (g)}$$

Initial: P_1 P_2 0

At equilibrium: $P_1 - \Delta P$ $P_2 - \Delta P$ $2\Delta P$

Since $P_1 = P_2$, $P_1 - \Delta P = P_2 - \Delta P$ $(P_1 - \Delta P)/(P_2 - \Delta P) = 1$ Answer(c)

11-23. For the following reaction ,

$$2NO_2 \text{ (g)} \rightleftharpoons 2NO \text{ (g)} + O_2 \text{ (g)} \qquad K_c = 7.4 \times 10^{-16}$$

If $(NO_2) = 0.10$ M, $(NO) = 0.01$ M, and $(O_2) = 1.0 \times 10^{-5}$, we can correctly predict:

   (a) That the reaction is at equilibrium
   (b) That the reaction must shift from right to left to reach equilibrium
   (c) That the reaction must shift from left to right to reach equilibrium
   (d) Nothing related to the question of which direction the reaction has to shift to reach equilibrium
   (e) Nothing related to the question of whether the reaction is at equilibrium

$Q_c = \{(NO)^2(O_2)\}/(NO_2)^2 = \{(0.01)^2 \times (1.0 \times 10^{-5})\}/(0.1)^2 = 1.0 \times 10^{-7}$

$Q_c > K_c$. So the reaction must shift from right to left to reach equilibrium.   Answer(b)

11-24. What is the relationship between the change in the concentrations of Cl_2 and F_2 as the

following reaction comes to equilibrium?

$$Cl_2 \text{ (g)} + 3F_2 \text{ (g)} \rightleftharpoons 2ClF_3 \text{ (g)}$$

For each mole of Cl_2 consumed, three moles of F_2 will be consumed. For each mole of

Cl_2 produced, three moles of F_2 will be produced. So if ΔC is the change in the concentration

of Cl_2 then the change in the concentration of F_2 will be $3\Delta C$.

11-25. What would happen if F_2 was removed from the following reaction while it was at

equilibrium?

$$Cl_2 \text{ (g)} + 3F_2 \text{ (g)} \rightleftharpoons 2ClF_3 \text{ (g)}$$

   (a) The concentrations of both Cl_2 and F_2 would increase as the reaction came back to equilibrium
   (b) The concentration of Cl_2 would decrease as the reaction came back to equilibrium
   (c) The concentrations of both Cl_2 and F_2 would decrease as the reaction came back to equilibrium
   (d) The concentrations of both Cl_2 and F_2 would remain the same because the reaction would be at
       equilibrium

To counteract this change the reaction will shift from right to left to reach equilibrium. So the concentration of both Cl_2 and F_2 will increase as the reaction comes back to equilibrium.

Answer(a)

11-26. For the reaction

$$2NOCl \,(g) \rightleftharpoons 2NO \,(g) + Cl_2 \,(g) \qquad K_c = 5.6 \times 10^{-6}$$

What is the relationship between the magnitude of the changes in the concentrations of NOCl and Cl_2 as this reaction comes to equilibrium?

For each mole of Cl_2 produced, two moles of NOCl will be consumed. For each mole of Cl_2 consumed, two moles of NOCl will be produced. So if ΔC is the change in the concentration of Cl_2, then the change in concentration of NOCl will be $2\Delta C$.

11-27. What is the relationship between the changes in the concentrations of NO_2 and O_2 as the following reaction comes to equilibrium?

$$2NO_2 \,(g) \rightleftharpoons 2NO \,(g) + O_2 \,(g) \qquad K_c = 7.4 \times 10^{-16} \text{ at } 25\ ^oC$$

For each mole of O_2 produced, two moles of NO_2 will be consumed. For each mole of O_2 consumed, two moles of NO_2 will be produced. So if ΔC is the change in the concentration of O_2, then the change in the concentration of NO_2 will be $2\Delta C$.

11-28. For the following reaction,

$$2C_2H_6 \,(g) + 7O_2 \,(g) \rightleftharpoons 4CO_2 \,(g) + 6H_2O \,(g)$$

If the reaction mixture initially contained all species at the same initial concentration, C , and if the concentartion of O_2 at equilibrium were expressed as $[O_2] = C + 7\Delta C$, then the concentration of H_2O at equilibrium would be expressed as $[H_2O] =$

   (a) $C + \Delta C$ (b) $C - \Delta C$ (c) $C + 6\Delta C$ (d) $C - 6\Delta C$ (e) None of the above

$$2C_2H_6 \,(g) + 7O_2 \,(g) \rightleftharpoons 4CO_2 \,(g) + 6H_2O \,(g)$$

Initial concentration:	C	C	C	C
At equilibrium:		$C + 7\Delta C$		?

This means that we are approaching the equilibrium from right to left. Then the concentration of H_2O at equilibrium will be $C - 6\Delta C$.

$$2C_2H_6 \text{ (g)} + 7O_2 \text{ (g)} \rightleftharpoons 4CO_2 \text{ (g)} + 6H_2O \text{ (g)}$$

Initial concentration:	C	C	C	C
At equilibrium:	$C + 2\Delta C$	$C + 7\Delta C$	$C - 4\Delta C$	$C - 6\Delta C$ Answer(d)

11-29. When solving a problem with the initial conditions outlined below, we rearrange the problem to give an intermediate set of conditions where the concentration of one of the reactants or products is equal to zero.

$$Fe(CO)_5 \rightleftharpoons Fe(CO)_4 + CO$$

Initial: 0.1 M   0.1 M   0.005 M

Our choice of whether to push the reaction all the way to the right or all the way to the left is determined by which of the following goals?

    (a) To make both Q_c and K_c large
    (b) To make both Q_c and K_c small
    (c) To bring Q_c as close as possible to K_c
    (d) To make the difference between Q_c and K_c as large as possible

K_c is constant. The intermediate condition should be such that it is not far from equilibrium, so that we can assume ΔC is small. This can be achieved by bringing Q_c as close as possible to K_c.   Answer(c)

11-30. Calculate the NO, NO_2, and O_2 concentrations in the following gas-phase reaction at equilibrium.

$$2NO_2 \text{ (g)} \rightleftharpoons 2NO \text{ (g)} + O_2 \text{ (g)} \quad K_c = 1.6 \times 10^{-8} \text{ (at 200 }^\circ C)$$

Initial: 0   0.1 M   0.1 M

When $K_c \ll 1$, we approach the equilibrium from left to right. We create an intermediate situation in which the concentration of one of the products is zero.

	$2NO_2$ (g) $\rightleftharpoons$	$2NO$ (g) +	O_2 (g)
Initial concentration:	0	0.1	0.1
Intermediate:	$0 + (2 \times 0.05)$	$0.1 - (2 \times 0.05)$	$0.1 - 0.05$
Intermediate:	0.1	0	0.05
At equilibrium:	$0.1 - 2\Delta C$	$2\Delta C$	$0.05 + \Delta C$

$K_c = \{(2\Delta C)^2 (0.05 + \Delta C)\}/(0.1 - 2\Delta C)^2 = 1.6 \times 10^{-8}$

Assumptions: $0.1 - 2\Delta C \approx 0.10$ and $0.05 + \Delta C \approx 0.05$

$\{(2\Delta C)^2 (0.05)\}/(0.1)^2 \approx 1.6 \times 10^{-8}$

$(\Delta C)^2 \approx 0.08 \times 10^{-8}$

$\Delta C \approx 2.8 \times 10^{-5}$

$\{2 \times (2.8 \times 10^{-5})/0.1\} \times 100 = 0.056\%$ (less than 5%)

$\{(2.8 \times 10^{-5})/0.05\} \times 100 = 0.056\%$ (less than 5%)

Assumptions $0.1 - 2\Delta C \approx 0.10$ and $0.05 + \Delta C \approx 0.05$ are valid.

Concentration of NO at equilibrium $\approx 2 \times 2.8 \times 10^{-5} \approx 5.6 \times 10^{-5}$ M

Concentration of O_2 at equilibrium ≈ 0.05 M

Concentration of NO_2 at equilibrium ≈ 0.10 M

11-31. Calculate the equilibrium concentrations of SO_3, SO_2, and O_2 if 0.10 mole of SO_3 in a

10-L flask decompose to form SO_2 and O_2 at 500K.

$$2SO_3 \text{ (g)} \rightleftarrows 2SO_2 \text{ (g)} + O_2 \text{ (g)} \quad K_c = 1.4 \times 10^{-11} \quad \text{(at 500K)}$$

When $K_c \ll 1$, we approach the equilibrium from left to right.

The initial concentration of SO_3 = 0.1/10 = 0.01 M (mole per liter)

	$2SO_3 \text{ (g)} \rightleftarrows$	$2SO_2 \text{ (g)} +$	$O_2 \text{ (g)}$
Initial concentration:	0.01 M	0	0
At equilibrium:	$0.01 - 2\Delta C$	$2\Delta C$	ΔC

$K_c = \{(2\Delta C)^2 \Delta C\}/(0.01 - 2\Delta C)^2 = 1.4 \times 10^{-11}$

Assumption: $0.01 - 2\Delta C \approx 0.01$

$4(\Delta C)^3 \approx 1.4 \times 10^{-15}$

$(\Delta C)^3 \approx 0.35 \times 10^{-15}$

$\Delta C \approx 7.0 \times 10^{-6}$

$\{2 \times (7.0 \times 10^{-6})/0.01\} \times 100 = 0.14\%$ (less than 5%)

Assumption $0.01 - 2\Delta C \approx 0.01$ is valid.

Concentration of O_2 at equilibrium $\approx 7.0 \times 10^{-6}$ M

Concentration of SO_2 at equilibrium $\approx 2 \times 7.0 \times 10^{-6} \approx 1.4 \times 10^{-5}$ M

Concentration of SO_3 at equilibrium ≈ 0.01 M

11-32. Calculate the concentration of all reactants and products when a mixture that is initially 0.10 M in N_2, H_2, and NH_3 comes to equilibrium at 600K.

$$N_2 \text{ (g)} + 3H_2 \text{ (g)} \rightleftharpoons 2NH_3 \text{ (g)} \quad K_c = 0.045 \text{ (at 600K)}$$

When $K_c \ll 1$, we approach the equilibrium from left to right. We create an intermediate situation in which the concentration of the product is zero.

	N_2 (g) +	$3H_2$ (g) $\rightleftharpoons$	$2NH_3$ (g)
Initial concentartion:	0.1	0.1	0.1
Intermediate:	0.1 + 0.05	0.1 + (3×0.05)	0.1 − (2×0.05)
Intermediate:	0.15	0.25	0
At equilibrium:	(0.15 − ΔC)	(0.25 − 3ΔC)	2ΔC

$K_c = (2\Delta C)^2 / \{(0.15 - \Delta C)(0.25 - 3\Delta C)^3\} = 0.045$

Assumptions: $(0.15 - \Delta C) \approx 0.15$ and $(0.25 - 3\Delta C) \approx 0.25$

$(2\Delta C)^2 / \{(0.15) \times (0.25)^3\} \approx 0.045 \qquad (\Delta C)^2 \approx 2.64 \times 10^{-5}$

$\Delta C \approx 5.1 \times 10^{-3}$

$\{3 \times (5.1 \times 10^{-3})/0.25\} \times 100 = 6.12\%$ (not less than 5%)

Assumption $(0.25 - 3\Delta C) \approx 0.25$ is not valid.

$\Delta C = 0.00465$ (by successive approximations).

Concentration of NH_3 at equilibrium = $2 \times 0.00465 = 9.3 \times 10^{-3}$ M

Concentration of N_2 at equilibrium = $(0.15 - 0.00465) = 0.145$ M

Concentration of H_2 at equilibrium = $0.25 - (3 \times 0.00465) = 0.236$ M

11-33. The equilibrium constant for the following reaction is $K_p = 4.1 \times 10^{-3}$:

$$COCl_2 \text{ (g)} \rightleftharpoons CO \text{ (g)} + Cl_2 \text{ (g)}$$

Calculate the equilibrium pressure of CO in a closed vessel that initially contained 0.30 atm of $COCl_2$ and no CO or Cl_2.

When $K_p \ll 1$, we approach the equilibrium from left to right.

	$COCl_2$ (g) $\rightleftharpoons$	CO (g) +	Cl_2 (g)
Initial:	0.3 atm	0	0
At equilibrium:	0.3 − ΔP	ΔP	ΔP

$K_p = (\Delta P)^2/(0.3 - \Delta P) = 4.1 \times 10^{-3}$

Assumption: $(0.3 - \Delta P) \approx 0.3$

$(\Delta P)^2/(0.3) \approx 4.1 \times 10^{-3}$

$(\Delta P)^2 \approx 1.23 \times 10^{-3}$

$\Delta P \approx 0.035$

$\{(0.035)/0.3\} \times 100 = 11.7\%$ (not less than 5%)

 Assumption: $(0.3 - \Delta P) \approx 0.3$ is not valid.

$\Delta P = 0.03295$ (by successive approximations)

Equilibrium pressure of CO = 0.03295 atm

Equilibrium pressure of Cl_2 = 0.03295 atm

Equilibrium pressure of $COCl_2$ = 0.3 − 0.03295 = 0.267 atm

11-34. The equilibrium constant for the following reaction is $K_p = 1.0 \times 10^4$:

$$2NO\ (g) + O_2\ (g) \rightleftharpoons 2NO_2\ (g)$$

Calculate the equilibrium pressure of oxygen in a system that initially contains 0.50 atm of NO, 0.10 atm of O_2, and 0.20 atm of NO_2.

When $K_p \gg 1$, we approach the equilibrium from right to left. We create an intermediate situation in which the concentration of one of the reactants is zero.

	2NO (g) +	O_2 (g) $\rightleftharpoons$	2NO$_2$ (g)
Initial:	0.5 atm	0.1 atm	0.2 atm
Intermediate:	0.5 − (2×0.1)	0.1 − 0.1	0.2 + (2×0.1)
Intermediate:	0.3	0	0.4
At equilibrium:	0.3 + 2ΔP	ΔP	0.4 − 2ΔP

$K_p = (0.4 - 2\Delta P)^2/\{(\Delta P)(0.3 + 2\Delta P)^2\} = 1.0 \times 10^4$

Assumptions: $(0.3 + 2\Delta P) \approx 0.3$ and $(0.4 - 2\Delta P) \approx 0.4$

$(0.4)^2/\{(\Delta P)(0.3)^2\} \approx 1.0 \times 10^4$

$\Delta P \approx 1.8 \times 10^{-4}$

$\{2 \times (1.8 \times 10^{-4})/0.3\} \times 100 = 0.12\%$ (less than 5%)

$\{2 \times (1.8 \times 10^{-4})/0.4\} \times 100 = 0.09\%$ (less than 5%)

Assumptions $(0.3 + 2\Delta P) \approx 0.3$ and $(0.4 - 2\Delta P) \approx 0.4$ are valid.

Equilibrium pressure of $O_2 \approx 1.8 \times 10^{-4}$ atm

Equilibrium pressure of NO ≈ 0.3 atm

Equilibrium pressure of $NO_2 \approx 0.4$ atm

11-35. $K_p = 2.1 \times 10^6$ for the following reaction at a given temperature:

$$2NH_3 \text{ (g)} \rightleftharpoons N_2 \text{ (g)} + 3H_2 \text{ (g)} \quad K_p = 2.1 \times 10^6$$

Starting with ammonia at 4.00 atm in a fixed-volume container, calculate the equilibrium pressures of all three gases.

When $K_p \gg 1$, we approach the equilibrium from right to left. We create an intermediate situation in which the concentration of the reactant is zero.

	$2NH_3 \text{ (g)} \rightleftharpoons$	$N_2 \text{ (g)} +$	$3H_2 \text{ (g)}$
Initial:	4 atm	0	0
Intermediate:	$4 - (2 \times 2)$	$0 + 2$	$0 + (3 \times 2)$
Intermediate:	0	2	6
At equilibrium:	$2\Delta P$	$2 - \Delta P$	$6 - (3\Delta P)$

$K_p = \{(2 - \Delta P)(6 - 3\Delta P)^3\}/(2\Delta P)^2 = 2.1 \times 10^6$

Assumptions: $(2 - \Delta P) \approx 2$ and $(6 - 3\Delta P) \approx 6$

$\{(2)(6)^3\}/(2\Delta P)^2 \approx 2.1 \times 10^6$

$(\Delta P)^2 \approx 5.14 \times 10^{-5}$

$\Delta P \approx 7.2 \times 10^{-3}$

$\{(7.2 \times 10^{-3})/2\} \times 100 = 0.36\%$ (less than 5%)

$\{3 \times (7.2 \times 10^{-3})/6\} \times 100 = 0.36\%$ (less than 5%)

Assumptions: $(2 - \Delta P) \approx 2$ and $(6 - 3\Delta P) \approx 6$ are valid.

Equilibrium pressure of $NH_3 \approx 2 \times 7.2 \times 10^{-3} \approx 1.44 \times 10^{-2}$ atm

Equilibrium pressure of $N_2 \approx 2$ atm

Equilibrium pressure of $H_2 \approx 6$ atm

11-36. The following equilibrium is established in the gas phase at 600K by heating phosgene, $COCl_2$:

$$COCl_2 (g) \rightleftharpoons CO (g) + Cl_2 (g) \quad K_p = 4.10 \times 10^{-3} \text{ atm}$$

If the initial pressure of phosgene is 0.120 atm, what is the equilibrium pressure of Cl_2?

When $K_p \ll 1$, we approach the equilibrium from left to right.

$$COCl_2 (g) \rightleftharpoons CO (g) + Cl_2 (g)$$

Initial:	0.120 atm	0	0
At equilibrium:	$0.12 - \Delta P$	ΔP	ΔP

$K_p = (\Delta P)^2/(0.12 - \Delta P) = 4.1 \times 10^{-3}$

Assumption: $\quad 0.12 - \Delta P \approx 0.12$

$(\Delta P)^2 \approx 4.92 \times 10^{-4}$

$\Delta P \approx 0.022$

$\{(0.022)/0.12\} \times 100 = 18.33\%$ (not less than 5%)

Assumption $\quad 0.12 - \Delta P \approx 0.12$ is not valid.

$\Delta P = 0.02025$ (by successive approximations)

Equilibrium pressure of $Cl_2 = 2.025 \times 10^{-2}$ atm

11-37. The equilibrium constant for the reaction of NO (g) with Cl_2 (g) to produce NOCl (g) is 52 atm^{-1}. If 0.5 atm of NO and 0.25 atm of Cl_2 react to equilibrium in a closed system that initially contained no NOCl, calculate the partial pressures of all three components at equilibrium.

When $K_p \gg 1$, we approach the equilibrium from right to left. In this case we create an intermediate situation in which the concentration of each reactant is zero.

$$2NO (g) + Cl_2 (g) \rightleftharpoons 2NOCl (g)$$

	2NO (g) +	Cl_2 (g)	2NOCl (g)
Initial:	0.5 atm	0.25 atm	0
Intermediate:	$0.5 - (2 \times 0.25)$	$0.25 - 0.25$	(2×0.25)
Intermediate:	0	0	0.5
At equilibrium:	$2\Delta P$	ΔP	$0.5 - 2\Delta P$

$K_p = (0.5 - 2\Delta P)^2/\{(2\Delta P)^2(\Delta P)\} = 52$

Assumption: $\quad 0.5 - 2\Delta P \approx 0.5$

$(\Delta P)^3 \approx (0.5)^2/(4 \times 52) \approx 1.2 \times 10^{-3}$

$\Delta P \approx 0.106$

$\{2 \times (0.106)/0.5\} \times 100 = 42.4\%$ (not less than 5%)

Assumption $0.5 - 2\Delta P \approx 0.5$ is not valid.

$\Delta P = 0.0815$ (by successive approximations)

Partial pressure of Cl_2 at equilibrium = 0.0815 atm

Partial pressure of NO at equilibrium = $2 \times 0.0815 = 0.163$ atm

Partial pressure of NOCl at equilibrium = $0.5 - (2 \times 0.0815) = 0.337$ atm

11-38. Assume an equilibrium constant $K_c = 1.7 \times 10^{-2}$ for the reaction:

$$N_2 \text{ (g)} + 3H_2 \text{ (g)} \rightleftharpoons 2NH_3 \text{ (g)}$$

Calculate the changes in the N_2 and H_2 concentrations and the equilibrium concentration of NH_3 when 0.1 M (mole per liter) each of N_2 and H_2 are mixed and allowed to react to equilibrium.

When $K_c \ll 1$, we approach the equilibrium from left to right.

$$N_2 \text{ (g)} + 3H_2 \text{ (g)} \rightleftharpoons 2NH_3 \text{ (g)}$$

Initial: 0.1 M   0.1 M   0

At equilibrium: $(0.1 - \Delta C)$ $(0.1 - 3\Delta C)$ $2\Delta C$

$K_c = (2\Delta C)^2/\{(0.1 - \Delta C) (0.1 - 3\Delta C)^3\} = 1.7 \times 10^{-2}$

Assumptions: $(0.1 - \Delta C) \approx 0.1$ and $(0.1 - 3\Delta C) \approx 0.1$

$(\Delta C)^2 \approx 4.25 \times 10^{-7}$

$\Delta C \approx 6.52 \times 10^{-4}$

$\{(6.52 \times 10^{-4})/0.1\} \times 100 = 0.652\%$ (less than 5%)

$\{3 \times (6.52 \times 10^{-4})/0.1\} \times 100 = 1.956\%$ (less than 5%)

Assumptions $(0.1 - \Delta C) \approx 0.1$ and $(0.1 - 3\Delta C) \approx 0.1$ are valid.

Change in the concentration of $H_2 \approx -3 \times 6.52 \times 10^{-4} = -1.96 \times 10^{-3}$ M

Change in the concentration of $N_2 \approx -6.52 \times 10^{-4}$ M

Equilibrium concentration of $NH_3 \approx 2 \times 6.52 \times 10^{-4} \approx 1.30 \times 10^{-3}$ M

11-39. Calculate the $COCl_2$, CO and Cl_2 concentrations when the following gas-phase reaction reaches equilibrium at 300 °C:

$$COCl_2 \text{ (g)} \rightleftharpoons CO \text{ (g)} + Cl_2 \text{ (g)} \qquad K_c = 4.0 \times 10^{-5}$$

Initial: 0.1 M   0   0

When $K_c \ll 1$, we approach the equilibrium from left to right.

$$COCl_2 \text{ (g)} \rightleftharpoons CO \text{ (g)} + Cl_2 \text{ (g)}$$

Initial :   0.1 M   0   0

At equilibrium: $0.1 - \Delta C$ ΔC ΔC

$K_c = (\Delta C)^2/(0.1 - \Delta C) = 4.0 \times 10^{-5}$

Assumption: $(0.1 - \Delta C) \approx 0.1$

$(\Delta C)^2/(0.1) \approx 4.0 \times 10^{-5}$

$(\Delta C)^2 \approx 4.0 \times 10^{-6}$

$\Delta C \approx 2.0 \times 10^{-3}$

$\{(2.0 \times 10^{-3})/0.1\} \times 100 = 2\%$ (less than 5%)

Assumption $(0.1 - \Delta C) \approx 0.1$ is valid.

Concentration of $COCl_2$ at equilibrium ≈ 0.1 M

Concentration of CO at equilibrium $\approx 2.0 \times 10^{-3}$ M

Concentration of Cl_2 at equilibrium $\approx 2.0 \times 10^{-3}$ M

11-40. Calculate the $COCl_2$, CO and Cl_2 concentrations when the following gas- phase reaction reaches equilibrium at 300 °C.

$$COCl_2 \text{ (g)} \rightleftharpoons CO \text{ (g)} + Cl_2 \text{ (g)} \quad K_c = 4.0 \times 10^{-5}$$

Initial: 0   0.1 M   0.5 M

When $K_c \ll 1$, we approach the equilibrium from left to right. We create an intermediate situation in which the concentration of one of the products is zero.

$$COCl_2 \text{ (g)} \rightleftharpoons CO \text{ (g)} + Cl_2 \text{ (g)}$$

Initial :   0   0.1   0.5

Intermediate: $0 + 0.1$ $0.1 - 0.1$ $0.5 - 0.1$

Intermediate: 0.1   0   0.4

At equilibrium: $0.1 - \Delta C$ ΔC $0.4 + \Delta C$

$K_c = (\Delta C)(0.4 + \Delta C)/(0.1 - \Delta C) = 4.0 \times 10^{-5}$

Assumptions: $(0.1 - \Delta C) \approx 0.1$ and $(0.4 + \Delta C) \approx 0.4$

$(\Delta C) \times 4 \approx 4.0 \times 10^{-5}$

$\Delta C \approx 1.0 \times 10^{-5}$

$\{(1.0 \times 10^{-5})/0.1\} \times 100 = 0.01\%$ (less than 5%)

$\{(1.0 \times 10^{-5})/0.4\} \times 100 = 1.25 \times 10^{-3}\%$ (less than 5%)

Assumptions $(0.1 - \Delta C) \approx 0.1$ and $(0.4 + \Delta C) \approx 0.4$ are valid.

Equilibrium concentration of $COCl_2 \approx 0.1$ M

Equilibrium concentration of $CO \approx 1.0 \times 10^{-5}$ M

Equilibrium concentration of $Cl_2 \approx 0.4$ M

11-41. Calculate the NO, NO_2, and O_2 concentrations in the following gas-phase reaction at equilibrium.

$$2NO_2 (g) \rightleftharpoons 2NO (g) + O_2 (g) \quad K_c = 1.6 \times 10^{-8}$$

Initial: 0.1 M   0   0

When $K_c \ll 1$, we approach the equilibrium from left to right.

$$2NO_2 (g) \rightleftharpoons 2NO (g) + O_2 (g)$$

Initial: 0.1 M   0   0

At equilibrium: $(0.1 - 2\Delta C)$ $2\Delta C$ ΔC

$K_c = (2\Delta C)^2 (\Delta C) / (0.1 - 2\Delta C)^2 = 1.6 \times 10^{-8}$

Assumption: $(0.1 - 2\Delta C) \approx 0.1$

$4(\Delta C)^3 \approx 1.6 \times 10^{-8} \times (0.1)^2 \approx 1.6 \times 10^{-10}$

$(\Delta C)^3 \approx 4 \times 10^{-11}$

$\Delta C \approx 3.4 \times 10^{-4}$

$\{2 \times (3.4 \times 10^{-4})/0.1\} \times 100 = 0.68\%$ (less than 5%)

Assumption $(0.1 - 2\Delta C) \approx 0.1$ is valid.

Equilibrium concentration of $O_2 \approx 3.4 \times 10^{-4}$ M

Equilibrium concentration of $NO \approx 2 \times 3.4 \times 10^{-4} \approx 6.8 \times 10^{-4}$ M

Equilibrium concentration of $NO_2 \approx 0.1$ M

11-42. What would be the effect of removing some NO_2 from the following system after the reaction reaches equilibrium?

$$2NO_2 \text{ (g)} \rightleftharpoons 2NO \text{ (g)} + O_2 \text{ (g)} \quad K_c = 7.4 \times 10^{-16} \text{ at } 25 \text{ °C}$$

(a) The NO and O_2 concentrations would increase
(b) The NO and O_2 concentrations would decrease.
(c) The NO concentration would increase but the concentration of O_2 would decrease
(d) The O_2 concentration would increase but the concentration of NO would decrease
(e) The concentrations of NO and O_2 would remain constant

To counteract this change, the equilibrium will be shifted to the left. NO_2 will be produced by consuming NO and O_2. So NO and O_2 concentrations will decrease.   Answer(b)

11-43. What would happen if Cl_2 is added to the following system at equilibrium?

$$H_2 \text{ (g)} + Cl_2 \text{ (g)} \rightleftharpoons 2HCl \text{ (g)}$$

(a) The partial pressures of both H_2 and HCl would increase
(b) The partial pressures of both H_2 and HCl would decrease
(c) The partial pressure of H_2 would increase, but the partial pressure of HCl would decrease
(d) The partial pressure of H_2 would decrease, but the partial pressure of HCl would increase

To counteract this change, the equilibrium will be shifted to the right. HCl will be produced by consuming H_2 and Cl_2. So the partial pressure of H_2 will decrease and the partial pressure of HCl will increase.   Answer(d)

11-44. What would happen if O_2 were removed from the following system at equilibrium?

$$2NO_2 \text{ (g)} \rightleftharpoons 2NO \text{ (g)} + O_2 \text{ (g)}$$

(a) The NO_2 and NO concentrations would increase
(b) The NO_2 and NO concentrations would decrease
(c) The NO_2 concentration would increase and the NO concentration would decrease
(d) The NO_2 concentration would decrease and the NO concentration would increase

To counteract this change, the equilibrium will be shifted to the right. NO_2 will be consumed to produce NO and O_2. So the concentration of NO_2 will decrease and the concentration of NO will increase.   Answer(d)

11-45. The addition of I_2 (g) to a fixed volume container with I_2 (g), H_2 (g), and HI (g) at equilibrium at a given temperature would cause:

(a) The partial pressures of both H_2 and HI to decrease
(b) The partial pressures of both H_2 and HI to increase
(c) The partial pressure of H_2 to increase and the partial pressure of HI to decrease
(d) The partial pressure of H_2 to decrease and the partial pressure of HI to increase

$$H_2 \text{ (g)} + I_2 \text{ (g)} \rightleftharpoons 2 HI \text{ (g)}$$

248

To counteract this change, HI will be produced by consuming H_2 and I_2. So the partial pressure of H_2 will decrease and the partial pressure of HI will increase. Answer(d)

11-46. The equilibrium constant for the following reaction becomes larger as the temperature at which the reaction is run increases.

$$2NOCl\ (g) \rightleftharpoons 2NO\ (g) + Cl_2\ (g) \quad K_c = 5.6 \times 10^{-6}\ \text{ at } 400K$$

What does this tell about the reaction?

(a) Heat is given off in the reaction
(b) Heat is absorbed in the reaction
(c) Heat is neither given off nor absorbed in the reaction
(d) Increasing the amount of NOCl would also increase the equilibrium constant for the reaction
(e) Increasing the amount of NO_2 would also increase the equilibrium constant for the reaction

Increasing the amount of NOCl will not change the value of the equilibrium constant for the reaction. Increasing the concentration of NO_2 will not change the value of the equilibrium constant for the reaction. The equilibrium constant is independent of concentration. Changing the temperature at which the reaction is run will change the equilibrium constant for the reaction.In an endothermic reaction increasing the the temperature will increase the equilibrium constant for the reaction. In an exothermic reaction increasing the temperature will decrease the equilibrium constant for the reaction. In the above reaction increasing the temperature increases the equilibrium constant. So the above reaction is an endothermic reaction and heat will be absorbed. Answer(b)

11-47. What is the effect of increasing the temperature on the following reaction?

$$2CO\ (g) + O_2\ (g) \rightleftharpoons 2CO_2\ (g) + heat$$

(a) The equilibrium constant will increase
(b) The equilibrium constant will decrease
(c) The equilibrium constant will not change

Changing the temperature at which the reaction is run will change the equilibrium constant for the reaction.In an endothermic reaction increasing the the temperature will increase the equilibrium constant for the reaction. In an exothermic reaction increasing the temperature will decrease the equilibrium constant for the reaction. In the above reaction heat is given off in the reaction. So it is an exothermic reaction and increasing the temperature will decrease the equilibrium constant for the reaction. Answer(b)

11-48. What would happen to the extent of the decomposition of PCl_5 at a given temperature if the pressure on the system were decreased?

$$PCl_5 (g) \rightleftharpoons PCl_3 (g) + Cl_2 (g)$$

(a) It would increase   (b) It would decrease   (c) It would remain constant

There is a net increase in the number of molecules in the system as the reactant is converted to products, which leads to an increase in the pressure of the system. To counteract the change (decrease in the pressure on the system), the reaction will be shifted towards the products, because this will increase the number of particles. So the extent of decomposition of PCl_5 will increase. Answer(a)

11-49. What would be the effect of decreasing the pressure on the following reaction after it reaches equilibrium?

$$2NO_2 (g) \rightleftharpoons 2NO (g) + O_2 (g)$$

(a) The NO and O_2 concentrations would increase
(b) The NO and O_2 concentrations would decrease
(c) The NO concentration would increase but the O_2 concentration would decrease
(d) The O_2 concentration would increase but the NO concentration would decrease
(e) The concentrations of NO and O_2 would remain constant

There is a net increase in the number of molecules in the system as the reactant is converted to products, which leads to an increase in the pressure of the system. To counteract the change (decreasing the pressure on the reaction), the reaction will be shifted towards the products, because this will increase the numebr of particles. So NO and O_2 concentrations will increase. Answer(a)

11-50. At a fixed temperature and 1.00 atm total pressure NO_2 gas and N_2O_4 gas are in equilibrium.

$$2NO_2 (g) \rightleftharpoons N_2O_4 (g)$$

What is the effect of increasing the total pressure to 10.0 atm at a constant temperature.

(a) K_p increases, but the partial pressure of N_2O_4 decreases
(b) K_p decreases, but the partial pressure of N_2O_4 increases
(c) Both K_p and the partial pressure of N_2O_4 decrease
(d) Both K_p and the partial pressure of N_2O_4 increase
(e) K_p remains the same and the partial pressure of N_2O_4 increases

The temperature of the reaction is constant. So K_p will remain the same. To counteract the change (increasing the total pressure to 10.0 atm), the reaction will be shifted towards the product, because this will reduce the number of particles. So the partial pressure of N_2O_4 will increase. Answer(e)

11-51. What would happen to the equilibrium constant for the following reaction if the pressure of the system were increased?

$$3O_2 \text{ (g)} \rightleftharpoons 2 O_3 \text{ (g)}$$

(a) It would increase (b) It would decrease (c) It would remain the same

If the temperature is constant, the equilibrium constant will remain the same. Answer(c)

11-52. One of the products of the reaction between NO_2 and CO is CO_2. By writing the equation for the reaction, decide whether an increase in pressure :

(a) Favors CO_2 formation
(b) Decreases CO_2 formation
(c) Has no effect on CO_2 formation
(d) Cannot tell from information supplied

$$NO_2 \text{ (g)} + CO \text{ (g)} \rightleftharpoons CO_2 \text{ (g)} + NO \text{ (g)}$$

Δn is the difference between the number of moles of products and the number of moles of reactants in the balanced equation.

$\Delta n = (1+1) - (1+1) = 0$

So this equilibrium will not be affected by increase in pressure. Answer(c)

11-53. Which of the following equilibria would not be affected by changes in pressure?

(a) $2NO \text{ (g)} + O_2 \text{ (g)} \rightleftharpoons 2NO_2 \text{ (g)}$
(b) $N_2O_4 \text{ (g)} \rightleftharpoons 2NO_2 \text{ (g)}$
(c) $4NH_3 \text{ (g)} + 5O_2 \text{ (g)} \rightleftharpoons 4NO \text{ (g)} + 6H_2O \text{ (g)}$
(d) $CO \text{ (g)} + H_2O \text{ (g)} \rightleftharpoons CO_2 \text{ (g)} + H_2 \text{ (g)}$

Δn is the difference between the number of moles of products and the number of moles of reactants in the balanced equation.

(a) $\Delta n = (2) - (2+1) = -1$

(b) $\Delta n = (2) - (1) = +1$

(c) $\Delta n = (4+6) - (5+4) = +1$

(d) $\Delta n = (1+1) - (1+1) = 0$

When Δn = 0, the equilibrium will not be affected by changes in pressure.

In the case of (d), Δn = 0. So (d) will not be affected by changes in pressure.   Answer(d)

CHAPTER 12

ACIDS AND BASES

INVALUABLE INFORMATION

TYPICAL BRØNSTED ACIDS AND THEIR CONJUGATE BASES

COMPOUND	K_a	pK_a	CONJUGATE BASE	K_b	pK_b
HI	3.0×10^9	-9.5	I^-	3.0×10^{-24}	23.5
HBr	1.0×10^9	-9.0	Br^-	1.0×10^{-23}	23
HCl	1.0×10^6	-6.0	Cl^-	1.0×10^{-20}	20
H_2SO_4	1.0×10^3	-3.0	HSO_4^-	1.0×10^{-17}	17
H_3O^+	55	-1.7	H_2O	1.8×10^{-16}	15.7
HNO_3	28	-1.4	NO_3^-	3.6×10^{-16}	15.4
PH_4^+	1	0	PH_3	1.0×10^{-14}	14.0
H_3PO_4	7.1×10^{-3}	2.1	$H_2PO_4^-$	1.4×10^{-12}	11.9
H_2Te	2.5×10^{-3}	2.6	HTe^-	4.0×10^{-12}	11.4
HF	7.2×10^{-4}	3.14	F^-	1.4×10^{-11}	10.86
H_2Se	2.0×10^{-4}	3.70	HSe^-	5.0×10^{-11}	10.30
CH_3CO_2H	1.8×10^{-5}	4.70	$CH_3CO_2^-$	5.6×10^{-10}	9.3
H_2S	1.0×10^{-7}	7.00	HS^-	1.0×10^{-7}	7.0
NH_4^+	5.6×10^{-10}	9.25	NH_3	1.8×10^{-5}	4.75
HS^-	1.3×10^{-13}	12.9	S^{2-}	8×10^{-2}	1.10
CH_3OH	3.16×10^{-16}	15.5	CH_3O^-	31.6	-1.5
H_2O	1.8×10^{-16}	15.74	OH^-	55	-1.74
AsH_3	1.0×10^{-23}	23	AsH_2^-	1.0×10^9	-9
$HC \equiv CH$	1.0×10^{-25}	25	$HC \equiv C^-$	1.0×10^{11}	-11
GeH_4	1.0×10^{-25}	25	GeH_3^-	1.0×10^{11}	-11
PH_3	1.0×10^{-27}	27	PH_2^-	1.0×10^{13}	-13
NH_3	1.0×10^{-33}	33	NH_2^-	1.0×10^{19}	-19

H_2	1.0×10^{-35}	35	H^-	1.0×10^{21}	-21
SiH_4	1.0×10^{-35}	35	SiH_3^-	1.0×10^{21}	-21
$H_2C=CH_2$	1.0×10^{-44}	44	$H_2C=C^-$	1.0×10^{30}	-30
CH_4	1.0×10^{-49}	49	CH_3^-	1.0×10^{35}	-35

12-1. Which of the following groups contains salts that all form basic solutions in H_2O?

(a) $NaNO_3$, NH_4CN, $NaOAc$, NH_4Cl (b) Na_2CO_3, KCl, $NaOAc$, NH_4Cl

(c) Na_2CO_3, NaF, $NaOAc$, $NaCN$ (d) $NaHCO_3$, NaF, NH_4Cl, Na_2SO_3

(e) None of the above.

(a): NH_4Cl is a salt of strong acid and a weak base. NH_4^+ will undergo hydrolysis to give H_3O^+ ions and reduce the pH to provide acidic solution.

$$NH_4^+ (aq) + H_2O (l) \rightleftharpoons NH_3 (aq) + H_3O^+ (aq)$$

(b): NH_4Cl will give acidic solution.

(d): NH_4Cl will acidic solution.

(c): Na_2CO_3: Salt of a strong base and a weak acid. CO_3^- will undergo hydrolysis to give basic solution.

$$CO_3^{2-} (aq) + H_2O (l) \rightleftharpoons HCO_3^- (aq) + OH^- (aq)$$

NaF: Salt of a strong base and a weak acid. F^- will undergo hydrolysis to give basic solution.

$$F^- (aq) + H_2O (l) \rightleftharpoons HF (aq) + OH^- (aq)$$

NaOAc: Salt of a strong base and a weak acid. OAc^- will undergo hydrolysis to give basic solution.

$$CH_3COO^- (aq) + H_2O (l) \rightleftharpoons CH_3COOH (aq) + OH^- (aq)$$

NaCN: Salt of a strong base and a weak acid. CN^- will undergo hydrolysis to give basic solution.

$$CN^- (aq) + H_2O (l) \rightleftharpoons HCN (aq) + OH^- (aq) \qquad \text{Answer(c)}$$

12-2. Which of the following compounds would give a 0.10 M solution with a pH less than 7?

(a) Na_2S (b) K_2CO_3 (c) NH_4Cl (d) Na_3PO_4

Na_2S: $S^{2-} (aq) + H_2O (l) \rightleftharpoons HS^- (aq) + OH^- (aq)$ $pH > 7$

K_2CO_3: $CO_3^{2-} (aq) + H_2O (l) \rightleftharpoons HCO_3^- (aq) + OH^- (aq)$ $pH > 7$

NH$_4$Cl: NH$_4^+$ (aq) + H$_2$O (l) $\rightleftharpoons$ NH$_3$ (aq) + H$_3$O$^+$ (aq) pH < 7

Na$_3$PO$_4$: PO$_4^{3-}$ (aq) + H$_2$O (l) $\rightleftharpoons$ HPO$_4^{2-}$ (aq) + OH$^-$ (aq) pH > 7 Answer(c)

12-3. The following compounds dissolve in H$_2$O to form solutions that turn litmus from red to blue. Which of these compounds satisfy the operational definition of a base?

(a) NaHCO$_3$ (b) CaO (c) CaH$_2$ (d) NH$_3$ (e) All of the above

Since all of them turn litmus from red to blue, all of them are bases. Answer(e)

12-4. What is the H$_3$O$^+$ concentration in a solution that has a pH of 5.75?

(a) 5.8×10^{-5} (b) 3.2×10^{-3} (c) 0.75×10^{-5} (d) 1.8×10^{-6}

pH = − log (H$_3$O$^+$) = 5.75 Log (H$_3$O$^+$) = − 5.75 (H$_3$O$^+$) = 1.8×10^{-6} Answer(d)

12-5. (i) Element X reacts with excess oxygen to form a basic oxide with the formula XO. In which Group of the periodic table should X be placed?

(a) Group 2 (b) Group 13 (c) Group 14 (d) Group 16 (e) Group 17

Compound XO: Oxidation state of X is +2. So it must be from Group 2. Answer(a)

(ii) If the element described in (i) were to react with hydrogen it should form:

(a) An acidic anhydride, XH (b) An acidic anhydride, XH$_2$

(c) A basic anhydride, XH (d) A basic anhydride, XH$_2$ (e) None of the above

Metal hydrides are bases. The oxidation state of the metal is +2. So a basic anhydride XH$_2$ will be formed. Answer(d)

12-6. Which Group 14 element is most likely to form an oxide with the formula XO$_2$ that dissolves in H$_2$O to form an acidic solution?

(a) C (b) Si (c) Ge (d) Sn (e) Pb

Non metal oxides form acidic solutions with H$_2$O. Sn and Pb are metals. Si and Ge are semimetals. C is a non-metal which is most likely to form an oxide (CO$_2$) that dissolves in H$_2$O to form an acidic solution. Answer(a)

12-7. Which of the following elements would be the most likely to form an acidic oxide with the formula XO$_2$ and an acidic anhydride with the formula XH$_2$?

(a) Na (b) Mg (c) Al (d) S (e) Cl

Non-metal oxides and hydrides are acidic. Na, Mg and Al are metals. Cl forms HCl.

Only S can form SO_2 (acidic oxide) and H_2S (acidic anhydride). Answer(d)

12-8. Which compound should dissolve in H_2O to form a 0.1 M solution with a pH of 8.9?

(a) HCl (b) NaOAc (c) HOAc (d) NaCl (e) KOH

(a): HCl is a strong acid and the pH will be < 7.

(c): CH_3COOH is a weak acid and the pH will be < 7.

(d): NaCl is a salt of a strong acid and a strong base. So it will not undergo hydrolysis in water and will not change the pH of the water.

(e): KOH is a strong base.

$[OH^-] = 0.1$

$[H^+][OH^-] = K_w = 10^{-14}$

$[H^+] = (10^{-14})/(0.1) = 10^{-13}$

pH = 13

(b): NaOAc is a salt of of a strong base and a weak acid. CH_3COO^- will undergo hydrolysis to give a pH > 7.

$$CH_3COO^- \text{ (aq)} + H_2O \text{ (l)} \rightleftharpoons CH_3COOH \text{ (aq)} + OH^- \text{ (aq)}$$

pH = ½ pK_w + ½ pK_a + ½ log C = 7.0 + 2.37 + ½ log (0.1) = 7.0 + 2.37 + (− 0.5) = 8.9

Answer(b)

12-9. Which of the following acid-base reactions should go more or less to completion?

(a) $HClO_4^-$ (aq) + OCl^- (aq) $\longrightarrow$ ClO_4^- (aq) + HOCl (aq)
(b) HF (aq) + I^- (aq) $\longrightarrow$ HI (aq) + F^- (aq)
(c) H_2SO_3 (aq) + SO_4^{2-} (aq) $\longrightarrow$ H_2SO_4 (aq) + SO_3^{2-} (aq)
(d) H_2S (aq) + $2Cl^-$ (aq) $\longrightarrow$ S^{2-} (aq) + 2HCl (aq)
(e) All of these reactions would go to completion, giving high yields of products.

HF, H_2SO_3, and H_2S are weak acids and HI, H_2SO_4, and HCl are strong acids. So, reactions

(b),(c), and (d) will not go to completion. $HClO_4$ is a strong acid; reaction (a) should go more

or less to completion. Answer(a)

12-10. Which of the following won't produce an acidic solution when dissolved in water?

(a) $FeCl_3$ (b) NH_4Cl (c) Aceticacid (CH_3COOH)

(d) Sodium acetate (CH_3COONa) (e) $HClO_4$

(a) $FeCl_3$ when dissolved in water will produce an acidic solution due to complex formation.

(b) NH_4Cl is a salt of a strong acid and a weak base. NH_4^+ will undergo hydrolysis to produce an acidic solution.

$$NH_4^+ (aq) + H_2O (l) \rightleftharpoons H_3O^+ (aq) + NH_3 (aq)$$

(c) Acetic acid is a weak acid and will produce an acidic solution in H_2O.

(d) Sodium acetate is a salt of a strong base and a weak acid. CH_3COO^- will undergo hydrolysis to give pH > 7.

$$CH_3COO^- (aq) + H_2O (l) \rightleftharpoons CH_3COOH (aq) + OH^- (aq)$$

(e) $HClO_4$ is a strong acid and will produce an acidic solution in H_2O.   Answer(d)

12-11. Which of the following statements is correct?

    (a) Nonmetals usually form basic oxides
    (b) Metals usually form basic hydroxides
    (c) Nonmetals have the lowest electronegativities
    (d) Nonmetals usually form ionic compounds with other nonmetals
    (e) All of the above are correct

Nonmetals usually form acidic oxides.e.g., CO_2

Metals usually form basic hydroxides. e.g., NaOH

Electronegativity increases from left to right in a period.

Nonmetals usually form covalent compounds with other nonmetals. e.g., CH_4 Answer(b)

12-12. Which of the following compounds exhibits the following behavior: It dissolves in water to give a colorless, odorless solution with a pH of 11. The solution reacts with HCl to form an odorless gas. When $AgNO_3$ is added to this solution, a yellow precipitate is formed.

    (a) H_2SO_4 (b) $Cu(OH)_2$ (c) Na_2CO_3 (d) Na_2S (e) NaCl

NaCl is a salt of a strong acid and a strong base. So it will not undergo hydrolysis and the pH will not be 11 when it dissolves in water. H_2SO_4 is a strong acid and the pH will not be 11. The reaction between $Cu(OH)_2$ and HCl does not form an odorless gas. The reaction between Na_2S and HCl produces H_2S which is not an odorless gas. Na_2CO_3 reacts with HCl to form CO_2 which is odorless. Na_2CO_3 reacts with $AgNO_3$ to form $AgCO_3$ a yellow precipitate.

Answer(c)

12-13. Which of the following gives a basic solution when dissolved in water?

    (a) CaO (b) CO_2 (c) Cl_2O (d) SO_2 (e) SeO_3

Metal oxides give basic solutions when dissolved in water and nonmetal oxides give acidic solutions when dissolved in water. Ca is a metal and C,Cl,S and Se are nonmetals.   Answer(a)

12-14. Which of the following compounds is not an acid in H_2O?

   (i)SO_2 (ii) HNO_3 (iii) SrO (iv) HI (v) K_2S

   (a) i and ii (b) ii and iii (c) iii and v (d) i, ii and iv

SO_2: Nonmetal oxide will produce acidic solution in water.

HNO_3: Strong acid will produce acidic solution in water.

SrO: Metal oxide will produce basic solution in water.

HI: Strong acid will produce acidic solution in water.

K_2S: Salt of a strong base and a weak acid will produce basic solution in water.   Answer(c)

12-15. Which of the following compounds isn't a base in water?

   (a) $Ca(OH)_2$ (b) CuO (c) NO_2 (d) LiO (e) CsOH

$Ca(OH)_2$: Metal hydroxide is a base in water.

CuO: Metal oxide is a base in water.

NO_2: Nonmetal oxide is an acid in water

LiO: Metal oxide is a base in water.

CsOH: Metal hydroxide is a base in water.   Answer(c)

12-16. Which of the following is most likely to be a white solid with a high melting point that dissolves in water to form a basic solution?

   (a) O_2 (b) CO_2 (c) Na_2O (d) P_4O_{10} (e) Cl_2O_7

O_2 and CO_2 are gases.

 Na_2O: Metal oxide produces a basic solution in water.

P_4O_{10} : Nonmetal oxide produces an acidic solution in water.

Cl_2O_7: Nonmetal oxide produces an acidic solution in water.   Answer(c)

12-17. Which of the following oxides would yield the most basic solution when dissolved in water?

   (a) ClO_2 (b) SiO_2 (c) SO_2 (d) Al_2O_3

Metal oxides will produce basic solutions in water. Nonmetal oxides will produce acidic solutions in water. Cl and S are nonmetals. Si is a semimetal. Al_2O_3 is an amphoteric oxide.

258

$Al(OH)_3$ can act as a base.

$$Al(OH)_3 \text{ (s)} + 3H^+ \text{ (aq)} \rightleftharpoons Al^{3+} \text{ (aq)} + 3H_2O \text{ (l)} \qquad \text{Answer(d)}$$

12-18. Consider the following reaction:

$$H_2PO_4^- \text{ (aq)} + HCO_3^- \text{ (aq)} \longrightarrow H_2CO_3 \text{ (aq)} + HPO_4^- \text{ (aq)}$$

Brønsted would identify the acidic species as:

(a) $H_2PO_4^-$ and H_2CO_3 (b) $H_2PO_4^-$ and HPO_4^{2-} (c) HCO_3^- and H_2CO_3

(d) HCO_3^- and HPO_4^{2-} (e) H_2CO_3 and HPO_4^{2-}

Brønsted acid is a proton (H^+ ion) donor. So $H_2PO_4^-$ and H_2CO_3 are Brønsted acids in the

above reaction. Answer(a)

12-19. Label the Brønsted acids and bases in the following reaction.

$$HSO_4^- \text{ (aq)} + H_2O \text{ (l)} \longrightarrow H_3O^+ \text{ (aq)} + SO_4^{2-} \text{ (aq)}$$

Brønsted acid is a proton (H^+ ion) donor and Brønsted base is a proton (H^+ ion) acceptor. So

HSO_4^- and H_3O^+ are Brønsted acids and H_2O and SO_4^{2-} are Brønsted bases in the above

reaction.

12-20. Which compound cannot act as a Brønsted base?

(a) NH_3 (b) H_2O (c) CH_4 (d) Cl^-

NH_3: Can accept a H^+ ion to form NH_4^+.

H_2O: Can accept a H^+ ion to form H_3O^+.

CH_4: Cannot accept a H^+ ion to form CH_5^+; does not have a pair of nonbonding valence

electrons which can accept a H^+ ion.

Cl^-: Can accept a H^+ ion to form HCl. Answer(c)

12-21. Which of the following compounds cannot be a Brønsted base?

(a) H_2O (b) NH_3 (c) CO_3^{2-} (d) OH^- (e) NH_4^+

H_2O: Can accept a H^+ ion to form H_3O^+.

NH_3: Can accept a H^+ ion to form NH_4^+.

CO_3^{2-}: Can accept a H^+ ion to form HCO_3^-.

OH^-: Can accept a H^+ ion to form H_2O.

NH_4^+: Cannot accept a H^+ ion; does not have a pair of nonbonding valence electrons which can accept a H$^+$ ion. Answer(e)

12-22. Which of the following compounds would not be a base in solution?

(a) NH$_3$ (b) CaO (c) Ca(OH)$_2$ (d) NaOH (e) NH$_4$NO$_3$

NH$_3$, CaO (metal oxide), Ca(OH)$_2$, and NaOH will be bases in solution.

NH$_4$NO$_3$: Salt of a strong acid and a weak base. NH$_4^+$ will undergo hydrolysis in water to give

acidic solution.

$$\text{NH}_4^+ \text{ (aq)} + \text{H}_2\text{O (l)} \rightleftharpoons \text{H}_3\text{O}^+ \text{ (aq)} + \text{NH}_3 \text{ (aq)}$$ Answer(e)

12-23. Which of the following compounds cannot be a Brønsted base?

(a) H$_2$O (b) MnO$_4^-$ (c) BH$_4^-$ (d) CN$^-$ (e) S^{2-}

H$_2$O: Can accept a H$^+$ ion to form H$_3$O$^+$.

MnO$_4^-$: Can accept a H$^+$ ion to form HMnO$_4$.

BH$_4^-$: Cannot accept a H$^+$ ion; does not have a pair of nonbonding valence electrons which can

accept a H$^+$ ion.

CN$^-$: Can accept a H$^+$ ion to form HCN.

S^{2-}: Can accept a H$^+$ ion to form HS$^-$. Answer(c)

12-24. Which of the following compounds cannot be a Brønsted base?

(a) O$_2$ (b) CO$_2$ (c) PH$_3$ (d) SF$_4$ (e) CH$_3^+$

O$_2$: Can accept a H$^+$ ion to form HO$_2^+$.

CO$_2$: Can accept a H$^+$ ion to form HCO$_2^+$.

PH$_3$: Can accept a H$^+$ ion to form PH$_4^+$.

SF$_4$: Can accept a H$^+$ ion to form HSF$_4^+$.

CH$_3^+$: Cannot accept a H$^+$ ion; does not have a pair of nonbonding valence electrons which can

accept a H$^+$ ion. Answer(e)

12-25. Which of the following compounds can act as either a Brønsted acid or a Brønsted base?

(i)   NaHCO$_3$ (ii) Na$_2$CO$_3$ (iii) H$_2$CO$_3$ (iv) CO$_2$ (v) H$_2$O

(a) i (b) i and ii (c) ii and iii (d) i and v (e) v

Na$_2$CO$_3$ and CO$_2$ cannot donate H$^+$ ions.

H$_2$CO$_3$ can donate H$^+$ ions but cannot accept H$^+$ ions.

NaHCO₃ can donate and accept H^+ ion.

H₂O can donate and accept H^+ ion.   Answer(d)

12-26. Which compound is most likely to be amphoteric?

    (a) Na_2O (b) CaO (c) Al_2O_3 (d) P_4O_{10} (e) Cl_2O_7

(a) and (b) are basic metal oxides.

(d) and (e) are acidic nonmetal oxides.

$$Al(OH)_3 \text{ (s)} + OH^- \text{ (aq)} \longrightarrow Al(OH)_4^- \text{ (aq)}$$

$$Al(OH)_3 \text{ (s)} + 3H^+ \text{ (aq)} \longrightarrow Al^{3+} \text{ (aq)} + 3H_2O \text{ (l)}$$

(c) can act both as an acid and a base. So it is amphoteric.   Answer(c)

12-27. Which of the following is the conjugate acid of the HPO_4^{2-} ion?

    (a) H_3PO_4 (b) $H_2PO_4^-$ (c) HPO_4^{2-} (d) PO_4^{3-} (e) H_3O^+

$$HPO_4^{2-} \text{ (aq)} + H_2O \text{ (l)} \rightleftharpoons H_2PO_4^- \text{ (aq)} + OH^- \text{ (aq)}$$

  Base   Acid   Conjugate   Conjugate

                                Acid   Base   Answer(b)

12-28. Which of the following is the conjugate base of the HPO_4^{2-} ion?

    (a) PO_4^{3-} (b) HPO_4^{2-} (c) $H_2PO_4^{2-}$ (d) H_3PO_4 (e) None of the above

$$HPO_4^{2-} \text{ (aq)} + H_2O \text{ (l)} \rightleftharpoons H_3O^+ \text{ (aq)} + PO_4^{3-} \text{ (aq)}$$

Acid   Base   Conjugate   Conjugate

                                  acid   base   Answer(a)

12-29. What is the conjugate base of HSO_4^-?

    (a) H_2O (b) H_2SO_4 (c) HSO_4^- (d) SO_3 (e) None of the above

$$HSO_4^- \text{ (aq)} + H_2O \text{ (l)} \rightleftharpoons SO_4^{2-} \text{ (aq)} + H_3O^+ \text{ (aq)}$$

Acid   Base   Conjugate   Conjugate

                                  base   acid   Answer(e)

12-30. Methanol is described by the following skeleton structure:

$$H - \overset{\displaystyle\overset{H}{\mid}}{\underset{\displaystyle\underset{H}{\mid}}{C}} - O - H$$

Which compound is formed when CH_3OH acts as a BrØnsted base?

(a) CH_3O^- (b) CH_3OH (c) $CH_3OH_2^+$ (d) H_3O^+

Brønsted base is a H^+ ion acceptor.

$$CH_3OH + H^+ \longrightarrow CH_3OH_2^+$$

CH_3OH will accept a H^+ ion to form $CH_3OH_2^+$.   Answer(c)

12-31. Which of the following is the conjugate base of a strong acid?

(a) OH^- (b) HSO_4^- (c) NH_2^- (d) S^{2-}

OH^- is the conjugate base of H_2O (a weak acid, $K_a = 1.8 \times 10^{-16}$).

HSO_4^- is conjugate base of H_2SO_4 (a strong acid, $K_a = 1.0 \times 10^3$).

NH_2^- is a conjugate base of NH_3 (a weak acid, $K_a = 1 \times 10^{-33}$).

S^{2-} is a conjugate base of HS^- (a weak acid, $K_a = 1.3 \times 10^{-13}$).   Answer(b)

12-32. Which of the following represents a conjugate acid/base pair?

(a) H^+, OH^- (b) H_3O^+, OH^- (c) NH_3, NH_4^+ (d) $NaOH, HCl$

OH^- is the conjugate base of H_2O

Cl^- is the conjugate base of HCl.

NH_3 is the conjugate base of NH_4^+.   Answer(c)

12-33. Which of the following isn't a conjugate acid/base pair?

(a) NH_4^+ (aq) and NH_3 (aq)   (b) H_2O (l) and OH^- (aq)

(c) H_3O^+ (aq) and OH^- (aq)   (d) NH_3 (aq) and NH_2^- (aq)

(e) All of these are conjugate acid/base pairs

(a) NH_4^+ (aq) + H_2O (l) $\rightleftharpoons$ H_3O^+ (aq) + NH_3 (aq)   (a) is a conjugate acid/ base pair.

(b) H_2O (l) + H_2O (l) $\rightleftharpoons$ H_3O^+ (aq) + OH^- (aq)   (b) is a conjugate acid/ base pair.

(c) H_2O (l) + H_2O (l) $\rightleftharpoons$ H_3O^+ (aq) + OH^- (aq)   (c) is not a conjugate acid/ base pair.

(d) NH_3 + NH_3 $\rightleftharpoons$ NH_4^+ + NH_2^- (d) is a conjugate acid/ base pair.

                    Answer(c)

12-34. Which of the following isn't a Brønsted conjugate acid/base pair?

(a) NH_3/NH_2^- (b) CH_3COOH/CH_3COO^- (c) H_2O/O^{2-} (d) H_3O^+/H_2O

(a) $NH_3 + NH_3 \rightleftharpoons NH_4^+ + NH_2^-$ (a) is a conjugate acid/ base pair.

(b) $CH_3COOH (aq) + H_2O (l) \rightleftharpoons CH_3COO^- (aq) + H_3O^+ (aq)$ (b)is a conjugate acid/ base pair.

(d) $H_2O (l) + H_2O (l) \rightleftharpoons H_3O^+ (aq) + OH^- (aq)$ (d) is a conjugate acid/ base pair.

H_2O (acid)/OH^- (conjugate base) is a conjugate acid/ base pair and

OH^- (acid)/O^{2-} (conjugate base) is a conjugate acid/ base pair.

H_2O/O^{2-} is not a conjugate acid/ base pair.   Answer(c)

12-35. Which of the following is the strongest Brønsted acid?

(a) NH_2^- (b) HS^- (c) H_2O (d) CH_4 (e) NH_4^+

NH_2^- cannot donate a H^+ ion but it can accept a H^+ ion to form NH_3. So it is a Brønsted base.

	K_a
HS^-	1.3×10^{-13}
CH_4	1×10^{-49}
H_2O	1.8×10^{-16}
NH_4^+	5.6×10^{-10}

NH_4^+ is the strongest Brønsted acid.   Answer(e)

12-36. Which of the following is the strongest Brønsted acid?

(a) CsH (b) H_2O (c) HF (d) H_2Te (e) HI

(a) CsH: metal hydride is a base.

	K_a
H_2O	1.8×10^{-16}
HF	7.2×10^{-4}
H_2Te	2.5×10^{-3}
HI	3.0×10^9

HI is the strongest Brønsted acid. Electronegativity of F (3.98) > Electronegativity of I (2.66).

But the increase in size from F to I decreases the bond dissociation enthalpy making HI the

strongest Brønsted acid.   Answer(e)

12-37. Which of the following would be the strongest Brønsted acid?

(a) H_3O^+ (b) HF   (c) NH_3 (d) HSO_4^- (e) NaOH

NaOH is a strong base.

Greater the negative charge, stronger the conjugate base. So H_2O will be the weakest conjugate base and H_3O^+ will be the strongest acid.

	K_a
NH_3	1.0×10^{-33}
HF	7.2×10^{-4}
HSO_4^-	1.0×10^{-2}
H_3O^+	55

H_3O^+ is the strongest Brønsted acid.   Answer(a)

12-38. Which of the following would be the weakest Brønsted acid?

(a) H_2Se (b) H_2S (c) H_2O (d) SH^- (e) OH^-

H_2Se	$K_{a1} = 1.3 \times 10^{-4}$
H_2S	$K_{a1} = 8.9 \times 10^{-8}$
H_2O	$K_{a1} = 1.8 \times 10^{-16}$
SH^-	$K_a = 1.3 \times 10^{-13}$
OH^-	$K_a < 1.0 \times 10^{-19}$

OH^- is the weakest Brønsted acid.   Answer(e)

12-39. Which of the following is the weakest acid?

(a) $HClO_4$ (b) HCl (c) HF (d) HI (e) HBr

	K_a
$HClO_4$	$K_a = 1.0 \times 10^3$
HCl	$K_a = 1.0 \times 10^6$
HF	$K_a = 7.2 \times 10^{-4}$
HI	$K_a = 3.9 \times 10^9$
HBr	$K_a = 1.0 \times 10^9$

The smaller size of F increases the bond dissociation enthalpy of HF making it the weakest acid. Answer(c)

12-40. Which of the following is the weakest Brønsted acid?

(a) H_2SO_3 (b) H_2CrO_4 (c) H_2BO_3 (d) C_6H_5OH

	K_a
H_2SO_3	1.4×10^{-2}
H_2CrO_4	9.6
H_2BO_3	7.3×10^{-10}
C_6O_5OH	1.0×10^{-10}

C_6O_5OH is the weakest Brønsted acid (lowest K_a value). Answer(d)

12-41. Which of the following is the strongest base?

(a) I^- (b) NO_3^- (c) HS^- (d) O^{2-} (e) OH^-

Base	Conjugate acid
I^-	HI ($K_a = 3.9\times10^9$)
NO_3^-	HNO_3 ($K_a = 28$)
HS^-	H_2S ($K_a = 1.0\times10^{-7}$)
O^{2-}	OH^- ($K_a < 10^{-19}$)
OH^-	H_2O ($K_a = 1.8\times10^{-16}$)

OH^- is the weakest conjugate acid. So O^{2-} is the strongest base. Answer(d)

12-42. Which of the following is the strongest base?

(a) NH_2^- (b) PH_2^- (c) CH_3^- (d) SiH_3^- (e) OH^-

Base	Conjugate acid
NH_2^-	NH_3 ($K_a = 1.0\times10^{-33}$)
PH_2^-	PH_3 ($K_a = 1.0\times10^{-27}$)
CH_3^-	CH_4 ($K_a = 1.0\times10^{-49}$)
SiH_3^-	SiH_4 ($K_a = 1.0\times10^{-7}$)
OH^-	H_2O ($K_a = 1.8\times10^{-16}$)

CH_4 is the weakest conjugate acid. So CH_3^- is the strongest base. Answer(c)

12-43. Which of the following is the strongest base?

(a) H_2O (b) OH^- (c) NH_2^- (d) NH_3 (e) CH_4

CH_4 does not have a pair of nonbonding valence electrons which can accept a H^+ ion. So it is not a base.

Base	Conjugate acid
H_2O	H_3O^+ ($K_a = 55$)
OH^-	H_2O ($K_a = 1.8 \times 10^{-16}$)
NH_2^-	NH_3 ($K_a = 1.0 \times 10^{-33}$)
NH_3	NH_4^+ ($K_a = 5.6 \times 10^{-10}$)

NH_3 is the weakest conjugate acid. So NH_2^- will be the strongest base.   Answer(c)

12-44. Which of the following is the strongest base?

(a) F^- (b) OH^- (c) H_2O (d) Cl^- (e) O^{2-}

Base	Conjugate acid
F^-	HF ($K_a = 7.2 \times 10^{-4}$)
OH^-	H_2O ($K_a = 1.8 \times 10^{-16}$)
H_2O	H_3O^+ ($K_a = 55$)
Cl^-	HCl ($K_a = 1.0 \times 10^6$)
O^{2-}	OH^- ($K_a < 10^{-19}$)

OH^- is the weakest conjugate acid. So O^{2-} is the strongest base.   Answer(e)

12-45. Which of the following would be the strongest Brønsted base?

(a) HSO_3^- (b) HSO_4^- (c) H_2SO_3 (d) SO_3^{2-} (e) SO_4^{2-}

H_2SO_3 is an acid.

Greater the negative charge, stronger the base. So SO_3^{2-} and SO_4^{2-} will be stronger bases. In the case of SO_3^{2-} the oxidation state of S in the conjugate acid HSO_3^- is +4 and in the case of SO_4^{2-} the oxidation state of S in the conjugate acid HSO_4^- is +6. Greater the oxidation state of the central atom, stronger the acid. So HSO_4^- is a stronger conjugate acid than HSO_3^- and SO_3^{2-} is a stronger base than SO_4^{2-}.

Base	Conjugate acid
HSO_3^-	H_2SO_3 ($K_a = 1.4 \times 10^{-2}$)

HSO_4^-	H_2SO_4 ($K_a = 1.0 \times 10^3$)
SO_3^{2-}	HSO_3^- ($K_a = 6.3 \times 10^{-8}$)
SO_4^{2-}	HSO_4^- ($K_a = 1.0 \times 10^{-2}$)

HSO_3^- is the weakest conjugate acid. So SO_3^{2-} is the strongest base.   Answer(d)

12-46. Which of the following would be the strongest Bronsted base?

   (a) H_2O (b) HS^- (c) OH^- (d) S^{2-} (e) O^{2-}

Base	Conjugate acid
H_2O	H_3O^+ ($K_a = 55$)
HS^-	H_2S ($K_a = 1.0 \times 10^{-7}$)
OH^-	H_2O ($K_a = 1.8 \times 10^{-16}$)
S^{2-}	HS^- ($K_a = 1.3 \times 10^{-13}$)
O^{2-}	OH^- ($K_a < 10^{-19}$)

OH^- is the weakest conjugate acid. So O^{2-} is the strongest base.   Answer(e)

12-47. Which oxide would dissolve in H_2O to produce the most basic solution?

   (a) CaO (b) Ga_2O_3 (c) GeO_2 (d) P_4O_{10} (e) SO_3

Metal oxides produce basic solutions in water and nonmetal oxides produce acidic solutions in water. Ca and Ga are metals, Ge is a semimetal. P and S are nonmetals. Metallic character decreases from left to right in a period. So CaO will produce the most basic solution in H_2O.

Answer(a)

12-48. Which of the following is a weak base?

   (a) $(CH_3)_3N$ (b) HF   (c) KOH   (d) H_2

(a) $(CH_3)_3N$ can accept a H^+ ion. It is a weak base ($K_b = 6.3 \times 10^{-5}$).

(b) HF is a weak acid.

(c) KOH is a strong base.

(d) H_2 does not have a pair of nonbonding valence electrons to accept a H^+ ion.   Answer(a)

12-49. Which of the following would have the strongest conjugate base?

   (a) H_2O (b) H_2S (c) NH_3 (d) PH_3 (e) CH_4

Conjugate base	Acid
OH^-	H_2O ($K_a = 1.8 \times 10^{-16}$)
HS^-	H_2S ($K_a = 1.0 \times 10^{-7}$)
NH_2^-	NH_3 ($K_a = 1.0 \times 10^{-33}$)
PH_2^-	PH_3 ($K_a = 1.0 \times 10^{-27}$)
CH_3^-	CH_4 ($K_a = 1.0 \times 10^{-49}$)

CH_4 is the weakest acid. So it will have the strongest conjugate base.   Answer(e)

12-50. What is the correct increasing base strength order?

(a) $NH_3 < PH_3 < H_2O < H_2S$ (b) $H_2S < H_2O < PH_3 < NH_3$

(c) $H_2O < H_2S < NH_3 < PH_3$ (d) $H_2O < H_2O < PH_3 < NH_3$

H_2S is a weak acid.

	K_b
NH_3	1.8×10^{-5}
PH_3	1.0×10^{-14}
H_2O	1.8×10^{-16}

The correct increasing base strength order is: $H_2S < H_2O < PH_3 < NH_3$ Answer(b)

12-51. Which of the following statements is true?

(a) H_2O is a stronger acid than HF(aq)
(b) H_2O is a stronger acid than H_2S(aq)
(c) H_2O is a stronger acid than H_3O^+ (aq) ion
(d) H_2O is a stronger acid than NH_3 (aq)
(e) None of the above is true

Acid	K_a
H_2O	1.8×10^{-16}
HF	7.2×10^{-4}
H_2S	1.0×10^{-7}
H_3O^+	55
NH_3	1.0×10^{-33}

The increasing acid strength order is: $NH_3 < H_2O < H_2S < HF < H_3O^+$ Answer(d)

12-52. Which of the following statements is true?

(a) H_2O is a weaker acid than NH_3
(b) PH_3 is a weaker acid than NH_3
(c) OH^- is a weaker base than H_2O
(d) NH_2^- is the conjugate acid of NH_3
(e) ClO^- is a stronger base than ClO_4^-

(a)H_2O ($K_a = 1.8\times10^{-16}$) is a stronger acid than NH_3 ($K_a = 1.0\times10^{-33}$).

(b)PH_3 ($K_a = 1.0\times10^{-27}$) is a stronger acid than NH_3 ($K_a = 1.0\times10^{-33}$).

(c)OH^- ($K_b = 55$) ia stronger base than H_2O ($K_b = 1.8\times10^{-16}$).

(d) NH_4^+ is the conjugate acid of NH_3.

(e) $HClO_4$ ($K_a = 1.0\times10^3$) is a stronger acid than $HOCl$ ($K_a = 2.9\times10^{-8}$). So ClO^- is a stronger

base than ClO_4^-. Answer(e)

12-53. What can be correctly concluded from the fact that the following acid-base reaction

proceeds to the right, as written?

$$NH_2^- \text{ (aq)} + HSO_4^- \text{ (aq)} \longrightarrow NH_3 \text{ (aq)} + SO_4^{2-} \text{ (aq)}$$

(a) NH_3 is a stronger base than NH_2^-
(b) NH_3 is a stronger acid than HSO_4^-
(c) NH_3 is a weaker acid than NH_2^-
(d) NH_3 is a weaker base than HSO_4^-
(e) NH_3 is a weaker acid than HSO_4^-

NH_3 is the conjugate acid of NH_2^- and SO_4^{2-} is a conjugate base of HSO_4^-. The equation

proceeds to the right. This means that NH_3 ($K_a = 1.0\times10^{-33}$) is a weak acid compared to HSO_4^-

($K_a = 1.2\times10^{-2}$). Answer(e)

12-54. $NaHCO_3$ can be used to neutralize strong bases, such as $NaOH$. What conclusion can be

drawn from the fact that the following acid base reaction proceeds to the right as written?

$$HCO_3^- \text{ (aq)} + OH^- \text{ (aq)} \longrightarrow CO_3^{2-} \text{ (aq)} + H_2O \text{ (l)}$$

(a) HCO_3^- is a stronger acid than H_2O
(b) HCO_3^- is a stronger base than CO_3^-
(c) HCO_3^- is a stronger base than OH^-
(d) CO_3^- is a stronger base than OH^-
(e) H_2O is a stronger acid than HCO_3^-

CO_3^{2-} is a conjugate base of HCO_3^- and H_2O is a conjugate acid of OH^-. The equation

proceeds to the right. This means that HCO_3^- ($K_a = 4.7\times10^{-11}$) is a stronger acid than H_2O (K_a

$= 1.8\times10^{-16}$). Answer(a)

12-55. What can we conclude from the fact that the following reaction proceeds as written?

$$NaNH_2 \text{ (s)} + H_2O \text{ (l)} \longrightarrow NH_3 \text{ (aq)} + OH^- \text{ (aq)}$$

(a) OH^- is a stronger base than NH_2^-
(b) NH_3 is a stronger acid than H_2O
(c) NH_2 is a stronger acid than H_2O
(d) OH^- is a weaker base than NH_2^-
(e) More than one of the above is true

NH_3 is the conjugate acid of NH_2^-. OH^- is the conjugate base of H_2O. The reaction proceeds to the right. This means that NH_2^- ($K_b = 1.0 \times 10^{19}$) is a stronger base than OH^- ($K_b = 55$).

Answer(d)

12-56. Which compound cannot accept a pair of electrons and function as a Lewis acid?

(a) SO_3 (b) BF_3 (c) Ag^+ (d) CH_4 (e) H^+

CH_4: The electron configuration of C: [He] $2s^2 2p^2$

The hybridisation is sp^3. There are no vacant orbitals to accept a pair of electrons. So CH_4 cannot accept a pair of electrons and act as a Lewis acid.

Answer(d)

12-57. Consider the following data:

HA: $K_a = 1.0 \times 10^{-4}$ HB : $K_a = 1.0 \times 10^{-7}$

HC : $K_a = 1.0 \times 10^{-10}$ HD: $K_a = 1.0 \times 10^{-11}$

   (i)   Solutions of each acid are prepared in which the initial concentration of the acid is

   0.10 M. Which of the four solutions will be the most acidic?

      (a) HA   (b) HB   (c) HC   (d)HD

      (e) Not enough information provided to answer the question

(a)   HA (aq) + H_2O (l) $\rightleftharpoons$ H_3O^+ (aq) + A^- (aq)

Initial concentration: 0.1 M   ≈0   0

At equilibrium: $0.1 - \Delta C$ ΔC ΔC

Assumption: $0.1 - \Delta C \approx 0.1$

$K_a = (\Delta C)^2/(0.1 - \Delta C)$

$(\Delta C)^2/(0.1) \approx 1.0 \times 10^{-4}$ $(\Delta C)^2 \approx 1.0 \times 10^{-5}$

$\Delta C \approx 3.16 \times 10^{-3}$

$\{(3.16 \times 10^{-3})/(0.1)\} \times 100 = 3.16\%$ (less than 5%)   Assumption $0.1 - \Delta C \approx 0.1$ is valid.

$[H_3O^+] \approx 3.16 \times 10^{-3}$ Log $[H_3O^+] \approx -2.5$ pH ≈ 2.5

(b) $HB \text{ (aq)} + H_2O \text{ (l)} \rightleftharpoons H_3O^+ \text{ (aq)} + B^- \text{ (aq)}$

Initial concentration: 0.1 M   ≈0   0

At equilibrium: $0.1 - \Delta C$ ΔC ΔC

Assumption: $0.1 - \Delta C \approx 0.1$

$K_a = (\Delta C)^2/(0.1 - \Delta C)$

$(\Delta C)^2/(0.1) \approx 1.0 \times 10^{-7}$

$(\Delta C)^2 \approx 1.0 \times 10^{-8}$

$\Delta C \approx 1.0 \times 10^{-4}$

$\{(1.0 \times 10^{-4})/(0.1)\} \times 100 = 0.1\%$ (less than 5%)   Assumption $0.1 - \Delta C \approx 0.1$ is valid.

$[H_3O^+] \approx 1.0 \times 10^{-4}$ Log $[H_3O^+] \approx -4.0$ pH ≈ 4.0

(c) $HC \text{ (aq)} + H_2O \text{ (l)} \rightleftharpoons H_3O^+ \text{ (aq)} + C^- \text{ (aq)}$

Initial concentration: 0.1 M   ≈0   0

At equilibrium: $0.1 - \Delta C$ ΔC ΔC

Assumption: $0.1 - \Delta C \approx 0.1$

$K_a = (\Delta C)^2/(0.1 - \Delta C)$

$(\Delta C)^2/(0.1) \approx 1.0 \times 10^{-10}$

$(\Delta C)^2 \approx 1.0 \times 10^{-11}$

$\Delta C \approx 3.16 \times 10^{-6}$

$\{(3.16 \times 10^{-6})/(0.1)\} \times 100 = 3.16 \times 10^{-3}\%$ (less than 5%)   Assumption $0.1 - \Delta C \approx 0.1$ is valid.

$[H_3O^+] \approx 3.16 \times 10^{-6}$ Log $[H_3O^+] \approx -5.5$ pH ≈ 5.5

(d) $HD \text{ (aq)} + H_2O \text{ (l)} \rightleftharpoons H_3O^+ \text{ (aq)} + D^- \text{ (aq)}$

Initial concentration: 0.1 M   ≈0   0

At equilibrium: $0.1 - \Delta C$ ΔC ΔC

Assumption: $0.1 - \Delta C \approx 0.1$

$K_a = (\Delta C)^2/(0.1 - \Delta C)$

$(\Delta C)^2/(0.1) \approx 1.0 \times 10^{-11}$

$(\Delta C)^2 \approx 1.0 \times 10^{-12}$

$\Delta C \approx 1.0 \times 10^{-6}$

$\{(1.0 \times 10^{-6})/(0.1)\} \times 100 = 1.0 \times 10^{-3}\%$ (less than 5%) Assumption $0.1 - \Delta C \approx 0.1$ is valid.

$[H_3O^+] \approx 1.0 \times 10^{-6}$ $Log\ [H_3O^+] \approx -6.0$ $pH \approx 6.0$

The acid solution with the largest K_a value is the most acidic solution.   Answer(a)

    (ii)   0.5 M solutions of NaA, NaB, NaC and NaD are prepared. Which solution is the

    most basic?

    (a) NaA   (b) NaB   (c) NaC   (d) NaD   (e) all of the solutions have the same pH.

(a) $A^-\ (aq) + H_2O\ (l) \rightleftharpoons HA\ (aq) + OH^-\ (aq)$

At equilibrium: $0.5 - \Delta C$ ΔC ΔC

Assumption: $(0.5 - \Delta C) \approx 0.5$

$K_b = (10^{-14})/(1.0 \times 10^{-4}) = 1.0 \times 10^{-10}$

$K_b = (\Delta C)^2/(0.5 - \Delta C)$

$(\Delta C)^2/(0.5) \approx 1.0 \times 10^{-10}$

$(\Delta C)^2 \approx 0.5 \times 10^{-10}$

$\Delta C \approx 7.1 \times 10^{-6}$

$\{(7.1 \times 10^{-6})/0.5\} \times 100 = 1.42 \times 10^{-3}\%$ (less than 5%)   Assumption $(0.5 - \Delta C) \approx 0.5$ is valid.

$[OH^-] \approx 7.1 \times 10^{-6}$

$Log\ [OH^-] \approx -5.15$

$pOH \approx 5.15$

(b) $B^-\ (aq) + H_2O\ (l) \rightleftharpoons HB\ (aq) + OH^-\ (aq)$

At equilibrium: $0.5 - \Delta C$ ΔC ΔC

Assumption: $(0.5 - \Delta C) \approx 0.5$

$K_b = (10^{-14})/(1.0 \times 10^{-7}) = 1.0 \times 10^{-7}$

$K_b = (\Delta C)^2/(0.5 - \Delta C)$

$(\Delta C)^2/(0.5) \approx 1.0 \times 10^{-7}$

$(\Delta C)^2 \approx (0.5) \times 10^{-7}$

$\Delta C \approx 2.24 \times 10^{-4}$

$\{(2.24 \times 10^{-4})/0.5\} \times 100 = 4.5 \times 10^{-2}\%$ (less than 5%)   Assumption $(0.5 - \Delta C) \approx 0.5$ is valid.

$[OH^-] \approx 2.24 \times 10^{-4}$

$\text{Log } [OH^-] \approx -3.65$

$pOH \approx 3.65$

(c) $\qquad\qquad\qquad\qquad C^- \text{ (aq)} + H_2O \text{ (l)} \rightleftharpoons HC \text{ (aq)} + OH^- \text{ (aq)}$

At equilibrium: $\qquad\qquad\qquad\qquad 0.5 - \Delta C \qquad\qquad\qquad \Delta C \qquad \Delta C$

Assumption: $(0.5 - \Delta C) \approx 0.5$

$K_b = (10^{-14}) / (1.0 \times 10^{-10}) = 1.0 \times 10^{-4}$

$K_b = (\Delta C)^2 / (0.5 - \Delta C)$

$(\Delta C)^2 / (0.5) \approx 1.0 \times 10^{-4}$

$(\Delta C)^2 \approx 0.5 \times 10^{-4}$

$\Delta C \approx 7.1 \times 10^{-3}$

$\{(7.1 \times 10^{-3})/0.5\} \times 100 = 1.42\%$ (less than 5%) Assumption $(0.5 - \Delta C) \approx 0.5$ is valid.

$[OH^-] \approx 7.1 \times 10^{-3}$

$\text{Log } [OH^-] \approx -2.15 \qquad\qquad\qquad pOH = 2.15$

(d) $\qquad\qquad\qquad\qquad D^- \text{ (aq)} + H_2O \text{ (l)} \rightleftharpoons HD \text{ (aq)} + OH^- \text{ (aq)}$

At equilibrium: $\qquad\qquad\qquad\qquad 0.5 - \Delta C \qquad\qquad\qquad \Delta C \qquad \Delta C$

Assumption: $(0.5 - \Delta C) \approx 0.5$

$K_b = (10^{-14}) / (1.0 \times 10^{-11}) = 1.0 \times 10^{-3}$

$K_b = (\Delta C)^2 / (0.5 - \Delta C)$

$(\Delta C)^2 / (0.5) \approx 1.0 \times 10^{-3}$

$(\Delta C)^2 \approx 0.5 \times 10^{-3}$

$\Delta C \approx 2.2 \times 10^{-2}$

$\{(2.2 \times 10^{-2})/0.5\} \times 100 = 4.4\%$ (less than 5%) Assumption $(0.5 - \Delta C) \approx 0.5$ is valid.

$[OH^-] \approx 2.2 \times 10^{-2}$

$\text{Log } [OH^-] \approx -1.7 \qquad\qquad\qquad pOH \approx 1.7$

The solution with the largest K_b value (smallest K_a value) is the most basic. $\qquad$ Answer(d)

(iii)   A solution is prepared with initial concentrations of HA and NaA of 0.5 M and 1.00

M respectively. What is the pH of the solution?

(a) 3.70   (b) 4.00   (c) 4.30   (d) 9.70   (e) 10.30

The solution is an acidic buffer solution.

pH = pK_a + log {[Salt]/[Acid]}

$pH = -\log 10^{-4} + \log \{(1)/(0.5)\} = 4 + \log 2 = 4.3$ Answer(c)

12-58. Which of the following solutions would have a pH greater than 7.0?

(a) NH4Cl   (b) NaCl   (c) HOAc   (d) NaOAc

(a) NH4Cl is a salt of strong acid and a weak base. NH_4^+ will undergo hydrolysis in H_2O to give H_3O^+ ions and reduce the pH to < 7.

$$NH_4^+ \text{ (aq)} + H_2O \text{ (l)} \rightleftharpoons NH_3 \text{ (aq)} + H_3O^+ \text{ (aq)}$$

(b)NaCl is a salt of a strong base and a strong acid. NaCl will not undergo hydrolysis in water and it will not change the pH of pure water.

(c)HOAc is a weak acid and the pH will be less than 7.

(d)NaOAc is a salt of a strong base and a weak acid. OAc^- will undergo hydrolysis in H_2O and the pH will be more than 7.

$$OAc^- \text{ (aq)} + H_2O \text{ (l)} \rightleftharpoons HOAc \text{ (aq)} + OH^- \text{ (aq)}$$ Answer(d)

12-59. Which of the following solutions has a pH above 7.0?

(a) 0.10 M NH4Cl   (b) 0.10 M NH4NO3   (c) 0.10 M HCN

(d) 0.10 M NaCN   (e) None of the above

(a) NH4Cl is a salt of strong acid and a weak base. NH_4^+ will undergo hydrolysis in H_2O to give H_3O^+ ions and reduce the pH to < 7.

$$NH_4^+ \text{ (aq)} + H_2O \text{ (l)} \rightleftharpoons NH_3 \text{ (aq)} + H_3O^+ \text{ (aq)}$$

(b) NH4NO3 is a salt of strong acid and a weak base. NH_4^+ will undergo hydrolysis in H_2O to give H_3O^+ ions and reduce the pH to < 7.

$$NH_4^+ \text{ (aq)} + H_2O \text{ (l)} \rightleftharpoons NH_3 \text{ (aq)} + H_3O^+ \text{ (aq)}$$

(c)HCN is a weak acid and the pH will be less than 7.

(d) NaCN is a salt of a strong base and a weak acid. CN^- will undergo hydrolysis in H_2O and the pH will be more than 7.

$$CN^- \text{ (aq)} + H_2O \text{ (l)} \rightleftharpoons HCN \text{ (aq)} + OH^- \text{ (aq)}$$ Answer(d)

12-60. Which of the following negative ions would be the strongest base?

(a) $Cl_2CHCO_2^-$ ($Cl_2CHCO_2H : K_a = 7.8 \times 10^{-3}$)
(b) $ClCH_2CO_2^-$ ($ClCH_2CO_2H : K_a = 1.4 \times 10^{-3}$)
(c) HCO_2^- ($HCO_2H : K_a = 1.8 \times 10^{-4}$)
(d) $CH_3CO_2^-$ ($CH_3COOH : K_a = 1.8 \times 10^{-5}$)

The negative ions will undergo hydrolysis in water to produce OH^- ion.

$$A^- (aq) + H_2O (l) \rightleftharpoons HA (aq) + OH^- (aq)$$

The ion with the weakest conjugate acid will be the strongest base (reference 12-57).

$CH_3CO_2^-$ will be the strongest base.   Answer(d)

12-61. Which of the following acids would have the strongest conjugate base?

(a) $HOCl$ ($K_a = 2.9 \times 10^{-8}$)   (b) $HOBr$ ($K_a = 2.4 \times 10^{-9}$)   (c) HOI ($K_a = 2.3 \times 10^{-11}$)

(d) H_2O_2 ($K_a = 2.2 \times 10^{-12}$)   (e) H_2O ($K_a = 1.8 \times 10^{-16}$)

Weakest acid will have the strongest conjugate base. So H_2O will have the strongest

conjugate base.   Answer(e)

12-62. Which of the following solutions would be the most acidic?

(a) 0.10 M CH_3COOH ($K_a = 1.8 \times 10^{-5}$)   (b) 0.10 M HCO_2H ($K_a = 1.8 \times 10^{-4}$)

(c) 0.10 M $ClCH_2COOH$ ($K_a = 1.4 \times 10^{-3}$)   (d) 0.10 M Cl_2CHCO_2H ($K_a = 5.1 \times 10^{-2}$)

(e) All of these would have the same pH.

The acidic solution with the largest K_a value will be the most acidic solution (reference: 12-57).

So (d) will be the most acidic solution.   Answer(d)

12-63. If the strength of the following acids increases from left to right

$$CH_3NH_2 < CH_3NH_3^+ < C_6H_5NH_3^+ < C_6H_5CO_2H < HCl$$

The order of increasing base strength, reading from left to right, must be

(a) $CH_3NH^- < CH_3NH_2 < C_6H_5NH_2 < C_6H_5CO_2^- < Cl^-$
(b) $Cl^- < C_6H_5CO_2^- < C_6H_5NH_2 < CH_3NH_2 < CH_3NH^-$
(c) $C_6H_5NH_2 < CH_3NH_2 < Cl^- < C_6H_5CO_2^- < CH_3NH^-$
(d) $CH_3NH^- < C_6H_5CO_2^- < Cl^- < CH_3NH_2 < C_6H_5NH_2$
(e) None of the above

The stronger the acid the weaker will be the conjugate base. So the order of increasing base

strength, reading from left to right is:

$$Cl^- < C_6H_5CO_2^- < C_6H_5NH_2 < CH_3NH_2 < CH_3NH^-$$ Answer(b)

12-64. What is the pH of a solution that is 1.7×10^{-4} M in H^+?

(a) 3.77   (b) 4.77   (c) 4.23   (d)10.23   (e) None of the above

$H^+ = 1.7 \times 10^{-4}$ $Log [H^+] = -3.77$ $pH = -log [H^+] = 3.77$ Answer(a)

12-65. What is the hydroxide ion concentration in a pH = 5.14 solution?

(a) 1.0×10^{-14} (b) 1.4×10^{-9} (c) 1.0×10^{-7} (d) 7.2×10^{-6} (e) 1.4×10^{-2}

pH = 5.14

pOH = 14 − 5.14 = 8.86 $- Log [OH^-] = 8.86$ $Log [OH^-] = -8.86$

$[OH^-] = 1.4 \times 10^{-9}$ Answer(b)

12-66. Which of the following solutions has the largest pH?

(a) 0.3 M Na_2CO_3 (b) 1 M HOAc   (c) 0.3 M NH_4Cl

(d)H_2O (e) 10^{-3} M HCl

(a): 0.3 M Na_2CO_3:

$$CO_3^{2-} \text{ (aq)} + H_2O \text{ (l)} \rightleftharpoons HCO_3^- \text{ (aq)} + OH^- \text{ (aq)}$$

At equilibrium: (0.3 − ΔC) M   ΔC   ΔC

Assumption: (0.3 − ΔC) ≈ 0.3

$K_b = [HCO_3^-] [OH^-]/[CO_3^{2-}]$

$K_b = (ΔC)^2/(0.3 − ΔC)$

$(ΔC)^2/(0.3) \approx 2.1 \times 10^{-4}$

$(ΔC)^2 \approx 0.63 \times 10^{-4}$

$ΔC \approx 8 \times 10^{-3}$

$\{(8 \times 10^{-3})/0.3\} \times 100 = 2.67\%$ (less than 5%)   Assumption (0.3 − ΔC) ≈ 0.3 is valid.

$[OH^-] \approx 8 \times 10^{-3}$ $Log [OH^-] \approx -2.1$ $pOH \approx 2.1$ $pH \approx 14 − 2.1 \approx 11.9$

(b) HOAc is a weak acid and the pH < 7.

(c) NH_4Cl is a salt of a strong acid and weak base. NH_4^+ will undergo hydrolysis in H_2O and the pH < 7.

(d) pH of pure water = 7.

(e) $[H^+] = 10^{-3}$ pH = 3.

Solution (a) has the largest pH.   Answer(a)

12-67. The addition of sodium formate (HCO_2Na) to a solution containing formic acid

(HCO_2H: $K_a = 1.8 \times 10^{-4}$) will cause

    (a) The pH to increase   (b) The pH to decrease   (c) No change in pH

    (d) A change in the pH, the direction of which cannot be predicted

$HCOOH$ (aq) + H_2O (l) $\rightleftharpoons$ H_3O^+ (aq) + $HCOO^-$ (aq)

$K_a = [H_3O^+][HCOO^-]/[HCOOH]$

$[H_3O^+] = K_a [HCOOH]/[HCOO^-]$

Addition of sodium formate will increase the $HCOO^-$ ion concentration and decrease the

$[H_3O^+]$ and increase the pH.   Answer(a)

12-68. Which equation best describes 0.1 M solution of a strong acid, such as hydrioiodic acid

(HI: $K_a = 3 \times 10^9$)?

    (a) $[H_3O]_T \approx C_{HI}$ (b) $[H_3O]_T \approx [OH^-]_W$ (c) $[H_3O]_T \approx [HI]$ (d) $[H_3O]_T \approx [H_2O]$

    (e) $[H_3O]_T \approx K_a C_a$

In the case of strong acid with a very high K_a value the dissociation will be > 99.99% and the

$[H_3O^+]_T$ will be almost equal to the initial concentration of the acid.   Answer(a)

12-69. A solution is known to have a hydronium ion concentration of 1.0×10^{-6} M. What

percentage of the total H_3O^+ ion concentration comes from the dissociation of H_2O?

    (a) 1% (b) 3% (c) 5% (d) 10% (e) 100%

$[H_3O^+]_T = [H_3O^+]_{HA} + [H_3O^+]_W$

$[H_3O^+]_T [OH^-] = K_w = 10^{-14}$

$[OH^-] = (10^{-14})/[H_3O^+]_T = (10^{-14})/(10^{-6}) = 10^{-8}$

$[OH^-] = [H_3O^+]_W = 10^{-8}$

Percentage from dissociation of H_2O = {$(10^{-8})/(10^{-6})$}×100 = 1%   Answer(a)

12-70. Assume a solution is prepared by adding 1.0×10^{-3} moles of a weak monoprotic acid

($K_a = 2.0 \times 10^{-4}$) to enough water to give a liter of solution. If you want to compute the H_3O^+ ion

concentration with an error of 5% or less, you can legitimately ignore:

    (a) Both the contribution of the dissociation of water and the additive ΔC term
    (b) Neither the dissociation of water nor the additive ΔC term
    (c) The dissociation of water but not the additive ΔC term

(d) The additive ΔC term but not the dissociation of water

(e) You can't compute the H_3O^+ concentration with an error of less than 5%

$$HA \text{ (aq)} + H_2 \text{ (l)} \rightleftharpoons H_3O^+ \text{ (aq)} + A^- \text{ (aq)}$$

At equilibrium: $(0.001 - \Delta C)$ M ΔC ΔC

$K_a = [H_3O^+] [A^-]/ [HA] = (\Delta C)^2/(0.001 - \Delta C) = 2\times10^{-4}$

Assumption: $0.001 - \Delta C \approx 0.001$

$(\Delta C)^2 \approx 2\times10^{-7}$

$\Delta C \approx 4.5\times10^{-4}$

$\{(4.5\times10^{-4})/0.001\}\times100 = 45\%$ (not less than 5%). So the assumption that $0.001 - \Delta C \approx 0.001$

is not valid. So ΔC can't be ignored in the calculation of $[H_3O^+]$.

$(\Delta C)^2/(0.001 - \Delta C) = 2\times10^{-4}$

$(\Delta C)^2 = (2\times10^{-4})\times (0.001-\Delta C) = (2\times10^{-7}) - \{(2\times10^{-4})\times\Delta C\}$

$\Delta C = 3.6\times10^{-4}$ (by quadratic formula)

$[H_3O^+]_T = 3.6\times10^{-4}$

$[OH^-] = (10^{-14})/ [H_3O^+]_T = (10^{-14})/ (3.6\times10^{-4}) = 2.8\times10^{-11}$

$[H_3O^+]_W = [OH^-] = 2.8\times10^{-11}$

Percentage contribution to the total H_3O^+ ion concentration from dissociation of

$H_2O = \{(2.8\times10^{-11})/ (3.6\times10^{-4})\}\times100 = 0.8\times10^{-5}$ % (less than 5%)

So the dissociation of H_2O can be ignored.   Answer(c)

12-71. For a particular weak acid, HA, in pure water, as the initial concentration of the weak

acid increases, the acid-dissociation equilibrium constant of that acid should:

(a) Increase   (b) Decrease   (c) Remain the same

(d) Increase until the ionization of water can be neglected

(e) Decrease until the ionization of water can be neglected

$$HA \text{ (aq)} + H_2O \text{ (l)} \rightleftharpoons H_3O^+ \text{ (aq)} + A^- \text{ (aq)}$$

$K_a = [H_3O^+] [A^-]/[HA]$ will be constant.

When the initial concentration of the weak acid increases, $[H_3O^+]$ and $[A^-]$ both will increase

and the K_a will not change.   Answer(c)

12-72. What would happen if more formic acid (HCOOH) was added to the following solution at equilibrium?

$$HCO_2H \text{ (aq)} + H_2O \text{ (l)} \rightleftharpoons H_3O^+ \text{ (aq)} + HCO_2^- \text{ (aq)}$$

(a) $[H_3O^+]$ should increase
(b) $[H_3O^+]$ and $[HCO_2^-]$ should both increase
(c) $[H_3O^+]$ and $[HCO_2^-]$ should both decrease
(d) $[H_3O^+]$ should increase but $[HCO_2^-]$ should decrease
(e) $[H_3O^+]$ should decrease but $[HCO_2^-]$ should increase

$$HCO_2H \text{ (aq)} + H_2O \text{ (l)} \rightleftharpoons H_3O^+ \text{ (aq)} + HCO_2^- \text{ (aq)}$$

$K_a = [H_3O^+] [HCO_2^-]/[HCO_2H]$

To counteract this change, as per Lechatelier's principle, the equilibrium will be shifted to the

right and both $[H_3O^+]$ and $[HCO_2^-]$ should increase.   Answer(b)

12-73. An 0.0024 M solution of boric acid (H_3BO_3: $K_a = 5.8 \times 10^{-10}$) is an example of which

kind of problem?

(a) Strong acid
(b) A weak acid where we can ignore the dissociation of water and assume that ΔC is small compared to C_a
(c) A weak acid where we can ignore the dissociation of water but we can't assume that ΔC is small compared to C_a
(d) A weak acid where we can't ignore the dissociation of water but we can assume that ΔC is small compared to C_a
(e) A weak acid where we can't ignore the dissociation of water and assume that ΔC is small compared to C_a

$$H_3BO_3 \text{ (aq)} + H_2O \text{ (l)} \rightleftharpoons H_3O^+ \text{ (aq)} + H_2BO_3^- \text{ (aq)}$$

Initial concentration:	0.0024 M	≈ 0	0
At equilibrium:	$(0.0024 - \Delta C)$	ΔC	ΔC

Assumption: $0.0024 - \Delta C \approx 0.0024$

$K_a = [H_3O^+]_T [H_2BO_3^{2-}]/[H_2BO_3]$

$(\Delta C)^2/(0.0024) \approx 5.8 \times 10^{-10}$

$(\Delta C)^2 = 1.4 \times 10^{-12}$

$\Delta C \approx 1.2 \times 10^{-6}$

$\{(1.2 \times 10^{-6})/0.0024\} \times 100 = 0.05\%$ (less than 5%)   Assumption $0.0024 - \Delta C \approx 0.0024$ is valid.

$[H_3O^+]_T \approx 1.2 \times 10^{-6}$

$[OH^-] = [H_3O^+]_W = 10^{-14}/[H_3O^+]_T \approx (10^{-14})/(1.2 \times 10^{-6}) \approx 8.3 \times 10^{-9}$

Percentage contribution to the total H_3O^+ ion concentration from dissociation of H_2O

$\approx \{(8.3\times10^{-9})/(1.2\times10^{-6})\}\times100 \approx 0.69\ \%$ (less than 5%)

So the dissociation of H_2O can be ignored.   Answer(b)

12-74. Which of the following equations is valid for a 0.10 M solution of formic acid (HCO_2H : $K_a = 1.8\times10^{-4}$)?

   (a) $[H_3O^+]_T \approx [OH^-]_W$ (b) $[H_3O^+]_W \approx [HCO_2^-]$ (c) $[H_3O^+]_T \approx [HCO_2^-]$

   (d) $[H_3O^+]_T \approx [H_2O]$ (e) $[OH^-] \approx [H_2O]$

$$HCO_2H\ (aq) + H_2O\ (l) \rightleftharpoons H_3O^+\ (aq) + HCO_2^-\ (aq)$$

Initial concentration:	0.1 M	≈ 0	0
At equilibrium:	$(0.1 - \Delta C)$	ΔC	ΔC

Assumption: $0.1 - \Delta C \approx 0.1$

$K_a = (\Delta C)^2/(0.1 - \Delta C)$

$(\Delta C)^2/(0.1) \approx 1.8\times10^{-4}$

$(\Delta C)^2 \approx 1.8\times10^{-5}$

$\Delta C \approx 4.24\times10^{-3}$

$\{(4.24\times10^{-3})/0.1\}\times100 = 4.24\%$ (less than 5%)

Assumption $0.1 - \Delta C \approx 0.1$ is valid.

$[H_3O^+]_T \approx 4.24\times10^{-3}$

$[OH^-] = [H_3O^+]_W \approx (10^{-14})/[H_3O^+]_T \approx (10^{-14})/(4.24\times10^{-3}) \approx 2.36\times10^{-12}$

$[H_3O^+]_W \approx 2.36\times10^{-12}$

Percentage contribution to the total H_3O^+ ion concentration from dissociation of H_2O

$\approx \{(2.36\times10^{-12})/(4.24\times10^{-3})\}\times100 \approx 5.6\times10^{-8}\ \%$ (less than 5%)

So $[H_3O^+]_T \approx [HCO_2^-]$ Answer(c)

12-75. Which of the following isn't true for the H_3O^+ ion concentration in an 0.01 M solution of acetic acid ($K_a = 1.8\times10^{-5}$)?

   (a) It is less than the H_3O^+ ion concentration in an 0.01 M solution of HCl
   (b) It is equal to the OAc^- concentration
   (c) It is equal to the OH^- concentration
   (d) The pH of the solution is less than 7
   (e) None of the above is true

$$\text{HOAc (aq)} + \text{H}_2\text{O (l)} \rightleftharpoons \text{H}_3\text{O}^+ \text{(aq)} + \text{OAc}^- \text{(aq)}$$

Initial concentration:	0.01 M	≈ 0	0
At equilibrium:	$(0.01 - \Delta C)$	ΔC	ΔC

Assumption: $0.01 - \Delta C \approx 0.01$

$K_a = (\Delta C)^2/(0.01 - \Delta C)$

$(\Delta C)^2/(0.01) \approx 1.8 \times 10^{-5}$

$(\Delta C)^2 \approx 1.8 \times 10^{-7}$

$\Delta C = 4.24 \times 10^{-4}$

$\{(4.24 \times 10^{-4})/0.01\} \times 100 = 4.24\%$ (less than 5%)   Assumption: $0.01 - \Delta C \approx 0.01$ is valid.

$[\text{H}_3\text{O}^+]_T \approx 4.24 \times 10^{-4}$

$[\text{OH}^-] = [\text{H}_3\text{O}^+]_W \approx (10^{-14})/(4.24 \times 10^{-4}) \approx 2.36 \times 10^{-11}$

$[\text{H}_3\text{O}^+]_W \approx 2.36 \times 10^{-11}$

Percentage contribution to the total H_3O^+ ion concentration from dissociation of H_2O

$\approx \{(2.36 \times 10^{-11})/(4.24 \times 10^{-4})\} \times 100 \approx 5.6 \times 10^{-6}$ % (less than 5%)

So $[\text{H}_3\text{O}^+]_T \approx [\text{OAc}^-]$

(a) $[\text{H}^+]$ in 0.01 M HCl = 0.01 M which is greater than 4.24×10^{-4} M. So the (a) is true.

(b) $[\text{H}_3\text{O}^+]_T \approx [\text{OAc}^-]$

(c) $[\text{H}_3\text{O}^+]_T \approx 4.24 \times 10^{-4}$ and $[\text{OH}^-] \approx (10^{-14})/(4.24 \times 10^{-4}) \approx 2.36 \times 10^{-11}$

So $[\text{H}_3\text{O}^+]_T$ is not equal to $[\text{OH}^-]$

(d) Acetic acid is a weak acid. So the pH will be less than 7.   Answer(c)

12-76. For a weak acid (where $C_a - \Delta C \approx C_a$) when can we ignore the contribution to the total H_3O^+ ion concentration from the dissociation of water?

$$\text{HA (aq)} + \text{H}_2\text{O (l)} \rightleftharpoons \text{H}_3\text{O}^+ \text{(aq)} + \text{A}^- \text{(aq)}$$

Initial concentration:	C_a	≈ 0	0
At equilibrium:	$C_a - \Delta C$	ΔC	ΔC

$K_a = (\Delta C)^2/(C_a - \Delta C)$

$(C_a - \Delta C) \approx C_a$

$K_a \approx (\Delta C)^2/(C_a)$

$$(\Delta C)^2 \approx K_a C_a$$

$$(\Delta C) \approx \sqrt{K_a C_a}$$

Case 1. $\qquad K_a C_a < 1.0 \times 10^{-13}$

$$K_a C_a = 0.9 \times 10^{-13}$$

$$(\Delta C) \approx \sqrt{0.9 \times 10^{-13}}$$

$$\approx 3 \times 10^{-7}$$

$$[H_3O^+]_T \approx 3 \times 10^{-7}$$

$$[OH^-] = [H_3O^+]_W \approx (10^{-14})/(3 \times 10^{-7}) \approx 3.33 \times 10^{-8}$$

Percentage contribution to the total H_3O^+ ion concentration from the dissociation of water

$$\approx \{(3.33 \times 10^{-8})/(3 \times 10^{-7})\} \times 100 \approx 11.1 \ \% \ (\text{more than } 5\%)$$

So the contribution due to the dissociation of water can't be ignored.

Case 2. $\qquad K_a C_a = 1.0 \times 10^{-13}$

$$(\Delta C) \approx \sqrt{1.0 \times 10^{-13}}$$

$$\approx 3.16 \times 10^{-7}$$

$$[H_3O^+]_T \approx 3.16 \times 10^{-7}$$

$$[OH^-] = [H_3O^+]_W \approx (10^{-14})/(3.16 \times 10^{-7}) \approx 3.16 \times 10^{-8}$$

Percentage contribution to the total H_3O^+ ion concentration from the dissociation of water

$$= \{(3.16 \times 10^{-8})/(3.16 \times 10^{-7})\} \times 100 = 10 \ \% \ (\text{more than } 5\%)$$

So the contribution due to the dissociation of water can't be ignored.

Case 3. $\qquad K_a C_a = 2.0 \times 10^{-13}$

$$(\Delta C) \approx \sqrt{2.0 \times 10^{-13}}$$

$$\approx 4.47 \times 10^{-7}$$

$$[H_3O^+]_T \approx 4.47 \times 10^{-7}$$

$$[OH^-] = [H_3O^+]_W \approx (10^{-14})/(4.47 \times 10^{-7}) = 2.24 \times 10^{-8}$$

Percentage contribution to the total H_3O^+ ion concentration from the dissociation of water

$$\approx \{(2.24 \times 10^{-8})/(4.47 \times 10^{-7})\} \times 100 \approx 5.01 \ \% \ (\text{more than } 5\%)$$

So the contribution due to the dissociation of water can't be ignored.

Case 4. $\qquad K_a C_a > 2.0 \times 10^{-13}$

$K_aC_a = 2.1 \times 10^{-13}$

$(\Delta C) \approx \sqrt{2.1 \times 10^{-13}}$

$\quad \approx 4.58 \times 10^{-7}$

$[H_3O^+]_T \approx 4.58 \times 10^{-7}$

$[OH^-] = [H_3O^+]_W \approx (10^{-14})/(4.58 \times 10^{-7}) \approx 2.18 \times 10^{-8}$

Percentage contribution to the total H_3O^+ ion concentration from the dissociation of water

$\approx \{(2.18 \times 10^{-8})/(4.58 \times 10^{-7})\} \times 100 \approx 4.8 \%$ (less than 5%)

So the contribution to the total H_3O^+ ion concentration from the dissociation of water can

be ignored when $K_aC_a > 2.0 \times 10^{-13}$.

12-77. Which of the following is not true for solutions of acetic acid ($K_a = 1.8 \times 10^{-5}$)?

   (a) The percent ionization increases as solutions are made more dilute
   (b) The $[H_3O^+]$ increases as solutions are made more dilute
   (c) In a 0.10 M of the acid, it is approximately 1% dissociated
   (d) The percent ionization is smaller in a 0.10 M solution containing sodium acetate than in pure water
   (e) The ionization constant is independent of concentration

(a) Case 1.   Initial concentration = 0.1 M

$$HOAc\ (aq) + H_2O\ (l) \rightleftharpoons H_3O^+\ (aq) + OAc^-\ (aq)$$

Initial concentration:	0.1 M	≈ 0	0
At equilibrium:	$(0.1 - \Delta C)$	ΔC	ΔC

Assumption: $(0.1 - \Delta C) \approx 0.1$

$K_a \approx (\Delta C)^2/(0.1)$

$(\Delta C)^2/(0.1) \approx 1.8 \times 10^{-5}$

$(\Delta C)^2 \approx 1.8 \times 10^{-6}$

$\Delta C \approx 1.34 \times 10^{-3}$

$\{(1.34 \times 10^{-3})/0.1\} \times 100 = 1.34\%$ (less than 5%)   Assumption: $(0.1 - \Delta C) \approx 0.1$ is valid.

$[H_3O^+]_T \approx 1.34 \times 10^{-3}$

$[OAc^-] \approx 1.34 \times 10^{-3}$

Percent ionization $\approx \{(1.34 \times 10^{-3})/(0.1)\} \times 100 \approx 1.34 \%$

   Case 2.   Initial concentration = 0.01 M

$$HOAc \text{ (aq)} + H_2O \text{ (l)} \rightleftharpoons H_3O^+ \text{ (aq)} + OAc^- \text{ (aq)}$$

Initial concentration: 0.01 M ≈ 0 0

At equilibrium: $(0.01 - \Delta C)$ ΔC ΔC

Assumption: $(0.01 - \Delta C) \approx 0.01$

$K_a \approx (\Delta C)^2/(0.01)$

$(\Delta C)^2/(0.01) \approx 1.8 \times 10^{-5}$

$(\Delta C)^2 \approx 1.8 \times 10^{-7}$

$\Delta C \approx 4.24 \times 10^{-4}$

$\{(4.24 \times 10^{-4})/0.01\} \times 100 = 4.24\%$ (less than 5%) Assumption: $(0.01 - \Delta C) \approx 0.01$ is valid.

$[H_3O^+]_T \approx 4.24 \times 10^{-4}$

$[OAc^-] \approx 4.24 \times 10^{-4}$

Percent ionization $\approx \{(4.24 \times 10^{-4})/(0.01)\} \times 100 \approx 4.24\%$

The percent ionization increases as solutions are made more dilute.

(b) Case 1.   Initial concentration = 0.1 M

$\Delta C \approx 1.34 \times 10^{-3}$

$[H_3O^+]_T \approx 1.34 \times 10^{-3}$

   Case 2.   Initial concentration = 0.01 M

$\Delta C \approx 4.24 \times 10^{-4}$

$[H_3O^+]_T \approx 4.24 \times 10^{-4}$

When solution is made more dilute the $[H_3O^+]$ decreases.

(c) In a 0.1 M acetic acid, the percent ionization $\approx 1.34\%$ ($\approx 1\%$)

(d) Case 1.   Initial concentration of HOAc = 0.1 M in pure water

Percent ionization $\approx 1.34\%$

   Case 2. $HOAc \text{ (aq)} + H_2O \text{ (l)} \rightleftharpoons H_3O^+ \text{ (aq)} + OAc^- \text{ (aq)}$

Initial concentration: 0.1 M ≈ 0 0.1

At equilibrium: $(0.1 - \Delta C)$ ΔC $0.1 + \Delta C$

Assumptions: $(0.1 + \Delta C) \approx 0.1$ and $(0.1 - \Delta C) \approx 0.1$

$K_a = (\Delta C)(0.1 + \Delta C)/(0.1 - \Delta C) \approx \Delta C$

$\Delta C \approx 1.8 \times 10^{-5}$

$\{(1.8 \times 10^{-5})/0.1\} \times 100 = 0.018\%$ (less than 5%)

Assumptions $(0.1 + \Delta C) \approx 0.1$ and $(0.1 - \Delta C) \approx 0.1$ are valid.

Percent ionization $\approx \{(1.8 \times 10^{-5})/(0.1)\} \times 100 \approx 0.018\%$

The percent ionization is smaller in a 0.100 M solution containing sodium acetate than in pure water.

(e) $HOAc\ (aq) + H_2O\ (l) \rightleftharpoons H_3O^+\ (aq) + OAc^-\ (aq)$

$K_a = [H_3O^+][OAc^-]/[HOAc]$

The ionization constant is independent of concentration. If initial concentration of HOAc is increased the equilibrium will be shifted to the right and the $[H_3O^+]$ and $[OAc^-]$ will increase and the K_a will not change. If initial concentration of HOAc is decreased the equilibrium will be shifted to the left and the $[H_3O^+]$ and $[OAc^-]$ will decrease and the K_a will not change.

Answer(b)

12-78. The value of equilibrium constant, K_a for the dissociation of formic acid in water,

$HCOOH\ (aq) + H_2O\ (l) \rightleftharpoons HCOO^-\ (aq) + H_3O^+\ (aq)$

would depend on

(a) The temperature   (b) The pressure   (c) The pH

(d) The concentration of HOAc   (e) The concentration of the OAc^- ion.

The value of equlibrium constant K_a depends on the temperature. In an endothermic reaction increasing the temperature will increase the K_a and in an exothermic reaction increasing the temperature will decrease the K_a. The dissociation of formic acid in water is an endothermic recation. So Increasing the temperature will increase the K_a.   Answer(a)

12-79. What is the percent ionization in 0.01 M HCN solution? (HCN: $K_a = 4 \times 10^{-10}$)

(a) 0.002%   (b) 0.02%   (c) 0.2%   (d) 2%   (e) 20%

$HCN\ (aq) + H_2O\ (l) \rightleftharpoons H_3O^+\ (aq) + CN^-\ (aq)$

At equilibrium: $0.01 - \Delta C$ ΔC ΔC

Assumption: $(0.01 - \Delta C) \approx 0.01$

$K_a = (\Delta C)^2/(0.01 - \Delta C) \approx (\Delta C)^2/(0.01)$

$(\Delta C)^2/(0.01) \approx 4 \times 10^{-10}$

$(\Delta C)^2 \approx 4 \times 10^{-12}$

$\Delta C \approx 2 \times 10^{-6}$

$\{(2 \times 10^{-6})/0.01\} \times 100 = 0.02\%$ (less than 5%)   Assumption $(0.01 - \Delta C) \approx 0.01$ is valid.

Percent ionization $\approx \{(2 \times 10^{-6})/(0.01)\} \times 100 \approx 0.02\%$ Answer(b)

12-80. Calculate the pH of a 0.10 M lactic acid solution. ($K_a = 8.4 \times 10^{-4}$)

   (a) ≈ 2 (b) ≈ 3 (c) ≈ 4 (d) ≈ 7 (e) ≈ 12

$$HA\ (aq) + H_2O\ (l) \rightleftharpoons H_3O^+\ (aq) + A^-\ (aq)$$

At equilibrium: $0.1 - \Delta C$ ΔC ΔC

Assumption: $(0.1 - \Delta C) \approx 0.1$

$K_a = (\Delta C)^2/(0.1 - \Delta C) \approx (\Delta C)^2/(0.1)$

$(\Delta C)^2/(0.1) \approx 8.4 \times 10^{-4}$

$(\Delta C)^2 \approx 8.4 \times 10^{-5}$

$(\Delta C) \approx 9.2 \times 10^{-3}$

$\{(9.2 \times 10^{-3})/(0.1)\} \times 100 = 9.2\%$ (not less than 5%) Assumption $(0.1 - \Delta C) \approx 0.1$ is not valid.

$(\Delta C)^2/(0.1 - \Delta C) = 8.4 \times 10^{-4}$

$\Delta C = 8.755 \times 10^{-3}$ (by quadratic formula)

$\text{Log}\ [H_3O^+] = -2.06$ $pH = 2.06 \approx 2$ Answer(a)

12-81. Calculate the pH of a 0.1 M solution of formic acid in water. (HCOOH: $K_a = 1.8 \times 10^{-4}$)

   (a) 2.4   (b) 3.7   (c) 4.7   (d) 7   (e) 11.6

$$HA\ (aq) + H_2O\ (l) \rightleftharpoons H_3O^+\ (aq) + A^-\ (aq)$$

At equilibrium: $0.1 - \Delta C$ ΔC ΔC

Assumption: $(0.1 - \Delta C) \approx 0.1$

$K_a = (\Delta C)^2/(0.1 - \Delta C) \approx (\Delta C)^2/(0.1)$

$(\Delta C)^2/(0.1) \approx 1.8 \times 10^{-4}$

$(\Delta C)^2 \approx 1.8 \times 10^{-5}$

$(\Delta C) \approx 4.24 \times 10^{-3}$

$\{(4.24 \times 10^{-3})/0.1\} \times 100 = 4.24\%$ (less than 5%)   Assumption $(0.1 - \Delta C) \approx 0.1$ is valid.

$[H_3O^+] \approx 4.24 \times 10^{-3}$ Log $[H_3O^+] \approx -2.37$ pH ≈ 2.4 Answer(a)

12-82. What is the pH of a 0.10 M solution of phenol (Phenol: $K_a = 1.0 \times 10^{-10}$)?

(a) 5.5 (b) 8.5 (c) 10 (d) 11 (e) None of the above

$$HOPh\ (aq) + H_2O\ (l) \rightleftharpoons H_3O^+\ (aq) + OPh^-\ (aq)$$

At equilibrium: $0.1 - \Delta C$ ΔC ΔC

Assumption: $(0.1 - \Delta C) \approx 0.1$

$K_a = (\Delta C)^2/(0.1 - \Delta C) \approx (\Delta C)^2/(0.1)$

$(\Delta C)^2/(0.1) \approx 1.0 \times 10^{-10}$

$(\Delta C)^2 \approx 1.0 \times 10^{-11}$

$(\Delta C) \approx 3.2 \times 10^{-6}$

$\{(3.2 \times 10^{-6})/0.1\} \times 100 = 3.2 \times 10^{-3}\%$ (less than 5%) Assumption $(0.1 - \Delta C) \approx 0.1$ is valid.

$[H_3O^+] \approx 3.2 \times 10^{-6}$ Log $[H_3O^+] \approx -5.49$ pH ≈ 5.5 Answer(a)

12-83. Hydrogen peroxide has been used as a bleach to change hair color, as a disinfectant to treat wounds, and as a rocket fuel. It is also a weak acid. Calculate the pH of 0.018 M H_2O_2 solution (H_2O_2: $K_a = 2.2 \times 10^{-12}$).

(a) 0.6 (b) 6.7 (c) 7.3 (d) 11.7 (e) 13.4

$$H_2O_2\ (aq) + H_2O\ (l) \rightleftharpoons H_3O^+\ (aq) + HO_2^-\ (aq)$$

At equilibrium: $0.018 - \Delta C$ ΔC ΔC

Assumption: $(0.018 - \Delta C) \approx 0.018$

$K_a = (\Delta C)^2/(0.018 - \Delta C) \approx (\Delta C)^2/(0.018)$

$(\Delta C)^2/(0.018) \approx 2.2 \times 10^{-12}$

$(\Delta C)^2 \approx 4 \times 10^{-14}$

$\Delta C \approx 2 \times 10^{-7}$

$\{(2 \times 10^{-7})/(0.018)\} \times 100 = 1.1 \times 10^{-3}\%$ (less than 5%) Assumption $(0.018 - \Delta C) \approx 0.018$ is valid.

$[H_3O^+] \approx 2 \times 10^{-7}$ Log $[H_3O^+] \approx -6.7$ pH ≈ 6.7 Answer(b)

12-84. A 0.10 M solution of a weak acid, HA, is found to be 1.50% ionized. Calculate the K_a for the acid.

(a) 1.5 (b) 2.3×10^{-1} (c) 2.3×10^{-5} (d) 2.3×10^{-6}

$$HA \ (aq) + H_2O \ (l) \rightleftharpoons H_3O^+ \ (aq) + A^- \ (aq)$$

At equilibrium: $0.1 - \Delta C$ ΔC ΔC

$\{(\Delta C)/(0.1)\} \times 100 =$ Percent ionization $= 1.5 \ \%$

$\Delta C = 1.5 \times 10^{-3}$

$\{(1.5 \times 10^{-3})/0.1\} \times 100 = 1.5\%$ (less than 5%)

$(0.1 - \Delta C) \approx 0.1$

$K_a = (\Delta C)^2/(0.1 - \Delta C) \approx (\Delta C)^2/(0.1) \approx (1.5 \times 10^{-3})^2/(0.1) \approx 2.3 \times 10^{-5}$ Answer(c)

12-85. A solution is prepared by dissolving 1.0×10^{-4} mole of HOBr in water and diluting to 1.00 L. What is the pH of the solution? (HOBr: $K_a = 2 \times 10^{-9}$)

    (a) 1.3   (b) 5.3   (c) 6.3   (d) 8.7   (e)12.7

$$HOBr \ (aq) + H_2O \ (l) \rightleftharpoons H_3O^+ \ (aq) + OBr^- \ (aq)$$

At equilibrium: $\{(1.0 \times 10^{-4}) - \Delta C\}$ ΔC ΔC

Assumption: $\{(1.0 \times 10^{-4}) - \Delta C\} \approx (1.0 \times 10^{-4})$

$K_a = (\Delta C)^2/\{(1.0 \times 10^{-4}) - \Delta C\} \approx (\Delta C)^2/(1.0 \times 10^{-4})$

$(\Delta C)^2/(1.0 \times 10^{-4}) \approx 2 \times 10^{-9}$

$(\Delta C)^2 \approx 2 \times 10^{-13}$

$(\Delta C) \approx 0.45 \times 10^{-6}$

$\{(0.45 \times 10^{-6})/(1.0 \times 10^{-4})\} \times 100 = 0.45\%$ (less than 5%)

Assumption $\{(1.0 \times 10^{-4}) - \Delta C\} \approx (1.0 \times 10^{-4})$ is valid.

$[H_3O^+] \approx 0.45 \times 10^{-6}$ Log $[H_3O^+] \approx -6.3$ pH ≈ 6.3 Answer(c)

12-86. What is the value of K_b for the formate ion if K_a for formic acid $= 1.8 \times 10^{-4}$?

    (a) $K_b = K_w K_a$ (b) $K_b = K_a/K_w$ (c) $K_b = K_w/K_a$ (d) $K_b = K_w + K_a$ (e) $K_b = K_w - K_a$

$$HCOOH \ (aq) + H_2O \ (l) \rightleftharpoons H_3O^+ \ (aq) + HCOO^- \ (aq)$$

$K_a = [H_3O^+] \ [HCOO^-]/[HCOOH]$

$$HCOO^- \ (aq) + H_2O \ (l) \rightleftharpoons HCOOH \ (aq) + OH^- \ (aq)$$

$K_b = [HCOOH] \ [OH^-]/[HCOO^-]$

$K_a \times K_b = \{[H_3O^+] \ [HCOO^-]/[HCOOH]\} \times \{ [HCOOH] \ [OH^-]/[HCOO^-]\} = [H_3O^+] \ [OH^-] = K_w$

$K_b = K_w/K_a = (10^{-14})/(1.8 \times 10^{-4}) = 5.56 \times 10^{-11}$ Answer(c)

12-87. Which is the correct relation between K_a for an acid and K_b for its conjugate base?

(a) $K_b = K_w K_a$ (b) $K_b = K_w/K_a$ (c) $K_b = -\log K_a$ (d) $K_b = 1/K_a$ (e) $K_b = K_a/K_w$

$$\text{Acid (aq)} + H_2O \text{ (l)} \rightleftharpoons \text{Congjugate base (aq)} + H_3O^+ \text{ (aq)}$$

$K_a = \text{[Conjugate base] } [H_3O^+]/\text{[Acid]}$

$$\text{Conjugate base (aq)} + H_2O \text{ (l)} \rightleftharpoons \text{Acid (aq)} + OH^- \text{ (aq)}$$

$K_b = \text{[Acid] } [OH^-]/\text{[Conjugate base]}$

$K_a K_b = \{\text{[Conjugate base] } [H_3O^+]/\text{[Acid]}\} \times \{\text{[Acid] } [OH^-]/\text{[Conjugate base]}\} = [H_3O^+] [OH^-]$

$\quad = K_w$

$K_b = (K_w)/(K_a)$ Answer(b)

12-88. Ammonia and the ammonium ion form a conjugate acid-base pair:

$$NH_3 \text{ (aq)} + H_2O \text{ (l)} \rightleftharpoons NH_4^+ \text{ (aq)} + OH^- \text{ (aq)}$$

Calculate the pH of 0.10 M NH_3 if K_a for the NH_4^+ ion is 5.6×10^{-10}.

  (a) 4.7   (b) 5.1   (c) 5.7   (d) 8.9   (e) 11.1

$K_a K_b = K_w$ $K_b = K_w/K_a = (10^{-14})/(5.6 \times 10^{-10}) = 0.18 \times 10^{-4}$

$$NH_3 \text{ (aq)} + H_2O \text{ (l)} \rightleftharpoons NH_4^+ \text{ (aq)} + OH^- \text{ (aq)}$$

At equilibrium: $0.1 - \Delta C$ ΔC ΔC

Assumption: $(0.1 - \Delta C) \approx 0.1$

$K_b = (\Delta C)^2/(0.1 - \Delta C) \approx (\Delta C)^2/(0.1)$

$(\Delta C)^2/(0.1) \approx 0.18 \times 10^{-4}$

$(\Delta C)^2 \approx 0.018 \times 10^{-4}$

$\Delta C \approx 0.13 \times 10^{-2}$

$\{(0.13 \times 10^{-2})/0.1\} \times 100 = 1.3\%$ (less than 5%) Assumption $(0.1 - \Delta C) \approx 0.1$ is valid.

$[OH^-] \approx 0.13 \times 10^{-2}$

Log $[OH^-] \approx -2.89$ pOH ≈ 2.89 pH $\approx 14 - 2.89 \approx 11.1$ Answer(e)

12-89. There are many ways of fluoridating water. One approach involves adding a salt of the fluoride ion, such as NaF. Calculate the pH of an 0.15 M NaF solution (HF: $K_a = 7.2 \times 10^{-4}$)

  (a) 2   (b) 5.8   (c) 8.16   (d) 10.9   (e) 12

$$H_2O$$

$$NaF \longrightarrow Na^+ \text{ (aq)} + F^- \text{ (aq)}$$

$$F^- \text{ (aq)} + H_2O \text{ (l)} \rightleftharpoons HF \text{ (aq)} + OH^- \text{ (aq)}$$

At equilibrium: $0.15 - \Delta C$ ΔC ΔC

Assumption: $(0.15 - \Delta C) \approx 0.15$

$K_b = (K_w)/(K_a) = (10^{-14})/(7.2 \times 10^{-4}) = 1.39 \times 10^{-11}$

$K_b = (\Delta C)^2 /(0.15 - \Delta C) \approx (\Delta C)^2 /(0.15)$

$(\Delta C)^2 /(0.15) \approx 1.39 \times 10^{-11}$

$(\Delta C)^2 \approx 0.15 \times 1.39 \times 10^{-11} \approx 2.1 \times 10^{-12}$ $\Delta C \approx 1.45 \times 10^{-6}$

$\{(1.45 \times 10^{-6})/0.15\} \times 100 = 9.6 \times 10^{-4}\%$ (less than 5%) Assumption $(0.15 - \Delta C) \approx 0.15$ is valid.

$[OH^-] \approx 1.45 \times 10^{-6}$ Log $[OH^-] \approx -5.84$ pOH ≈ 5.84 pH ≈ 8.16 Answer(c)

12-90. Calculate the pH of a solution prepared by dissolving 1.00×10^{-2} mole of sodium hypochlorite (NaOCl) in enough water to produce a liter of solution. Hypochlorous acid is a weak monoprotic acid with $K_a = 3.2 \times 10^{-8}$.

    (a) 4.3   (b) 4.7   (c) 7.5   (d) 9.3   (e) 9.7

$$H_2O$$

$$NaOCl \longrightarrow Na^+ \text{ (aq)} + OCl^- \text{ (aq)}$$

$$OCl^- \text{ (aq)} + H_2O \text{ (l)} \rightleftharpoons HOCl \text{ (aq)} + OH^- \text{ (aq)}$$

At equilibrium: $(1 \times 10^{-2}) - \Delta C$ ΔC ΔC

Assumption: $\{(1 \times 10^{-2}) - \Delta C\} \approx (1 \times 10^{-2})$

$K_b = (K_w)/(K_a) = (10^{-14})/(3.2 \times 10^{-8}) = 3.125 \times 10^{-7}$

$K_b = (\Delta C)^2 /\{(1 \times 10^{-2}) - \Delta C\} \approx (\Delta C)^2 /(1 \times 10^{-2})$

$(\Delta C)^2 /(1 \times 10^{-2}) \approx 3.125 \times 10^{-7}$

$(\Delta C)^2 \approx (1 \times 10^{-2}) \times (3.125 \times 10^{-7}) \approx 0.3 \times 10^{-8}$

$\Delta C \approx 0.55 \times 10^{-4}$

$\{(0.55 \times 10^{-4})/ (1 \times 10^{-2})\} \times 100 = 0.55\%$ (less than 5%)

Assumption $\{(1 \times 10^{-2}) - \Delta C\} \approx (1 \times 10^{-2})$ is valid.

$[OH^-] \approx 0.55 \times 10^{-4}$ log $[OH^-] \approx -4.26$ pOH ≈ 4.26

$pH \approx 14 - 4.26 \approx 9.74 \approx 9.7$ Answer(e)

12-91. What is the OH^- concentration in a 0.2 M NaOAc solution? (HOAc: $K_a = 1.8 \times 10^{-5}$)

(a) 1.1×10^{-10} (b) 5.7×10^{-10} (c) 1.8×10^{-6} (d) 1.1×10^{-5} (e) 1.3×10^{-3}

$$\overset{H_2O}{NaOAc \longrightarrow Na^+ (aq) + OAc^- (aq)}$$

$$OAc^- (aq) + H_2O (l) \rightleftharpoons HOAc (aq) + OH^- (aq)$$

At equilibrium: $\quad\quad\quad\quad\quad\quad 0.2 - \Delta C \quad\quad\quad\quad\quad\quad \Delta C \quad\quad \Delta C$

Assumption: $(0.2 - \Delta C) \approx (0.2)$

$K_b = (K_w)/(K_a) = (10^{-14})/(1.8 \times 10^{-5}) = 5.55 \times 10^{-10}$

$K_b = (\Delta C)^2/(0.2 - \Delta C) \approx (\Delta C)^2/(0.2)$

$(\Delta C)^2/(0.2) \approx 5.55 \times 10^{-10}$

$(\Delta C)^2 \approx (0.2) \times (5.55 \times 10^{-10}) \approx 1.11 \times 10^{-10}$

$\Delta C \approx 1.1 \times 10^{-5}$

$\{(1.1 \times 10^{-5})/0.2\} \times 100 = 5.5 \times 10^{-3}\%$ (less than 5%) $\quad$ Assumption $(0.2 - \Delta C) \approx (0.2)$ is valid.

$[OH^-] \approx 1.1 \times 10^{-5}$ Answer(d)

12-92. What is the H_3O^+ ion concentration in a 0.1 M NH_3 solution ($K_b = 1.8 \times 10^{-5}$)?

(a) 7.5×10^{-12} (b) 3.0×10^{-10} (c) 1.8×10^{-6} (d) 1.3×10^{-3}

$$NH_3 (aq) + H_2O (l) \rightleftharpoons NH_4^+ (aq) + OH^- (aq)$$

At equilibrium: $\quad\quad\quad\quad 0.1 - \Delta C \quad\quad\quad\quad\quad\quad \Delta C \quad\quad \Delta C$

Assumption: $(0.1 - \Delta C) \approx 0.1$

$K_b = (\Delta C)^2/(0.1 - \Delta C) \approx (\Delta C)^2/(0.1)$

$(\Delta C)^2/(0.1) \approx 1.8 \times 10^{-5}$

$(\Delta C)^2 \approx 1.8 \times 10^{-6}$

$\Delta C \approx 1.34 \times 10^{-3}$

$\{(1.34 \times 10^{-3})/0.1\} \times 100 = 1.34\%$ (less than 5%) $\quad$ Assumption $(0.1 - \Delta C) \approx 0.1$ is valid.

$[H_3O^+] = (10^{-14})/[OH^-] \approx (10^{-14})/(1.34 \times 10^{-3}) \approx 7.5 \times 10^{-12}$ Answer(a)

12-93. HA is a weak monoprotic acid with $K_a = 2.4 \times 10^{-6}$. What is $[OH^-]$ in a 0.30 M solution

of the salt NaA?

(a) 5.8×10^{-12} (b) 2.4×10^{-10} (c) 3.5×10^{-5} (d) 6.5×10^{-5} (e) 1.6×10^{-3}

$$NaA \xrightarrow{H_2O} Na^+(aq) + A^-(aq)$$

$$A^-(aq) + H_2O(l) \rightleftharpoons HA(aq) + OH^-(aq)$$

At equilibrium: $\quad\quad\quad\quad 0.3 - \Delta C \quad\quad\quad\quad\quad\quad \Delta C \quad\quad \Delta C$

Assumption: $\quad (0.3 - \Delta C) \approx 0.3$

$K_b = (K_w)/(K_a) = (10^{-14})/(2.4 \times 10^{-6}) = 4.17 \times 10^{-9}$

$K_b = (\Delta C)^2/(0.3 - \Delta C) \approx (\Delta C)^2/(0.3)$

$(\Delta C)^2/(0.3) \approx 4.17 \times 10^{-9}$

$(\Delta C)^2 \approx 0.3 \times 4.17 \times 10^{-9} \approx 1.25 \times 10^{-9}$

$\Delta C \approx 3.5 \times 10^{-5}$

$\{(3.5 \times 10^{-5})/0.3\} \times 100 = 0.12\%$ (less than 5%) Assumption $(0.3 - \Delta C) \approx 0.3$ is valid.

$[OH^-] \approx 3.5 \times 10^{-5}$ $\quad\quad\quad\quad\quad\quad\quad\quad\quad\quad\quad\quad\quad\quad\quad$ Answer(c)

12-94. Which of the following would make the best buffer?

   (a) A mixture of HCl and NaCl   (b) A mixture of NaOAc and NH_3

   (c) A mixture of HOAc and NH_4Cl (d) A mixture of NaOAc and NH_4Cl

   (e) A mixture of NH_3 and NH_4Cl

Either a mixture of a weak acid and its conjugate base (provided by a salt) or a mixture of

a weak base and its conjugate acid (provided by a salt) will make the best buffer.

(a): Not a buffer solution.

(b): NH_3 is not the conjugate acid of NaOAc.

(c): NH_4Cl is not the conjugate base of HOAc.

(d) NH_4Cl is not the conjugate acid of NaOAc.

(e) NH_3 is a weak base and NH_4^+ is its conjugate acid. So a mixture of NH_3 and NH_4Cl will be

the best buffer. $\quad\quad\quad\quad\quad\quad\quad\quad\quad\quad\quad\quad\quad\quad\quad\quad\quad\quad\quad$ Answer(e)

12-95. Which of the following would be acidic buffer?

   (a) 0.10 M HCl and 0.01 M NaOH
   (b) 0.10 M HCl and 0.10 M NaCl
   (c) 0.10 M HCOOH and 0.10 M HCOONa

(d) 0.10 NH$_3$ and 0.10 M NH$_4$Cl

(e) None of the above would be buffers.

(a)Not a buffer solution.

(b)Not a buffer solution.

(c)HCOOH: K$_a$ = 1.8×10^{-4} and HCOO$^-$: K$_b$ = 5.56×10^{-11}

K$_a$ > K$_b$ and K$_a$ > 1.0×10^{-7}. So this is an acidic buffer.

(d)NH$_3$: K$_b$ = 1.8×10^{-5} and NH$_4^+$: K$_a$ = 5.56×10^{-10}

K$_b$ > K$_a$ and K$_b$ > 1.0×10^{-7}. So this is a basic buffer.   Answer(c)

12-96. When NH$_4$Cl is added to 0.1 M NH$_3$ solution, the pH of this solution

(a) Increases   (b) Decreases   (c) Remains the same

(d)Increases at first and then decreases   (e) Decreases at first and then increases

$$NH_3 \text{ (aq)} + H_2O \text{ (l)} \rightleftharpoons NH_4^+ \text{ (aq)} + OH^- \text{ (aq)}$$

When NH$_4$Cl is added to 0.10 M NH$_3$ solution, according to LeChatelier's principle, to

counteract this the equilibrium will be shifted to the left. This means that the [OH$^-$] will

decrease and the pH will decrease.   Answer(b)

12-97. When would be pH of a solution prepared by adding sodium formate to formic acid be

equal to the pK$_a$ of formic acid?

(a) When [HCOOH] < [HCOO$^-$]   (b) When [HCOOH] = [HCOO$^-$]

(c)When [HCOOH] > [HCOO$^-$]

(d)The pH of this buffer will never be equal to the pK$_a$ of formic acid

pH = pK$_a$

$-$ log [H$^+$] = $-$ log K$_a$

[H$^+$] = K$_a$

[H$^+$]/K$_a$ = 1

$$HCOOH \text{ (aq)} + H_2O \text{ (l)} \rightleftharpoons H_3O^+ \text{ (aq)} + HCOO^- \text{ (aq)}$$

K$_a$ = [H$_3$O$^+$] [HCOO$^-$]/[HCOOH]

[H$_3$O$^+$]/K$_a$ = [HCOOH]/[HCOO$^-$] = 1

So [HCOOH] = [HCOO$^-$]   Answer(b)

12-98. Which of the following would be the best buffer when a small amount of NaOH is added to the solution?

    (a) A 1.0 L solution that contains 0.1 mole of HCOOH and 0.1 mole of HCOONa.
    (b) A 1.0 L solution that contains 0.2 mole of HCOOH and 0.1 mole of HCOONa.
    (c) A 1.0 L solution that contains 0.1 mole of HCOOH and 0.2 mole of HCOONa.
    (d) A 1.0 L solution that contains 0.2 mole of HCOOH and 0.2 mole of HCOONa.

$$HCOOH\ (aq) + H_2O\ (l) \rightleftharpoons H_3O^+\ (aq) + HCOO^-\ (aq)$$

$K_a = [H_3O^+]\ [HCOO^-]/[HCOOH]$

$[H_3O^+] = K_a\ [HCOOH]/\ [HCOO^-]$

$pH = pK_a + \log\ [HCOO^-]/[HCOOH]$

(a) $[HCOO^-] = [HCOOH] = 0.1\ M$

$pH = pK_a$

Add 10 ml of 1 M NaOH (the change in volume due to the addition of NaOH can be neglected). It will react with HCOOH and reduce the HCOOH concentration and increase the salt concentration.

$[HCOOH] = 0.1 - 0.01 = 0.09$ $[HCOO^-] = 0.1 + 0.01 = 0.11$

$pH = pK_a + \log\ \{(0.11)/(0.09)\} = pK_a + \log 1.22 = pK_a + 0.09$

The change in pH = 0.09

(b) $[HCOO^-] = 0.1\ M$ $[HCOOH] = 0.2\ M$

$pH = pK_a + \log\ \{(0.1)/(0.2)\} = pK_a + \log\ (0.5) = pK_a - 0.30$

Add 10 ml of 1 M NaOH (the change in volume due to the addition of NaOH can be neglected). It will react with HCOOH and reduce the HCOOH concentration and increase the salt concentration.

$[HCOOH] = 0.20 - 0.01 = 0.19$ $[HCOO^-] = 0.10 + 0.01 = 0.11$

$pH = pK_a + \log\ \{(0.11)/(0.19)\} = pK_a + \log 0.58 = pK_a - 0.24$

The change in pH = 0.06

(c) $[HCOO^-] = 0.2\ M$ $[HCOOH] = 0.1\ M$

$pH = pK_a + \log\ \{(0.2)/(0.1)\} = pK_a + \log\ (2.0) = pK_a + 0.30$

Add 10 ml of 1 M NaOH (the change in volume due to the addition of NaOH can be neglected). It will react with HCOOH and reduce the HCOOH concentration and increase the salt concentration.

$[HCOOH] = 0.10 - 0.01 = 0.09$ $\qquad$ $[HCOO^-] = 0.20 + 0.01 = 0.21$

$pH = pK_a + log \{(0.21)/(0.09)\} = pK_a + log 2.33 = pK_a + 0.37$

The change in pH = 0.07

(d)) $[HCOO^-] = 0.2 M$ $\qquad$ $[HCOOH] = 0.2 M$

$pH = pK_a + log \{(0.2)/(0.2)\} = pK_a + log (1.0) = pK_a$

Add 10 ml of 1 M NaOH (the change in volume due to the addition of NaOH can be neglected). It will react with HCOOH and reduce the HCOOH concentration and increase the salt concentration.

$[HCOOH] = 0.20 - 0.01 = 0.19$ $\qquad$ $[HCOO^-] = 0.20 + 0.01 = 0.21$

$pH = pK_a + log \{(0.21)/(0.19)\} = pK_a + log 1.1 = pK_a + 0.04$

The change in pH = 0.04

The change in pH is lowest in the case of (d). So (d) is the best buffer. $\qquad$ Answer(d)

12-99. Which of the following would be the best buffer when a small amount of NaOH is added to the solution?

  (a) A 1.0 L solution prepared by adding 0.10 mole of carbonic acid and 0.10 mole of sodiumbicarbonate
  (b) A 1.0 L solution prepared by adding 0.20 mole of carbonic acid and 0.20 mole of sodiumbicarbonate
  (c) A 1.0 L solution prepared by adding 0.10 mole of HCl and 0.10 mole of NaOH
  (d) A 1.0 L solution prepared by adding 0.20 mole of HCl and 0.20 mole of NaOH
  (e) All of these solutions would have the same buffer capacity

(c) and (d) are not buffer solutions.

(a) $\qquad$ $H_2CO_3 (aq) + H_2O (l) \rightleftharpoons H_3O^+ (aq) + HCO_3^- (aq)$

$K_a = [H_3O^+] [HCO_3^-]/[H_2CO_3]$

$[H_3O^+] = K_a [H_2CO_3]/ [HCO_3^-]$

$pH = pK_a + log [HCO_3^-]/[H_2CO_3]$

$[HCO_3^-] = [H_2CO_3] = 0.1$

So initially the pH = pK_a

Add 10 ml of 1 M of NaOH (the change in volume due to the addition of NaOH can be neglected). It will react with carbonic acid and reduce the carbonic acid concentration and increase the salt concentration.

$[H_2CO_3] = 0.1 - 0.01 = 0.09$ mole $\qquad$ $[HCO_3^-] = 0.1 + 0.01 = 0.11$ mole

$pH = pK_a + \log [HCO_3^-]/[H_2CO_3] = pK_a + \log (0.11)/(0.09) = pK_a + \log 1.22 = pK_a + 0.09$

The change in pH = 0.09

(b) $[HCO_3^-] = [H_2CO_3] = 0.2$

So initially the $pH = pK_a$

Add 10 ml of 1 M NaOH (the change in volume due to the addition of NaOH can be neglected). It will react with carbonic acid and reduce the carbonic acid concentration and increase the salt concentration.

$[H_2CO_3] = 0.2 - 0.01 = 0.19$ mole $\qquad$ $[HCO_3^-] = 0.2 + 0.01 = 0.21$ mole

$pH = pK_a + \log [HCO_3^-]/[H_2CO_3] = pK_a + \log (0.21)/(0.19) = pK_a + \log 1.10 = pK_a + 0.04$

The change in pH = 0.04

In the case of (b) the change in pH is only 0.04 {in the case of (a) the change in pH is 0.09}.

So (b) is the best buffer. $\qquad$ Answer(b)

12-100. Which of the following would be the best choice for preparing a pH = 7.46 buffer solution?

    (a) A mixture of formic acid ($K_a = 1.8 \times 10^{-4}$) and the formate ion.
    (b) A mixture of acetic acid ($K_a = 1.8 \times 10^{-5}$) and the acetate ion
    (c) A mixture of hypochlorous acid ($K_a = 3.5 \times 10^{-8}$) and hypochlorite ion
    (d) A mixture of boric acid ($K_a = 5.8 \times 10^{-10}$) and its conjugate base
    (e) All of these

Acidic buffer: $\quad pH = pK_a + \log \{[Salt]/[Acid]\}$

Let us assume [Salt] = [Acid] = 0.20 M in all cases. $pH = pK_a$

	K_a	$\log K_a$	pK_a	pH
(a)	1.8×10^{-4}	-3.75	3.75	3.75
(b)	1.8×10^{-5}	-4.75	4.75	4.75
(c)	3.5×10^{-8}	-7.46	7.46	7.46
(d)	5.8×10^{-10}	-9.24	9.24	9.24

So (c) will be the best choice. $\qquad$ Answer(c)

12-101. Which of the following would have the smallest pH? (HCOOH: $K_a = 1.8 \times 10^{-4}$;

NH$_4^+$: $K_a = 5.6 \times 10^{-10}$)

   (a) 0.10 M HCOOH   (b) 0.10M HCOONa   (c) 0.10 M NH$_4^+$

   (d) A solution that is 0.10 M in both HCOOH and HCOONa

   (e) A solution that is 0.10 M in both NH$_4^+$ and NH$_3$

(a)   HCOOH (aq) + H$_2$O (l) $\rightleftharpoons$ H$_3$O$^+$ (aq) + HCOO$^-$ (aq)

At equilibrium: $(0.1 - \Delta C)$ ΔC ΔC

Assumption: $(0.1 - \Delta C) \approx 0.1$

$K_a = (\Delta C)^2/(0.1 - \Delta C) \approx (\Delta C)^2/(0.1)$

$(\Delta C)^2/(0.1) \approx 1.8 \times 10^{-4}$

$(\Delta C)^2 \approx 1.8 \times 10^{-5}$

$(\Delta C) \approx 4.2 \times 10^{-3}$

$\{(4.2 \times 10^{-3})/0.1\} \times 100 = 4.2\%$ (less than 5%) Assumption $(0.1 - \Delta C) \approx 0.1$ is valid.

 [H$_3$O$^+$] $\approx 4.2 \times 10^{-3}$ log [H$_3$O$^+$] ≈ -2.38 pH ≈ 2.38

(b) HCOONa is a salt of a strong base and a weak acid.

pH = (1/2) pK_w + (1/2) pK_a + (1/2) log c

pK_a = $-$ log K_a = $-$ log (1.8×10^{-4}) = 3.74

pH = (1/2) 14 + (1/2) 3.74 + (1/2) log (0.1) = 7 + 1.87 $-$ 0.5 = 8.37

(c)   NH$_4^+$ (aq) + H$_2$O (l) $\rightleftharpoons$ H$_3$O$^+$ (aq) + NH$_3$ (aq)

At equilibrium: $(0.1 - \Delta C)$ ΔC ΔC

Assumption: $(0.1 - \Delta C) \approx 0.1$

$K_a = (\Delta C)^2/(0.1 - \Delta C) \approx (\Delta C)^2/(0.1)$

$(\Delta C)^2/(0.1) \approx 5.6 \times 10^{-10}$

$(\Delta C)^2 \approx 5.6 \times 10^{-11}$

$(\Delta C) \approx 7.5 \times 10^{-6}$

$\{(7.5 \times 10^{-6})/0.1\} \times 100 = 7.5 \times 10^{-3}\%$ (less than 5%)   Assumption $(0.1 - \Delta C) \approx 0.1$ is valid.

[H$_3$O$^+$] $\approx 7.5 \times 10^{-6}$ log [H$_3$O$^+$] ≈ -5.12 pH ≈ 5.12

(d)   Acidic buffer: pH = pK_a + log $\{$ [OAc$^-$]/[HOAc]$\}$

pH = pK_a = $-$ log K_a = $-$ log (1.8×10^{-4}) = 3.74

(e)   Basic buffer: $pOH = pK_b + \log \{[NH_4^+]/[NH_3]\}$

   $pOH = pK_b$

$K_b = (10^{-14})/K_a = (10^{-14})/(5.6 \times 10^{-10}) = 1.8 \times 10^{-5}$

$pK_b = -\log K_b = -\log (1.8 \times 10^{-5}) = 4.74$

$pOH = 4.74$ $pH = 14 - 4.74 = 9.26$

(a) has the smallest pH.   Answer(a)

12-102. Which of the following solutions will show the largest pH change when 10 ml of 1 M NaOH is added to it?

   (a) 10 ml of 1 M NaOH   (b) 100 ml of water

   (c) 10 ml of 1 M NaNO$_3$ (d) 10 ml of 0.01 HOAc

   (e) 100 ml of 0.1 M HOAc and 0.1 M NaOAc

(a) Addition of 10 ml of 1 M NaOH to 10 ml of 1 M NaOH will not change the pH.

Change in pH = 0.0

(b) Initial pH of water = 7.0

10 ml of 1M NaOH contains 0.01 mole of NaOH.

Concentration of OH$^-$ after addition = $\{(0.01)/110\} \times 1000 = 0.09$ mole per liter

$\log (0.09) = -1.04$ $pOH = 1.04$ $pH = 14 - 1.04 = 12.96$

Change in pH = 12.96 − 7.0 = 5.96

(c) NaNO$_3$ is a salt of a strong acid and a strong base. It will not undergo hydrolysis in water.

Initial pH = 7.0

Concentration of OH$^-$ after addition : $\{(0.01)/20\} \times 1000 = 0.5$ mole per liter

$\log (OH^-) = \log (0.5) = -0.3$ $pOH = 0.3$ $pH = 14 - 0.3 = 13.7$

Change in pH = 13.7 − 7.0 = 6.7

(d) $HOAc\ (aq) + H_2O\ (l) \rightleftharpoons H_3O^+\ (aq) + OAc^-\ (aq)$

At equilibrium: $0.01 - \Delta C$ ΔC ΔC

Assumption: $(0.01 - \Delta C) \approx 0.01$

$K_a = (\Delta C)^2 /(0.01 - \Delta C) \approx (\Delta C)^2 /(0.01)$

$(\Delta C)^2 /(0.01) \approx 1.8 \times 10^{-5}$

$(\Delta C)^2 \approx 1.8 \times 10^{-7}$

$\Delta C \approx 4.24 \times 10^{-4}$

$\{(4.24 \times 10^{-4})/0.01\} \times 100 = 4.24\%$ (less than 5%)   Assumption $(0.01 - \Delta C) \approx 0.01$ is valid

$[H_3O^+] \approx 4.24 \times 10^{-4}$ $\log [H_3O^+] \approx -3.37$ $pH \approx 3.37$

Initial $pH \approx 3.37$

Amount of HOAc = $\{(0.01)/1000\} \times 10 = 0.0001$ mole

Amount of NaOH = 0.01 mole

0.01 mole of NaOH will react with 0.0001 mole of HOAc to produce 0.0001 mole of NaOAC.

Amount of NaOH after the addition = $0.01 - 0.0001 = 0.0099$ mole

NaOH concentration after addition will be = 0.0099 mole/ 20 ml = 0.495 mole/liter.

Concentration of NaOAC = 0.0001 mole /20 ml = 0.005 mole per liter

$$OAc^- (aq) + H_2O (l) \rightleftharpoons HOAc (aq) + OH^- (aq)$$

At equilibrium: $0.005 - \Delta C$ ΔC $0.495 + \Delta C$

Assumptions: $(0.005 - \Delta C) \approx 0.005$ and $(0.495 - \Delta C) \approx 0.495$

$K_b = (\Delta C)(0.495 + \Delta C)/(0.005 - \Delta C) \approx \Delta C (0.495)/(0.005)$

$\Delta C (0.495)/(0.005) \approx 5.56 \times 10^{-10}$

$\Delta C \approx 5.61 \times 10^{-12}$

$\{(5.56 \times 10^{-12})/0.005\} \times 100 = 1.1 \times 10^{-7}$ % (less than 5%)

$\{(5.56 \times 10^{-12})/0.495\} \times 100 = 1.1 \times 10^{-9}$ % (less than 5%)

Assumptions $(0.005 - \Delta C) \approx 0.005$ and $(0.495 - \Delta C) \approx 0.495$ are valid.

$[OH^-] \approx 0.495 + 5.61 \times 10^{-12} \approx 0.495$ mole per liter $\log [OH^-] \approx -0.30$ $pOH \approx 0.30$

$pH \approx 13.7$

Change in $pH \approx 13.7 - 3.37 \approx 10.33$

(e) The solution is an acidic buffer.

$pH = pK_a + \log [OAc^-]/[HOAc] = pK_a = 4.74$

Initial pH = 4.74

The solution contains 0.01 mole of HOAc and 0.01 mole of HOAC.

0.01 mole of NaOH will react with 0.01 mole of HOAc to produce 0.01 mole of NaOAc.

(NaOAc) after addition = {(0.02)/110}×1000 = 0.18 mole per liter.

$$OAc^- (aq) + H_2O (l) \rightleftharpoons HOAc (aq) + OH^- (aq)$$

At equilibrium: $0.18 - \Delta C$ ΔC ΔC

Assumption: $(0.18 - \Delta C) \approx 0.18$

$K_b = (\Delta C)^2/(0.18 - \Delta C) \approx (\Delta C)^2/(0.18)$

$(\Delta C)^2/(0.18) \approx 5.6×10^{-10}$ $(\Delta C)^2 \approx 1.008×10^{-10}$

$(\Delta C) \approx 1.00×10^{-5}$

$\{(1.00×10^{-5})/0.18\}×100 = 5.56×10^{-3}\%$ (less than 5%)

Assumption: $(0.18 - \Delta C) \approx 0.18$ is valid.

$[OH^-] \approx 1.00×10^{-5}$ $Log [OH^-] \approx -5$ $pOH \approx 5$ $pH \approx 14 - 5 \approx 9$

Change in pH $\approx 9 - 4.74 \approx 4.26$ Answer(d)

12-103. Calculate the initial concentration of HCOONa you would have to add to 0.10 M

HCOOH to make a buffer with a pH of 4.00? (HCOOH: $K_a = 1.8×10^{-4}$)

   (a) 0.36 M   (b) 0.98 M   (c) 0.18 M   (d) 0.27 M

$$HCOOH (aq) + H_2O (l) \rightleftharpoons H_3O^+ (aq) + HCOO^- (aq)$$

$K_a = [H_3O^+][HCOO^-]/[HCOOH]$

The solution is an acidic buffer.

$pH = pK_a + log \{[HCOO^-]/[HCOOH]\}$

$4.0 = - log (1.8×10^{-4}) + log \{[HCOO^-]/[HCOOH]\}$

$4.0 = 3.74 + log \{[HCOO^-]/[HCOOH]\}$

$log \{[HCOO^-]/[HCOOH]\} = 4.0 - 3.74 = 0.26$

$[HCOO^-]/[HCOOH] = 1.82$

$[HCOO^-]/(0.10) = 1.82$ $[HCOO^-] = 0.18 M$ Answer(c)

12-104. What is the pH of a solution that is simultaneously 0.10 M in both NH$_3$ and NH$_4$ ion?

(NH$_4^+$: $K_a = 5.6×10^{-10}$)

   (a) 4.75   (b) 7.00   (c) 9.26   (d) The pH of this solution is impossible to predict

   (e) None of the above

$$NH_4^+ (aq) + H_2O (l) \rightleftharpoons H_3O^+ (aq) + NH_3 (aq)$$

The solution is a basic buffer.

$pOH = pK_b + \log\{[NH_4^+]/[NH_3]\} = pK_b$

$K_b = (10^{-14})/(5.6\times10^{-10}) = 1.8\times10^{-5}$

$pK_b = -\log K_b = -\log (1.8\times10^{-5}) = 4.74$

$pOH = 4.74$ $pH = 14 - 4.74 = 9.26$ Answer(c)

12-105. The pH of a solution that is simultaneously 0.10 M in both NH_3 and NH_4 ion is 9.26.

What would be the effect of doubling the $[NH_3]$ in the solution on the pH? (NH_4^+:

$K_a = 5.6\times10^{-10}$)

$$NH_4^+ (aq) + H_2O (l) \rightleftharpoons H_3O^+ (aq) + NH_3 (aq)$$

The solution is a basic buffer.

The increase in the $[NH_4^+]$ due to increase in the concentration of NH_3 will be very

small compared to 0.1 and can be neglected.

$pOH = pK_b + \log\{[NH_4^+]/[NH_3]\} = pK_b + \log \{(0.1)/(0.2)\} = 4.74 - 0.3 = 4.44$

$pH = 14 - 4.44 = 9.56$

The pH increases from 9.26 to 9.56.

12-106. How much HCOONa would you have to add to 0.10 M HCOOH to get a buffer

solution with a pH of 3.4? (HCOOH: $K_a = 1.8\times10^{-4}$)

     (a) 0.010 M HCOONa   (b) 0.046 M HCOONa   (c) 0.10 M HCOONa

     (d)0.20 M HCOONa   (e) None of the above

$$HCOOH (aq) + H_2O (l) \rightleftharpoons H_3O^+ (aq) + HCOO^- (aq)$$

$K_a = [H_3O^+][HCOO^-]/[HCOOH]$ The solution is an acidic buffer.

$pH = pK_a + \log \{[HCOO^-]/[HCOOH]\}$

$3.4 = -\log (1.8\times10^{-4}) + \log \{[HCOO^-]/[HCOOH]\} = 3.74 + \log \{[HCOO^-]/[HCOOH]\}$

$\log \{[HCOO^-]/[HCOOH]\} = 3.4 - 3.74 = -0.34$

$[HCOO^-]/[HCOOH] = 0.46$ $[HCOO^-]/(0.10) = 0.46$ $[HCOO^-] = 0.046 M$ Answer(b)

12-107. How much NaOAc would have to be added to 1.00 L of 0.10 M HOAc

to make a pH of 3.00 buffer? (HOAc: $K_a = 1.8\times10^{-5}$)

(a) 0.18 M   (b) 0.36 M   (c) 0.0018 M   (d) 0.09 M

$$CH_3COOH \ (aq) + H_2O \ (l) \rightleftharpoons CH_3COO^- \ (aq) + H_3O^+ \ (aq)$$

$K_a = [H_3O^+][CH_3COO^-]/[CH_3COOH]$

The solution is an acidic buffer.

$pH = pK_a + \log \{[CH_3COO^-]/[CH_3COOH]\}$

$3.0 = - \log (1.8 \times 10^{-5}) + \log \{[CH_3COO^-]/[CH_3COOH]\}$

$\quad = 4.74 + \log \{[CH_3COO^-]/[CH_3COOH]\}$

$\log \{[CH_3COO^-]/[CH_3COOH]\} = 3.0 - 4.74 = -1.74$

$[CH_3COO^-]/[CH_3COOH] = 0.018 \quad [CH_3COO^-]/(0.10) = 0.018 \quad [CH_3COO^-] = 0.0018 \ M$

Answer(c)

12-108. How many moles of sodium acetate (NaOAC) must be added to 500 ml of 0.25 acetic acid (HOAc) to give a pH 4.9 buffer solution? (HOAC: $K_a = 1.8 \times 10^{-5}$; assume no volume change)

(a) 0.36 mole   (b) 0.18 mole   (c) 0.27 mole   (d) 0.09 mole

$$CH_3COOH \ (aq) + H_2O \ (l) \rightleftharpoons CH_3COO^- \ (aq) + H_3O^+ \ (aq)$$

$K_a = [H_3O^+][CH_3COO^-]/[CH_3COOH]$

The solution is an acidic buffer.

$pH = pK_a + \log \{[CH_3COO^-]/[CH_3COOH]\}$

$4.9 = - \log (1.8 \times 10^{-5}) + \log \{[CH_3COO^-]/[CH_3COOH]\}$

$\quad = 4.74 + \log \{[CH_3COO^-]/[CH_3COOH]\}$

$\log \{[CH_3COO^-]/[CH_3COOH]\} = 4.9 - 4.74 = 0.16$

$[CH_3COO^-]/[CH_3COOH] = 1.45 \quad [CH_3COO^-]/(0.25) = 1.45 \quad [CH_3COO^-] = 0.36 \ M$

For 500 ml, 0.18 mole of NaOAc should be added.   Answer(b)

12-109. What is the [HOAc]/[OAc⁻] ratio in an acetic acid/sodium acetate buffer at pH = 4.9? (HOAC: $K_a = 1.8 \times 10^{-5}$)

(a) 0.31   (b) 0.70   (c) 1.40   (d) 2.4   (e) 4.9

$$CH_3COOH \ (aq) + H_2O \ (l) \rightleftharpoons CH_3COO^- \ (aq) + H_3O^+ \ (aq)$$

$K_a = [H_3O^+][CH_3COO^-]/[CH_3COOH]$

The solution is an acidic buffer.

$pH = pK_a + \log \{[CH_3COO^-]/[CH_3COOH]\}$

$4.9 = -\log (1.8 \times 10^{-5}) + \log \{[CH_3COO^-]/[CH_3COOH]\}$

$= 4.74 + \log \{[CH_3COO^-]/[CH_3COOH]\}$

$\log \{[CH_3COO^-]/[CH_3COOH]\} = 4.9 - 4.74 = 0.16$

$[CH_3COO^-]/[CH_3COOH] = 1.45$

$[CH_3COOH]/[CH_3COO^-] = (1)/(1.45) = 0.7$ Answer(b)

12-110. Calculate the pH of a buffer solution prepared by mixing 0.1 mole of sodium formate and 0.05 mol of formic acid in 1.0 L of solution? (HCOOH: $K_a = 1.8 \times 10^{-4}$)

   (a) 1.8×10^{-4} (b) 3.44 (c) 4.04 (d) 5.31 (e) None of the above

$$HCOOH (aq) + H_2O (l) \rightleftharpoons H_3O^+ (aq) + HCOO^- (aq)$$

$K_a = [H_3O^+][HCOO^-]/[HCOOH]$

The solution is an acidic buffer.

$pH = pK_a + \log \{[HCOO^-]/[HCOOH]\}$

$pH = -\log (1.8 \times 10^{-4}) + \log \{[HCOO^-]/[HCOOH]\} = 3.74 + \log \{[HCOO^-]/[HCOOH]\}$

$\log \{[HCOO^-]/[HCOOH]\} = \log \{(0.1)/(0.05)\} = 0.30$

$pH = 3.74 + 0.30 = 4.04$ Answer(c)

12-111. What is the pH of a solution that is 0.01 M CH_3NH_2 and 0.1 M in $CH_3NH_3^+$?

$$CH_3NH_2 (aq) + H_2O (l) \rightleftharpoons CH_3NH_3^+ (aq) + OH^- (aq) \quad K_b = 4.4 \times 10^{-4}$$

   (a) 2.36 (b) 3.36 (c) 9.64 (d) 10.64 (e) 11.64

$K_b = [CH_3NH_3^+] [OH^-]/[CH_3NH_2]$

The solution ia a basic buffer.

$pOH = pK_b + \log\{[CH_3NH_3^+]/[CH_3NH_2]\} = -\log (4.4 \times 10^{-4}) + \log \{(0.1)/(0.01)\}$

$$= 3.36 + 1.0 = 4.36$$

$pH = 14 - 4.36 = 9.64$ Answer(c)

12-112. If equimolar amounts of NH_3 and NH_4Cl are dissolved in 1.0 L of water, what is the H_3O^+ ion concentration in this solution? ($NH_3 : K_b = 1.8 \times 10^{-5}$)

   (a) 1.0×10^{-14} M (b) 5.5×10^{-10} M (c) 1.0×10^{-7} M (d) 1.8×10^{-5} M (e)It can not be calculated from the information given.

$$NH_3 \text{ (aq)} + H_2O \text{ (l)} \rightleftharpoons NH_4^+ \text{ (aq)} + OH^- \text{ (aq)}$$

$K_b = [NH_4^+] [OH^-]/[NH_3]$

The solution ia a basic buffer.

$pOH = pK_b + \log\{[NH_4^+]/[NH_3]\} = -\log (1.8 \times 10^{-5}) + \log 1.0 = 4.74 + 0.0 = 4.74$

$pH = 14 - 4.74 = 9.26$ $-\log [H_3O^+] = 9.26$ $[H_3O^+] = 5.5 \times 10^{-10}$ M   Answer(b)

12-113. What is the pH of a 0.10 M succinic acid solution if we assume that H_2Sc is a weak acid that undegoes stepwise dissociation?

$$H_2Sc \text{ (aq)} + H_2O \text{ (l)} \rightleftharpoons H_3O^+ \text{ (aq)} + HSc^- \text{ (aq)} \quad K_{a1} = 6.9 \times 10^{-5}$$

$$HSc^- \text{ (aq)} + H_2O \text{ (l)} \rightleftharpoons H_3O^+ \text{ (aq)} + Sc^{2-} \text{ (aq)} \quad K_{a2} = 2.5 \times 10^{-6}$$

(a) 2.1   (b) 2.6   (c) 3.3   (d) 4.2   (e) 5.6

$$H_2Sc \text{ (aq)} + H_2O \text{ (l)} \rightleftharpoons H_3O^+ \text{ (aq)} + HSc^- \text{ (aq)}$$

At equilibrium: $0.1 - \Delta C$ ΔC ΔC

Assumption: $(0.1 - \Delta C) \approx 0.1$

$K_{a1} = [H_3O^+] [HSc^-]/[H_2Sc]$

$K_{a1} = (\Delta C)^2/ (0.1 - \Delta C) \approx (\Delta C)^2/ (0.1)$

$(\Delta C)^2/ (0.1) \approx 6.9 \times 10^{-5}$

$(\Delta C)^2 \approx 6.9 \times 10^{-6}$

$\Delta C \approx 2.63 \times 10^{-3}$

$\{(2.63 \times 10^{-3})/0.1\} \times 100 = 2.63\%$ (less than 5%) Assumption $(0.1 - \Delta C) \approx 0.1$ is valid.

$[H_3O^+] \approx 2.63 \times 10^{-3}$

$[HSc^-] \approx 2.63 \times 10^{-3}$

$$HSc^- \text{ (aq)} + H_2O \text{ (l)} \rightleftharpoons H_3O^+ \text{ (aq)} + Sc^{2-} \text{ (aq)}$$

At equilibrium: $\{(2.63 \times 10^{-3}) - \Delta C\}$ $\{(2.63 \times 10^{-3}) + \Delta C\}$ ΔC

Assumptions: $(2.63 \times 10^{-3}) - \Delta C \approx 2.63 \times 10^{-3}$ and $(2.63 \times 10^{-3}) + \Delta C \approx (2.63 \times 10^{-3})$

$K_{a2} = [H_3O^+] [Sc^{2-}]/[HSc^-]$

$K_{a2} = \{(2.63 \times 10^{-3}) + \Delta C\} \Delta C/\{(2.63 \times 10^{-3}) - \Delta C) \approx \Delta C$

$\Delta C \approx 2.5 \times 10^{-6}$

$\{(2.5 \times 10^{-6})/(2.63 \times 10^{-3})\} \times 100 = 9.73 \times 10^{-2}\%$ (less than 5%)

Assumptions $(2.63 \times 10^{-3}) - \Delta C \approx 2.63 \times 10^{-3}$ and $(2.63 \times 10^{-3}) + \Delta C \approx (2.63 \times 10^{-3})$ are valid.

$[HSc^-] \approx (2.63 \times 10^{-3}) - (2.5 \times 10^{-6}) \approx 2.6275 \times 10^{-3}$

$[H_3O^+] \approx (2.63 \times 10^{-3}) + (2.5 \times 10^{-6}) \approx 2.6325 \times 10^{-3}$

$Log\ [H_3O^+] \approx -2.58$ \qquad pH ≈ 2.6 \hfill Answer(b)

12-114. What fraction of the HSc⁻ formed in the first step of the dissociation of succinic acid

goes on to dissociate to form Sc^{2-} ions in the second step?

    (a) None of the HSc⁻ ions dissociate further
    (b) Less than 1% of the HSc⁻ ions dissociate
    (c) Between 1 and 5 % of the HSc⁻ ions dissociate
    (d) Between 5 and 20 % of the HSc⁻ ions dissociate
    (e) Between 5 and 20 % of the HSc⁻ ions dissociate

$$HSc^- \text{ (aq)} + H_2O \text{ (l)} \rightleftharpoons H_3O^+ \text{ (aq)} + Sc^{2-} \text{ (aq)}$$

$K_{a2} = [H_3O^+] [Sc^{2-}]/[HSc^-] = 2.5 \times 10^{-6}$

Initial $[HSc^-] \approx 2.63 \times 10^{-3}$ (reference 12-113)

$[HSc^-]$ at equilibrium $\approx (2.63 \times 10^{-3}) - (2.5 \times 10^{-6}) \approx 2.6275 \times 10^{-3}$ (reference 12-113)

Percentage dissociation of $[HSc^-] \approx \{(2.5 \times 10^{-6})/(2.63 \times 10^{-3})\} \times 100 \approx 0.095\ \%\ (< 1\%\ )$

\hfill Answer(b)

12-115. Which of the following equations accurately describes a 0.10 M H_2Gly

solution? ($H_2Gly : K_{a1} = 4.5 \times 10^{-3}$, $K_{a2} = 2.5 \times 10^{-10}$)

    (a) $[H_3O^+] \approx [H_2Gly]$ (b) $[H_3O^+] > [H_2Gly]$ (c) $[H_3O^+] < [HGly^-]$

    (d) $[H_3O^+] \approx [HGly^-]$ (e)) $[H_3O^+] > [HGly^-]$

$$H_2Gly \text{ (aq)} + H_2O \text{ (l)} \rightleftharpoons H_3O^+ \text{ (aq)} + HGly^- \text{ (aq)}$$

At equilibrium: \qquad\qquad\qquad\qquad $0.1 - \Delta C$ \qquad\qquad ΔC \qquad ΔC

Assumption: $(0.1 - \Delta C) \approx 0.1$

$K_{a1} = [H_3O^+] [HGly^-]/[H_2Gly] = (\Delta C)^2/ (0.1 - \Delta C) \approx (\Delta C)^2/ (0.1)$

$(\Delta C)^2/ (0.1) \approx 4.5 \times 10^{-3}$

$(\Delta C)^2 \approx 4.5 \times 10^{-4}$

$(\Delta C) \approx 2.12 \times 10^{-2}$

$\{(2.12 \times 10^{-2})/0.1\} \times 100 = 21.2\%$ (not less than 5%)   Assumption $(0.1 - \Delta C) \approx 0.1$ is not valid.

$(\Delta C) = 1.91 \times 10^{-2}$ (by quadratic formula)

$[H_3O^+] = [HGly^-] = 1.91 \times 10^{-2}$

$$HGly^- \ (aq) + H_2O \ (l) \ \rightleftharpoons \ H_3O^+ \ (aq) \qquad + \ Gly^{2-} \ (aq)$$

At equilibrium: $\qquad (1.91 \times 10^{-2}) - \Delta C \qquad\qquad (1.91 \times 10^{-2}) + \Delta C \qquad \Delta C$

Assumptions: $(1.91 \times 10^{-2}) - \Delta C \approx (1.91 \times 10^{-2})$ and $(1.91 \times 10^{-2}) + \Delta C \approx (1.91 \times 10^{-2})$

$K_{a2} = \{(1.91 \times 10^{-2}) + \Delta C\} \ \Delta C / \{(1.91 \times 10^{-2}) - \Delta C\} \approx \Delta C$

$\Delta C \approx 2.5 \times 10^{-10}$

$\{(2.5 \times 10^{-10}) / (1.91 \times 10^{-2})\} \times 100 = 1.31 \times 10^{-6}\%$ (less than 5%)

Assumptions $(1.91 \times 10^{-2}) - \Delta C \approx (1.91 \times 10^{-2})$ and $(1.91 \times 10^{-2}) + \Delta C \approx (1.91 \times 10^{-2})$ are

valid.

$[H_3O^+] = (1.91 \times 10^{-2}) + \Delta C \approx (1.91 \times 10^{-2}) + (2.5 \times 10^{-10}) \approx (1.91 \times 10^{-2})$

$[HGly^-] = (1.91 \times 10^{-2}) - \Delta C) \approx (1.91 \times 10^{-2}) - (2.5 \times 10^{-10}) \approx (1.91 \times 10^{-2})$

So $[H_3O^+] \approx [HGly^-]$ for H_2Gly solution where K_{a2} is very small. $\qquad$ Answer(d)

12-116. For a weak diprotic acid such as glycine, it can be shown that:

(a) $[Gly^{2-}] \approx$ Initial concentration of H_2Gly (b) $[Gly^{2-}] \approx K_{a2}$ (c) $[Gly^{2-}] \approx [HGly^-]$

(d) $[Gly^{2-}] > [HGly^-]$ (e) $[Gly^{2-}] \approx [H_3O^+]$

$$HGly^- \ (aq) + H_2O \ (l) \rightleftharpoons H_3O^+ \ (aq) + Gly^{2-} \ (aq)$$

$K_{a2} = [H_3O^+] \ [Gly^{2-}] / [HGly^-]$

For H_2Gly solution $[H_3O^+] \approx [HGly^-]$ (reference: 12-115)

So $[Gly^{2-}] \approx K_{a2}$ $\qquad\qquad$ Answer(b)

12-117. Sulfurous acid (H_2SO_3) is an acid and sodium sulfite is a base. Which equation

correctly describes the relationship between the equlibrium constants of these compounds?

(a) $K_{a1} \times K_{b2} = K_w$ (b) $K_{a1} \times K_{a2} = K_w$ (c) $K_{a2} \times K_{b2} = K_w$

(d) $K_{a1} \times K_{a2} = 1$ (e) $K_{a1} \times K_{b2} = 1$

$$H_2SO_3 \ (aq) + H_2O \ (l) \ \rightleftharpoons \ H_3O^+ \ (aq) + HSO_3^- \ (aq) \quad K_{a1} = \{[H_3O^+] \ [HSO_3^-]\}/[H_2SO_3]$$

$$HSO_3^- \ (aq) + H_2O \ (l) \ \rightleftharpoons \ H_3O^+ \ (aq) + SO_3^{2-} \ (aq) \quad K_{a2} = \{[H_3O^+] \ [SO_3^{2-}]\}/[HSO_3^-]$$

$$H_2O$$

$$Na_2SO_3 \longrightarrow 2\,Na^+\,(aq) + SO_3^{2-}\,(aq)$$

$SO_3^{2-}\,(aq) + H_2O\,(l) \rightleftharpoons HSO_3^-\,(aq) + OH^-\,(aq)$ $K_{b1} = \{[HSO_3^-][OH^-]\}/[SO_3^{2-}]$

$HSO_3^-\,(aq) + H_2O\,(l) \rightleftharpoons H_2SO_3\,(aq) + OH^-\,(aq)$ $K_{b2} = \{[H_2SO_3][OH^-]\}/[HSO_3^-]$

$K_{a1} = \{[H_3O^+][HSO_3^-]\}/[H_2SO_3]$

$\quad = [H_3O^+][OH^-] \times \{[HSO_3^-]/[H_2SO_3][OH^-]\}$

$\quad = K_w \times \{(1)/(K_{b2})\} = (K_w)/(K_{b2})$

$K_{a1} \times K_{b2} = K_w$ Answer(a)

12-118. Use the relationship between K_{a1} and K_{a2} for H_2SO_3 and K_{b1} and K_{b2} for the SO_3^{2-} ion

to predict whether a 0.10 M solution of the HSO_3^- (bisulfite) ion would be acidic, basic or

neutral.

(H_2SO_3: $K_{a1} = 1.7 \times 10^{-2}$, $K_{a2} = 6.4 \times 10^{-8}$)

   (a) The solution would be strongly acidic, pH < 2
   (b) The solution would be moderately acidic, pH $\approx$ 4
   (c) The solution would be very close to neutral, pH $\approx$ 7
   (d) The solution would be slightly basic, pH $\approx$ 8
   (e) The solution would be strongly basic, pH > 12

$H_2SO_3\,(aq) + H_2O\,(l) \rightleftharpoons H_3O^+\,(aq) + HSO_3^-\,(aq)$ $K_{a1} = \{[H_3O^+][HSO_3^-]\}/[H_2SO_3]$

$HSO_3^-\,(aq) + H_2O\,(l) \rightleftharpoons H_3O^+\,(aq) + SO_3^{2-}\,(aq)$ $K_{a2} = \{[H_3O^+][SO_3^{2-}]\}/[HSO_3^-]$

$SO_3^{2-}\,(aq) + H_2O\,(l) \rightleftharpoons HSO_3^-\,(aq) + OH^-\,(aq)$ $K_{b1} = \{[HSO_3^-][OH^-]\}/[SO_3^{2-}]$

$HSO_3^-\,(aq) + H_2O\,(l) \rightleftharpoons H_2SO_3\,(aq) + OH^-\,(aq)$ $K_{b2} = \{[H_2SO_3][OH^-]\}/[HSO_3^-]$

$K_{a1} = \{[H_3O^+][HSO_3^-]\}/[H_2SO_3] = [H_3O^+][OH^-] \times \{[HSO_3^-]/[H_2SO_3][OH^-]$

$\quad = K_w \times \{(1)/(K_{b2})\} = (K_w)/(K_{b2})$

$K_{a1} \times K_{b2} = K_w$

$K_{a2} = 6.4 \times 10^{-8}$

$K_{b2} = K_w/K_{a1} = (10^{-14})/(1.7 \times 10^{-2}) = 5.88 \times 10^{-13}$

$HSO_3^-\,(aq) + H_2O\,(l) \rightleftharpoons H_3O^+\,(aq) + SO_3^{2-}\,(aq)$ $K_{a2} = 6.4 \times 10^{-8}$

$HSO_3^-\,(aq) + H_2O\,(l) \rightleftharpoons H_2SO_3\,(aq) + OH^-\,(aq)$ $K_{b2} = 5.88 \times 10^{-13}$

Since the $K_{a2} > K_{b2}$ we expect that the solution will be acidic.

$$HSO_3^- \text{ (aq)} + H_2O \text{ (l)} \rightleftharpoons H_2SO_3 \text{ (aq)} + OH^- \text{ (aq)}$$

At equilibrium: $0.1 - \Delta C$ ΔC ΔC

Assumption: $(0.1 - \Delta C) \approx 0.1$

$K_{b2} = (\Delta C)^2/(0.1 - \Delta C) \approx (\Delta C)^2/(0.1)$

$(\Delta C)^2/(0.1) \approx 5.88 \times 10^{-13}$

$(\Delta C)^2 \approx 5.88 \times 10^{-14}$

$\Delta C \approx 2.42 \times 10^{-7}$

$\{(2.42 \times 10^{-7})/0.1\} \times 100 = 2.42 \times 10^{-4}\%$ (less than 5%) Assumption $(0.1 - \Delta C) \approx 0.1$ is valid.

$[OH^-] \approx 2.42 \times 10^{-7}$

$$HSO_3^- \text{ (aq)} + H_2O \text{ (l)} \rightleftharpoons H_3O^+ \text{ (aq)} + SO_3^{2-} \text{ (aq)}$$

At equilibrium: $0.1 - \Delta C$ ΔC ΔC

Assumption: $(0.1 - \Delta C) \approx 0.1$

$K_{b2} = (\Delta C)^2/(0.1 - \Delta C) \approx (\Delta C)^2/(0.1)$

$(\Delta C)^2/(0.1) \approx 6.4 \times 10^{-8}$

$(\Delta C)^2 \approx 6.4 \times 10^{-9}$

$\Delta C \approx 8.0 \times 10^{-5}$

$\{(8.0 \times 10^{-5})/0.1\} \times 100 = 8.0 \times 10^{-2}\%$ (less than 5%)   Assumption $(0.1 - \Delta C) \approx 0.1$ is valid.

$[H_3O^+] \approx 8.0 \times 10^{-5}$

The solution will be moderately acidic , pH ≈ 4.   Answer(b)

12-119. What is the pH of a 0.1 M H_2CO_3 solution? (H_2CO_3 : $K_{a1} = 4.2 \times 10^{-7}$, $K_{a2} = 4.7 \times 10^{-11}$)

   (a) 3.1   (b) 3.7   (c) 6.3   (d) 6.8   (e) None of the above

$$H_2CO_3 \text{ (aq)} + H_2O \text{ (l)} \rightleftharpoons H_3O^+ \text{ (aq)} + HCO_3^- \text{ (aq)}$$

At equilibrium: $0.1 - \Delta C$ ΔC ΔC

Assumption: $(0.1 - \Delta C) \approx 0.1$

$K_{a1} = [H_3O^+] [HCO_3^-]/[H_2CO_3]$

$K_{a1} = (\Delta C)^2/ (0.1 - \Delta C) \approx (\Delta C)^2/ (0.1)$

$(\Delta C)^2/ (0.1) \approx 4.2 \times 10^{-7}$

$(\Delta C)^2 \approx 4.2 \times 10^{-8}$

$\Delta C \approx 2.05 \times 10^{-4}$

$\{(2.05 \times 10^{-4})/0.1\} \times 100 = 0.205\%$ (less than 5%)   Assumption $(0.1 - \Delta C) \approx 0.1$ is valid.

$[H_3O^+] \approx 2.05 \times 10^{-4}$

$[HCO_3^-] \approx 2.05 \times 10^{-4}$

$$HCO_3^- \ (aq) + H_2O \ (l) \rightleftharpoons H_3O^+ \ (aq) \quad + \quad CO_3^{2-} \ (aq)$$

At equilibrium: $(2.05 \times 10^{-4}) - \Delta C$ $(2.05 \times 10^{-4}) + \Delta C$ ΔC

Assumptions: $(2.05 \times 10^{-4}) - \Delta C \approx (2.05 \times 10^{-4})$ and $(2.05 \times 10^{-4}) + \Delta C \approx (2.05 \times 10^{-4})$

$K_{a2} = \{(2.05 \times 10^{-4}) + \Delta C\} \Delta C / \{(2.05 \times 10^{-4}) - \Delta C\} \approx \Delta C$

$\Delta C \approx 4.7 \times 10^{-11}$

$\{(4.7 \times 10^{-11})/(2.05 \times 10^{-4})\} \times 100 = 2.29 \times 10^{-5}\%$ (less than 5%)

Assumptions $(2.05 \times 10^{-4}) - \Delta C \approx (2.05 \times 10^{-4})$ and $(2.05 \times 10^{-4}) + \Delta C \approx (2.05 \times 10^{-4})$ are

valid.

$[H_3O^+] \approx (2.05 \times 10^{-4}) + \Delta C \approx \{(2.05 \times 10^{-4}) + (4.7 \times 10^{-11})\} \approx (2.05 \times 10^{-4})$

$Log \ [H_3O^+] = -3.7$ $pH = 3.7$ Answer(b)

12-120. A diprotic acid, H_2A has the following dissociation constants: $K_{a1} = 2.1 \times 10^{-7}$ and

$K_{a2} = 4.3 \times 10^{-13}$. The H_3O^+ ion concentration when this acid is dissolved in water is:

   (a) Roughly equal to the H_2A concentration
   (b) Much larger than the H_2A concentration
   (c) Much larger than the HA^- concentration
   (d) Approximately equal to the HA^- concentration
   (e) Much less than the HA^- concentration

For example let us consider a 0.10 M H_2A solution.

$$H_2A \ (aq) + H_2O \ (l) \rightleftharpoons H_3O^+ \ (aq) + HA^- \ (aq)$$

At equilibrium: $0.1 - \Delta C$ ΔC ΔC

Assumption: $(0.1 - \Delta C) \approx 0.1$

$K_{a1} = [H_3O^+] [HA^-]/[H_2A] = (\Delta C)^2/ (0.1 - \Delta C) \approx (\Delta C)^2/ (0.1)$

$(\Delta C)^2/ (0.1) \approx 2.1 \times 10^{-7}$

$(\Delta C)^2 \approx 2.1 \times 10^{-8}$

$(\Delta C) \approx 1.45 \times 10^{-4}$

$\{(1.45 \times 10^{-4})/0.1\} \times 100 = 0.145\%$ (less than 5%)   Assumption $(0.1 - \Delta C) \approx 0.1$ is valid.

$[H_3O^+] \approx 1.45 \times 10^{-4}$

$[HA^-] \approx 1.45 \times 10^{-4}$

$$HA^- (aq) + H_2O (l) \rightleftharpoons H_3O^+ (aq) + A^{2-} (aq)$$

At equilibrium: $\quad\quad\quad (1.45 \times 10^{-4}) - \Delta C \quad\quad\quad\quad (1.45 \times 10^{-4}) + \Delta C \quad \Delta C$

Assumptions: $(1.45 \times 10^{-4}) - \Delta C \approx (1.45 \times 10^{-4})$ and $(1.45 \times 10^{-4}) + \Delta C \approx (1.45 \times 10^{-4})$

$K_{a2} = \{(1.45 \times 10^{-4}) + \Delta C\} \Delta C / \{(1.45 \times 10^{-4}) - \Delta C\} \approx \Delta C$

$\Delta C \approx 4.3 \times 10^{-13}$

$\{(4.3 \times 10^{-13})/(1.45 \times 10^{-4})\} \times 100 = 2.97 \times 10^{-7}$ (less than 5%)

Assumptions $(1.45 \times 10^{-4}) - \Delta C \approx (1.45 \times 10^{-4})$ and $(1.45 \times 10^{-4}) + \Delta C \approx (1.45 \times 10^{-4})$ are

valid.

$[H_3O^+] \approx (1.45 \times 10^{-4}) + \Delta C \approx \{(1.45 \times 10^{-4}) + (4.3 \times 10^{-13})\} \approx (1.45 \times 10^{-4})$

$[HA^-] \approx (1.45 \times 10^{-4}) - \Delta C \approx \{(1.45 \times 10^{-4}) - (4.3 \times 10^{-13})\} \approx (1.45 \times 10^{-4})$

So for this diprotic acid $[H_3O^+] \approx [HA^-]$ $\quad\quad\quad\quad\quad\quad\quad\quad\quad$ Answer(d)

12-121. For a diprotic acid, H_2A for which $K_{a1} = 2.1 \times 10^{-7}$ and $K_{a2} = 4.3 \times 10^{-13}$, the $[A^{2-}]$ at

equilibrium will be:

  (a) Approximately equal to the initial concentration of H_2A
  (b) Roughly equal to K_{a2}
  (c) Roughly equal to the HA^- concentration
  (d) Much larger than the HA^- concentration
  (e) Approximately equal to the H_3O^+ concentration

$K_{a1} = 2.1 \times 10^{-7}$ and $K_{a2} = 4.3 \times 10^{-13}$

For this diprotic acid $[H_3O^+] \approx [HA^-]$ $\quad$ (Reference: 12-120)

$$HA^- (aq) + H_2O (l) \rightleftharpoons H_3O^+ (aq) + A^{2-} (aq)$$

$K_{a2} = [H_3O^+] [A^{2-}]/[HA^-] \approx [A^{2-}]$

$[A^{2-}] \approx K_{a2}$ $\quad\quad\quad\quad\quad\quad\quad\quad\quad\quad\quad\quad\quad\quad\quad\quad\quad$ Answer(b)

12-122. Which of the following equations can be used to calculate K_{b1} and K_{b2} for Na_2Gly from

K_{a1} and K_{a2} for H_2Gly?

  (a) $K_{b1} = K_w \times K_{a1}$ and $K_{b2} = K_w \times K_{a2}$
  (b) $K_{b1} = K_w/K_{a1}$ and $K_{b2} = K_w/K_{a2}$
  (c) $K_{b1} = K_{a1}/K_w$ and $K_{b2} = K_{a1}/K_w$

(d) $K_{b1} = K_w \times K_{a2}$ and $K_{b2} = K_w \times K_{a1}$

(e) $K_{b1} = K_w/K_{a2}$ and $K_{b2} = K_w/K_{a1}$

$H_2Gly \ (aq) + H_2O \ (l) \rightleftharpoons H_3O^+ \ (aq) + HGly^- \ (aq) \quad K_{a1} = [H_3O^+] \ [HGly^-]/[H_2Gly]$

$HGly^- \ (aq) + H_2O \ (l) \rightleftharpoons H_3O^+ \ (aq) + Gly^{2-} \ (aq) \quad K_{a2} = [H_3O^+] \ [Gly^{2-}]/[HGly^-]$

$$H_2O$$

$$Na_2Gly \longrightarrow 2 \ Na^+ \ (aq) + Gly^{2-} \ (aq)$$

$Gly^{2-} \ (aq) + H_2O \ (l) \rightleftharpoons HGly^- \ (aq) + OH^- \ (aq) \qquad K_{b1} = [HGly^-] \ [OH^-]/[Gly^{2-}]$

$HGly^- \ (aq) + H_2O \ (l) \rightleftharpoons H_2Gly \ (aq) + OH^- \ (aq) \qquad K_{b2} = [H_2Gly] \ [OH^-]/[HGly^-]$

$K_{a1} \times K_{b2} = \{[H_3O^+] \ [HGly^-]/[H_2Gly]\} \times \{[H_2Gly] \ [OH^-]/[HGly^-]\} = [H_3O^+] \ [OH^-] = K_w$

$K_{b2} = K_w/K_{a1}$

$K_{a2} \times K_{b1} = \{[H_3O^+] \ [Gly^{2-}]/[HGly^-]\} \times \{[HGly^-] \ [OH^-]/[Gly^{2-}]\} = [H_3O^+] \ [OH^-] = K_w$

$K_{b1} = K_w/K_{a2}$ Answer(e)

12-123. For H_2CO_3, $K_{a1} = 4.2 \times 10^{-7}$, $K_{a2} = 4.7 \times 10^{-11}$. Which of the following numbers is closest to the pH of a 0.200 M Na_2CO_3 solution?

(a) 5.9×10^{-3} (b) 2.2 (c) 2.48 (d) 11.6 (e) 11.8

$H_2CO_3 \ (aq) + H_2O \ (l) \rightleftharpoons H_3O^+ \ (aq) + HCO_3^- \ (aq) \quad K_{a1} = [H_3O^+] \ [HCO_3^-]/[H_2CO_3]$

$HCO_3^- \ (aq) + H_2O \ (l) \rightleftharpoons H_3O^+ \ (aq) + CO_3^{2-} \ (aq) \quad K_{a2} = [H_3O^+] \ [CO_3^{2-}]/[HCO_3^-]$

$$H_2O$$

$$Na_2CO_3 \longrightarrow 2 \ Na^+ \ (aq) + CO_3^{2-} \ (aq)$$

$CO_3^{2-} \ (aq) + H_2O \ (l) \rightleftharpoons HCO_3^- \ (aq) + OH^- \ (aq) \qquad K_{b1} = [HCO_3^-] \ [OH^-]/[CO_3^{2-}]$

$HCO_3^- \ (aq) + H_2O \ (l) \rightleftharpoons H_2CO_3 \ (aq) + OH^- \ (aq) \qquad K_{b2} = [H_2CO_3] \ [OH^-]/[HCO_3^-]$

$K_{b1} = K_w/K_{a2} = (10^{-14})/(4.7 \times 10^{-11}) = 2.13 \times 10^{-4}$

$K_{b2} = K_w/K_{a1} = (10^{-14})/(4.2 \times 10^{-7}) = 2.38 \times 10^{-8}$

$$CO_3^{2-} \ (aq) + H_2O \ (l) \rightleftharpoons HCO_3^- \ (aq) + OH^- \ (aq)$$

At equilibrium: $\qquad\qquad\qquad 0.2 - \Delta C \qquad\qquad\qquad \Delta C \qquad \Delta C$

Assumption: $(0.2 - \Delta C) \approx 0.2$

$K_{b1} = (\Delta C)^2/(0.2 - \Delta C) \approx (\Delta C)^2/(0.2)$

$(\Delta C)^2/(0.2) \approx 2.13 \times 10^{-4}$

$(\Delta C)^2 \approx 0.43 \times 10^{-4}$

$\Delta C \approx 6.56 \times 10^{-3}$

$\{(6.56 \times 10^{-3})/(0.2)\} \times 100 = 3.28 \%$ (less than 5%) Assumption $(0.2 - \Delta C) \approx 0.2$ is valid.

$[OH^-] \approx 6.56 \times 10^{-3}$

$[HCO_3^-] \approx 6.56 \times 10^{-3}$

$$HCO_3^- \text{ (aq)} + H_2O \text{ (l)} \rightleftharpoons H_2CO_3 \text{ (aq)} + OH^- \text{ (aq)}$$

At equilibrium: $\quad\quad\quad (6.56 \times 10^{-3}) - \Delta C \quad\quad\quad\quad \Delta C \quad\quad\quad\quad (6.56 \times 10^{-4}) + \Delta C$

Assumptions: $(6.56 \times 10^{-3}) - \Delta C \approx (6.56 \times 10^{-3})$ and $(6.56 \times 10^{-3}) + \Delta C \approx (6.56 \times 10^{-3})$

$K_{a2} = \{(6.56 \times 10^{-3}) + \Delta C\} \Delta C / \{(6.56 \times 10^{-3}) - \Delta C\} \approx \Delta C$

$\Delta C \approx 2.38 \times 10^{-8}$

$\{(2.38 \times 10^{-8})/(6.56 \times 10^{-3})\} \times 100 = 3.62 \times 10^{-4} \%$ (less than 5%)

Assumptions $(6.56 \times 10^{-3}) - \Delta C \approx (6.56 \times 10^{-3})$ and $(6.56 \times 10^{-3}) + \Delta C \approx (6.56 \times 10^{-3})$ are

valid. $[OH^-] \approx (6.56 \times 10^{-3}) + \Delta C \approx \{(6.56 \times 10^{-3}) + (2.38 \times 10^{-8})\} \approx (6.56 \times 10^{-3})$

$Log [OH^-] \approx -2.18 \quad pOH \approx 2.18 \quad pH \approx 11.82 \approx 11.8 \quad\quad\quad\quad\quad$ Answer(e)

12-124. In a 0.10 M solution of H_2Se, the Se^{2-} concentration is (H_2Se : $K_{a1} = 1.7 \times 10^{-4}$,

$K_{a2} = 1.0 \times 10^{-10}$):

   (a) $[Se^{2-}] = 0.10 \times K_{a2} = 1.0 \times 10^{-11}$ (b) $[Se^{2-}] \approx 1.0 \times 10^{-10}$

   (c) $[Se^{2-}] = (0.10 \times K_{a2})^{\frac{1}{2}} = 3.2 \times 10^{-6}$ (d) $[Se^{2-}] = (0.10 \times K_{a1})^{\frac{1}{2}} = 4.1 \times 10^{-3}$

   (e) None of the above

$$H_2Se \text{ (aq)} + H_2O \text{ (l)} \rightleftharpoons H_3O^+ \text{ (aq)} + HSe^- \text{ (aq)}$$

At equilibrium: $\quad\quad\quad\quad\quad\quad\quad\quad 0.1 - \Delta C \quad\quad\quad\quad \Delta C \quad\quad \Delta C$

Assumption: $(0.1 - \Delta C) \approx 0.1$

$K_{a1} = [H_3O^+] [HSe^-]/[H_2Se]$

$K_{a1} = (\Delta C)^2 / (0.1 - \Delta C) \approx (\Delta C)^2 / (0.1)$

$(\Delta C)^2 / (0.1) \approx 1.7 \times 10^{-4}$

$(\Delta C)^2 \approx 1.7 \times 10^{-5}$

$(\Delta C) \approx 4.12 \times 10^{-3}$

$\{(4.12 \times 10^{-3})/0.1\} \times 100 = 4.12\%$ (less than 5%) $\quad\quad$ Assumption $(0.1 - \Delta C) \approx 0.1$ is valid.

$[H_3O^+] \approx 4.12 \times 10^{-3}$

$[HSe^-] \approx 4.12 \times 10^{-3}$

$$HSe^- \text{ (aq)} + H_2O \text{ (l)} \rightleftharpoons H_3O^+ \text{ (aq)} + Se^{2-} \text{ (aq)}$$

At equilibrium: $(4.12 \times 10^{-3}) - \Delta C$ $(4.12 \times 10^{-3}) + \Delta C$ ΔC

Assumptions: $(4.12 \times 10^{-3}) - \Delta C \approx (4.12 \times 10^{-3})$ and $(4.12 \times 10^{-3}) + \Delta C \approx (4.12 \times 10^{-3})$

$K_{a2} = \{(4.12 \times 10^{-3}) + \Delta C)\} \Delta C / \{(4.12 \times 10^{-3}) - \Delta C\} \approx \Delta C$

$\Delta C \approx 1.0 \times 10^{-10}$

$\{(1.0 \times 10^{-10})/(4.12 \times 10^{-3})\} \times 100 = 2.43 \times 10^{-6}$ % (less than 5%)

Assumptions $(4.12 \times 10^{-3}) - \Delta C \approx (4.12 \times 10^{-3})$ and $(4.12 \times 10^{-3}) + \Delta C \approx (4.12 \times 10^{-3})$ are

valid. $[Se^{2-}] \approx 1.0 \times 10^{-10}$ Answer(b)

12-125. If 0.2 moles of sodium selenide is dissolved in 1.0 L of water, the base ionization

constant for the Se^{2-} would be $(H_2Se : K_{a1} = 1.7 \times 10^{-4}, K_{a2} = 1.0 \times 10^{-10})$:

   (a) $K_{b1} = K_{a1} = 1.7 \times 10^{-4}$ (b) $K_{b1} = K_{a2} = 1.0 \times 10^{-10}$ (c) $K_{b1} = K_w/K_{a1} = 5.9 \times 10^{-11}$

   (d) $K_{b1} = K_w/K_{a2} = 1.0 \times 10^{-4}$ (e) None of the above

$$H_2Se \text{ (aq)} + H_2O \text{ (l)} \rightleftharpoons H_3O^+ \text{ (aq)} + HSe^- \text{ (aq)} \quad K_{a1} = [H_3O^+][HSe^-]/[H_2Se]$$

$$HSe^- \text{ (aq)} + H_2O \text{ (l)} \rightleftharpoons H_3O^+ \text{ (aq)} + Se^{2-} \text{ (aq)} \quad K_{a2} = [H_3O^+][Se^{2-}]/[HSe^-]$$

$$H_2O$$

$$Na_2Se \longrightarrow 2 Na^+ \text{ (aq)} + Se^{2-} \text{ (aq)}$$

$$Se^{2-} \text{ (aq)} + H_2O \text{ (l)} \rightleftharpoons HSe^- \text{ (aq)} + OH^- \text{ (aq)} \quad K_{b1} = [HSe^-][OH^-]/[Se^{2-}]$$

$$HSe^- \text{ (aq)} + H_2O \text{ (l)} \rightleftharpoons H_2Se \text{ (aq)} + OH^- \text{ (aq)} \quad K_{b2} = [H_2Se][OH^-]/[HSe^-]$$

K_{b1} is the base ionization constant of Se^{2-} ion.

$K_{b1} \times K_{a2} = \{[HSe^-][OH^-]/[Se^{2-}]\} \times \{[H_3O^+][Se^{2-}]/[HSe^-]\} = [H_3O^+][OH^-] = K_w$

$K_{b1} = K_w/K_{a2} = (10^{-14})/(1.0 \times 10^{-10}) = 10^{-4}$ Answer(d)

12-126. If sodium selenide is added to water, we would expect the pH to $(H_2Se :$

$K_{a1} = 1.7 \times 10^{-4}, K_{a2} = 1.0 \times 10^{-10})$:

   (a) Increase   (b) Decrease   (c) Remain the same

   (d) Change in a direction that can't be predicted from this information

$$Se^{2-} \text{ (aq)} + H_2O \text{ (l)} \rightleftharpoons HSe^- \text{ (aq)} + OH^- \text{ (aq)} \quad K_{b1} = [HSe^-][OH^-]/[Se^{2-}]$$

$HSe^- (aq) + H_2O (l) \rightleftharpoons H_2Se (aq) + OH^- (aq)$ $K_{b2} = [H_2Se] [OH^-]/[HSe^-]$

$K_{b1} = K_w/K_{a2} = (10^{-14})/(1.0 \times 10^{-10}) = 1.0 \times 10^{-4}$

$K_{b2} = K_w/K_{a1} = (10^{-14})/(1.7 \times 10^{-4}) = 5.88 \times 10^{-11}$

For example let us add 0.1 mole of sodium selenide to 1.0 L of water.

$$Se^{2-} (aq) + H_2O (l) \rightleftharpoons HSe^- (aq) + OH^- (aq)$$

At equilibrium: $0.1 - \Delta C$ ΔC ΔC

Assumption: $(0.1 - \Delta C) \approx 0.1$

$K_{b1} = (\Delta C)^2/(0.1 - \Delta C) \approx (\Delta C)^2/(0.1)$

$(\Delta C)^2/(0.1) \approx 1.0 \times 10^{-4}$

$(\Delta C)^2 \approx 1.0 \times 10^{-5}$

$\Delta C \approx 3.16 \times 10^{-3}$

$\{(3.16 \times 10^{-3})/(0.1)\} \times 100 = 3.16 \%$ (less than 5%) Assumption $(0.1 - \Delta C) \approx 0.1$ is valid.

$[OH^-] \approx 3.16 \times 10^{-3}$

$[HSe^-] \approx 3.16 \times 10^{-3}$

$$HSe^- (aq) + H_2O (l) \rightleftharpoons H_2Se (aq) + OH^- (aq)$$

At equilibrium: $(3.16 \times 10^{-3}) - \Delta C$ ΔC $(3.16 \times 10^{-3}) + \Delta C$

Assumptions: $(3.16 \times 10^{-3}) - \Delta C \approx (3.16 \times 10^{-3})$ and $(3.16 \times 10^{-3}) + \Delta C \approx (3.16 \times 10^{-3})$

$K_{b2} = \{(3.16 \times 10^{-3}) + \Delta C\} \Delta C/\{(3.16 \times 10^{-3}) - \Delta C\} \approx \Delta C$

$\Delta C \approx 5.88 \times 10^{-11}$

$\{(5.88 \times 10^{-11})/(3.16 \times 10^{-3})\} \times 100 = 1.86 \times 10^{-6} \%$ (less than 5%)

Assumptions $(3.16 \times 10^{-3}) - \Delta C \approx (3.16 \times 10^{-3})$ and $(3.16 \times 10^{-3}) + \Delta C \approx (3.16 \times 10^{-3})$ are

valid.

$[OH^-] \approx (3.16 \times 10^{-3}) + \Delta C \approx \{(3.16 \times 10^{-3}) + (5.88 \times 10^{-11})\} \approx (3.16 \times 10^{-3})$

$Log [OH^-] \approx -2.5$ $pOH \approx 2.5$ $pH \approx 11.5$

Initial pH = 7.0

pH after addition of sodium selenide ≈ 11.5

Addition of sodium selenide will increase the pH of water.   Answer(a)

12-127. Hydrogen sulfide (H_2S) is an acid, and sodium sulfide (Na_aS) is a base. H_2S:

$K_{a1} = 1.0 \times 10^{-7}$, $K_{a2} = 1.3 \times 10^{-13}$

Which of the following statements concerning NaHS is correct?

  (a) NaHS is an acid because K_{a1} for H_2S is much larger than K_{a2}
  (b) NaHS is an acid because K_{a1} for H_2S is smaller than K_{b1} for Na_2S
  (c) NaHS is a base because K_{b1} for Na_2S is much larger than K_{b2}
  (d) NaHS is a base because K_{b2} for Na_2S is larger than K_{a2} for H_2S
  (e) It is impossible to tell whether NaHS is an acid or a base from these data

$H_2S \text{ (aq)} + H_2O \text{ (l)} \rightleftharpoons H_3O^+ \text{ (aq)} + HS^- \text{ (aq)}$ $K_{a1} = [H_3O^+][HS^-]/[H_2S]$

$HS^- \text{ (aq)} + H_2O \text{ (l)} \rightleftharpoons H_3O^+ \text{ (aq)} + S^{2-} \text{ (aq)}$ $K_{a2} = [H_3O^+][S^{2-}]/[HS^-]$

$$Na_2S \xrightarrow{H_2O} 2\,Na^+ \text{ (aq)} + S^{2-} \text{ (aq)}$$

$S^{2-} \text{ (aq)} + H_2O \text{ (l)} \rightleftharpoons HS^- \text{ (aq)} + OH^- \text{ (aq)}$ $K_{b1} = [HS^-][OH^-]/[S^{2-}]$

$HS^- \text{ (aq)} + H_2O \text{ (l)} \rightleftharpoons H_2S \text{ (aq)} + OH^- \text{ (aq)}$ $K_{b2} = [H_2S][OH^-]/[HS^-]$

$K_{b2} = K_w/K_{a1} = (10^{-14})/(1.0 \times 10^{-7}) = 1.0 \times 10^{-7}$

NaHS can act as an acid or a base.

$$NaHS \xrightarrow{H_2O} Na^+ \text{ (aq)} + HS^- \text{ (aq)}$$

$HS^- \text{ (aq)} + H_2O \text{ (l)} \rightleftharpoons H_3O^+ \text{ (aq)} + S^{2-} \text{ (aq)}$ $K_{a2} = 1.3 \times 10^{-13}$

$HS^- \text{ (aq)} + H_2O \text{ (l)} \rightleftharpoons H_2S \text{ (aq)} + OH^- \text{ (aq)}$ $K_{b2} = 1.0 \times 10^{-7}$

NaHS will be a base because $K_{b2} > K_{a2}$.   Answer(d)

12-128. H_2Gly is an acid and Na_2Gly ia a base. On the basis of the values of K_{a1} and K_{a2} for H_2Gly and the values of K_{b1} and K_{b2} for Na_2Gly, would an 0.10 M NaHGly solution be acidic, basic, or neutral? (H_2Gly: $K_{a1} = 4.5 \times 10^{-3}$, $K_{a2} = 2.5 \times 10^{-10}$)

  (a) Slightly acidic   (b) Slightly basic   (c) Neutral (pH = 7.0)

  (d)There is no way to predict whether NaHGly is an acid or base

$H_2Gly \text{ (aq)} + H_2O \text{ (l)} \rightleftharpoons H_3O^+ \text{ (aq)} + HGly^- \text{ (aq)}$ $K_{a1} = [H_3O^+][HGly^-]/[H_2Gly]$

$HGly^- \text{ (aq)} + H_2O \text{ (l)} \rightleftharpoons H_3O^+ \text{ (aq)} + Gly^{2-} \text{ (aq)}$ $K_{a2} = [H_3O^+][Gly^{2-}]/[HGly^-]$

$$Na_2Gly \xrightarrow{H_2O} 2\,Na^+ \text{ (aq)} + Gly^{2-} \text{ (aq)}$$

Gly^{2-} (aq) + H_2O (l) $\rightleftharpoons$ $HGly^-$ (aq) + OH^- (aq) $K_{b1} = [HGly^-][OH^-]/[Gly^{2-}]$

$HGly^-$ (aq) + H_2O (l) $\rightleftharpoons$ H_2Gly (aq) + OH^- (aq) $K_{b2} = [H_2Gly][OH^-]/[HGly^-]$

$K_{a1} \times K_{b2} = \{[H_3O^+][HGly^-]/[H_2Gly]\} \times \{[H_2Gly][OH^-]/[HGly^-]\} = [H_3O^+][OH^-] = K_w$

$K_{b2} = K_w/K_{a1} = (10^{-14})/(4.5 \times 10^{-3}) = 2.2 \times 10^{-12}$

$K_{a2} \times K_{b1} = \{[H_3O^+][Gly^{2-}]/[HGly^-]\} \times \{[HGly^-][OH^-]/[Gly^{2-}]\} = [H_3O^+][OH^-] = K_w$

$K_{b1} = K_w/K_{a2} = (10^{-14})/(2.5 \times 10^{-10}) = 4.0 \times 10^{-5}$

NaHGly can act as an acid or a base.

$$H_2O$$

NaHGly $\longrightarrow$ Na^+ (aq) + $HGly^-$ (aq)

$HGly^-$ (aq) + H_2O (l) $\rightleftharpoons$ H_3O^+ (aq) + Gly^{2-} (aq) $K_{a2} = 2.5 \times 10^{-10}$

$HGly^-$ (aq) + H_2O (l) $\rightleftharpoons$ H_2Gly (aq) + OH^- (aq) $K_{b2} = 2.2 \times 10^{-12}$

0.10 M NaHGly solution will be slightly acidic because $K_{a2} > K_{b2}$.   Answer(a)

CHAPTER 13

SOLUBILITY AND COMPLEX ION EQUILIBRIA

13-1. Calculate the solubility of CaF_2 in pure water. K_{sp} $(CaF_2) = 4.0 \times 10^{-11}$

Let 'Y' be the solubility of CaF_2 in mole per liter.

$$CaF_2 (S) \rightleftharpoons Ca^{2+} (aq) + 2F^- (aq)$$

Concentration: Y   Y   2Y

$K_{sp} = [Ca^{2+}] [F^-]^2 = (Y) (2Y)^2 = 4Y^3 = 4.0 \times 10^{-11}$

$Y^3 = 10^{-11}$

$Y = 2.15 \times 10^{-4}$ mole per liter

Molecular weight of $CaF_2 = 40.08 + (19 \times 2) = 78.08$

Solubility of $CaF_2 = 2.15 \times 10^{-4} \times 78.08$ gram per liter $= 0.01679$ gram per liter

13-2. Calculate the solubility in mole per liter of silver sulfide in order to decide whether it is accurately labelled when described as an insoluble salt. K_{sp} $(Ag_2S) = 6.3 \times 10^{-50}$

Ler 'Y' be the solubility of Ag_2S in moles per liter.

$$Ag_2S (s) \rightleftharpoons 2Ag^+ (aq) + S^{2-} (aq)$$

Concentration: Y   2Y   Y

$K_{sp} = [Ag^+]^2 [S^{2-}] = (2Y)^2(Y) = 4Y^3 = 6.3 \times 10^{-50}$

$Y^3 = 1.575 \times 10^{-50}$

$Y = 2.5 \times 10^{-7}$ mole per liter

$Y < 0.001$ M. It is correct to say that the salt is an insoluble salt.

13-3. Determine which salt – $CaCO_3$ or Ag_2CO_3 – is more soluble in water in units of mole per liter . K_{sp} $(CaCO_3) = 2.8 \times 10^{-9}$ and K_{sp} $(Ag_2CO_3) = 8.1 \times 10^{-12}$ $CaCO_3$:

Let 'Y' be the solubility of $CaCO_3$ in mole per liter.

$$CaCO_3 (s) \rightleftharpoons Ca^{2+} (aq) + CO_3^{2-} (aq)$$

Concentration: Y   Y   Y

$K_{sp} = [Ca^{2+}] [CO_3^{2-}] = Y^2 = 2.8 \times 10^{-9}$

$Y = 5.3 \times 10^{-5}$ mole per liter

Ag_2CO_3:

Let 'Z' be the solubility of Ag_2CO_3 in moles per liter.

$$Ag_2CO_3 \text{ (s)} \rightleftharpoons 2Ag^+ \text{ (aq)} + CO_3^{2-} \text{ (aq)}$$

Concentration: Z   2Z   Z

$K_{sp} = [Ag^+]^2[CO_3^{2-}] = (2Z)^2(Z) = 4Z^3 = 8.1 \times 10^{-12}$

$Z^3 = 2.025 \times 10^{-12}$

$Z = 1.265 \times 10^{-4}$ mole per liter

Ag_2CO_3 is more soluble than $CaCO_3$ in units of mole per liter

13-4. Calculate the solubility of AgCl in pure water. K_{sp} (AgCl) $= 1.77 \times 10^{-10}$

Let 'Y' be the solubility of AgCl in moles per liter.

$$AgCl \text{ (s)} \rightleftharpoons Ag^+ \text{ (aq)} + Cl^- \text{ (aq)}$$

Concentration: Y   Y   Y

$K_{sp} = [Ag^+] [Cl^-] = Y^2 = 1.77 \times 10^{-10}$

$Y = 1.33 \times 10^{-5}$ mole per liter

Solubility of AgCl in pure water $= 1.33 \times 10^{-5}$ mole per liter

13-5. Calculate the solubility of AgCl in each situation :

(i)   In 0.5 M NaCl, ignoring the formation of any complex ions
(ii)   In 0.5 M NaCl, taking the formation of complex ions into account, assuming that $[AgCl_2]^-$ is the only complex ion that forms in significant concentration

K_{sp} (AgCl) $= 1.77 \times 10^{-10}$ and K_f for $[AgCl_2]^- = 1.1 \times 10^5$

(i)

$K_{sp} = [Ag^+] [Cl^-]$

$[Cl^-] = 0.5$ M (The contribution due to AgCl may be neglected in comparison with 0.5 M).

$[Ag^+] = (K_{sp})/[Cl^-] = 1.77 \times 10^{-10} / [Cl^-]$

$[Ag^+] = (1.77 \times 10^{-10}) / (0.5) = 3.54 \times 10^{-10}$ mole per liter

Solubility of AgCl in 0.5 M NaCl $= 3.54 \times 10^{-10}$ mole per liter

(ii)

$$AgCl\ (s) \rightleftharpoons Ag^+\ (aq) + Cl^-\ (aq) \qquad K_{sp} = 1.77 \times 10^{-10}$$

$$\underline{Ag^+\ (aq) + 2Cl^-\ (aq) \rightleftharpoons [AgCl_2]^-\ (aq) \qquad K_f = 1.1 \times 10^5}$$

$$AgCl\ (s) + Cl^-\ (aq) \rightleftharpoons [AgCl_2]^-\ (aq) \qquad K = K_{sp} \times K_f = 1.95 \times 10^{-5}$$

Let Y be the solubility of AgCl in 0.5 M NaCl solution.

$$AgCl\ (s) + Cl^- \rightleftharpoons [AgCl_2]^-$$

At equilibrium: $\qquad\qquad\qquad\qquad\qquad\qquad\qquad 0.5 - Y \qquad\qquad Y$

$K = Y\ /(0.5 - Y) = 1.95 \times 10^{-5}$

$Y = 9.75 \times 10^{-6}\ M$

Solubility of AgCl in 0.5 M NaCl (taking the formation of complex ions into account,

assuming that $[AgCl_2]^-$ is the only complex ion that forms in significant concentration)

$= 9.75 \times 10^{-6}$ mole per liter

13-6. Calculate the pH at which $Cr(OH)_3$ starts to precipitate from a 0.10 M Cr^{3+} solution. Use

the results of this calculation to explain why it is impossible to prepare a 0.10 M Cr^{3+} solution

at neutral pH. $K_{sp}\ \{Cr(OH)_3\} = 6.3 \times 10^{-31}$

$$Cr(OH)_3\ (s) \rightleftharpoons Cr^{3+}\ (aq) + 3(OH)^-\ (aq)$$

$K_{sp} = [Cr^{3+}]\ [OH^-]^3$

$[OH^-]^3 = (\ K_{sp})/\ (Cr^{3+})$

$[Cr^{3+}] = 0.10\ M$

$[OH^-]^3 = (6.3 \times 10^{-31})\ /\ (0.10)$

$[OH^-] = 1.85 \times 10^{-10}$ mole per liter

$[H^+]\ [OH^-] = 10^{-14}$

$[H^+] = (10^{-14})/\ (1.85 \times 10^{-10}) = 0.54 \times 10^{-4}$

$pH = -\log\ (H^+) = 4.27$

When the concentration of $OH^- > 1.85 \times 10^{-10}$ mole per liter (pH > 4.27), $Q_{sp} > K_{sp}$ and $Cr(OH)_3$

will precipitate from solution.

At pH = 7, the concentration of $OH^- = 10^{-7}\ (> 1.85 \times 10^{-10})$. So it is impossible to prepare a

0.10 M Cr^{3+} solution at neutral pH.

13-7. Calculate the S^{2-} ion concentration at which MnS will begin to precipitate from a 0.10 M

Mn^{2+} solution. K_{sp} (MnS) $= 3\times10^{-13}$

$$MnS~(s) \rightleftharpoons Mn^{2+}~(aq) + S^{2-}~(aq)$$

$K_{sp} = [Mn^{2+}]~[S^{2-}]$

$[S^{2-}] = (K_{sp})~/~[Mn^{2+}] = (3\times10^{-13})~/[Mn^{2+}]$

$[Mn^{2+}] = 0.1$ M

$[S^{2-}] = (3\times10^{-13})/(0.1) = 3\times10^{-12}$ mole per liter

We can keep MnS from precipitating from a 0.10 M Mn^{2+} solution if we keep the

S^{2-} concentration $\leq 3\times10^{-12}$ mole per liter.

13-8. Calculate the pH at which a saturated solution of H_2S in water must be adjusted for the

S^{2-} concentration to be 3×10^{-12} M.

$H_2S~(aq) + H_2O~(l) \rightleftharpoons H_3O^+~(aq) + HS^-~(aq)$ $K_{a1} = 1.0\times10^{-7}$

$HS^-~(aq) + H_2O~(l) \rightleftharpoons H_3O^+~(aq) + S^{2-}~(aq)$ $K_{a2} = 1.3\times10^{-13}$

$K_{a1} = [H_3O^+]~[HS^-]/~[H_2S]$

$K_{a2} = [H_3O^+]~[S^{2-}]/~[HS^-]$

A saturated solution of H_2S in water has a concentration of 0.1 M at room temperature. In pure

water, H_2S dissociates by losing one proton at a time. But when we add strong acid or strong

base we can treat H_2S as if it dissociates two protons in a step.

$$H_2S~(aq) + 2H_2O~(l) \rightleftharpoons 2H_3O^+~(aq) + S^{2-}~(aq)$$

$K = [H_3O^+]^2~[S^{2-}]/~[H_2S] = \{[H_3O^+]~[HS^-]/~[H_2S]\} \times\{~[H_3O^+]~[S^{2-}]/~[HS^-]\}$

$= K_{a1}\times K_{a2} = (~1.0\times10^{-7})\times (1.3\times10^{-13}) = 1.3\times10^{-20}$

$$H_2S~(aq) + 2H_2O~(l) \rightleftharpoons 2H_3O^+~(aq) + S^{2-}~(aq)$$

Initial concentration: 0.1 M   0   0

At equilibrium: 0.1$-\Delta$C   ? ΔC

Assumption: $0.1 - \Delta C \approx 0.1$

$\Delta C = 3\times10^{-12}$ M

$\{(3\times10^{-12})/0.1\}\times100 = 3\times10^{-9}$% (less than 5%) Assumption $0.1 - \Delta C \approx 0.1$ is valid.

$$[H_3O^+]^2~[S^{2-}]/~[H_2S] = 1.3\times10^{-20}$$

$$[H_3O^+]^2 \approx (1.3\times10^{-20})\times(0.1)/~[S^{2-}] \approx (1.3\times10^{-21})/~[S^{2-}]$$

$$[H_3O^+]^2 \approx (1.3\times10^{-21})/ (3\times10^{-12})$$

$$[H_3O^+] \approx 2.08\times10^{-5}$$

Log $(H^+) \approx - 4.7$ pH ≈ 4.7

13-9. Calculate complex formation equilibrium constant for the two-coordinate $Fe(SCN)_2^+$ complex from the following data.

Fe^{3+} (aq) + SCN^- (aq) $\rightleftharpoons$ $Fe(SCN)^{2+}$ (aq) $K_{f1} = 890$

$Fe(SCN)^{2+}$ (aq) + SCN^- (aq) $\rightleftharpoons$ $Fe(SCN)_2^+$ (aq) $K_{f2} = 2.6$

Fe^{3+} (aq) +2(SCN)$^-$ (aq) $\rightleftharpoons$ $Fe(SCN)_2^+$ (aq)

$K_f = [Fe(SCN)_2^+]/[Fe^{3+}][SCN^-]^2$

$K_f = \{[Fe(SCN)_2^+]/[Fe(SCN)^{2+}]\ [SCN^-]\}\times\{\ [Fe(SCN)^{2+}]/[Fe^{3+}][SCN^-]\} = K_{f2}\times K_{f1} = 2.6\times890$

$= 2.314\times10^3$

13-10. Calculate the complex dissociation constant for the $Cu(NH_3)_4^{2+}$ ion from the value of K_f for this complex.

Cu^{2+} (aq) + $4NH_3$ (aq) $\rightleftharpoons$ $Cu(NH_3)_4^{2+}$ (aq) $K_f = 2.1\times10^{13}$

$Cu(NH_3)_4^{2+}$ (aq) $\rightleftharpoons$ Cu^{2+} (aq) + $4NH_3$ (aq) $K_d = ?$

$K_d = \{[Cu^{2+}][NH_3]^4\}/[Cu(NH_3)_4^{2+}]$

$K_f = [Cu(NH_3)_4^{2+}]/\{[Cu^{2+}][NH_3]^4\}$

$K_d = 1/ K_f = 1/(2.1\times10^{13}) = 4.76\times10^{-14}$

13-11. Calculate the concentration of Cu^{2+} ion in a solution that is initially 0.10 M Cu^{2+} and 1.0 M NH_3.

Cu^{2+} (aq) + $4NH_3$ (aq) $\rightleftharpoons$ $Cu(NH_3)_4^{2+}$ (aq) $K_f = 2.1\times10^{13}$

$K_f \gg 1$. So we approach the equilibrium from right to left. We create an intermediate situation in which the concentration of one of the reactants is zero.

	Cu^{2+} (aq) +	$4NH_3$ (aq) $\rightleftharpoons$	$Cu(NH_3)_4^{2+}$
Initial concentration:	0.1 M	1.0M	0
Intermediate:	(0.1 − 0.1)	{1.0 − (4×0.1)}	0.1
Intermediate:	0	0.6	0.1
At equilibrium:	ΔC	0.6 + 4ΔC	0.1 − ΔC

$K_f = (0.1 - \Delta C)/(\Delta C)(0.6 + 4\Delta C)^4 = 2.1 \times 10^{13}$

Assumptions: $0.1 - \Delta C \approx 0.1$ and $0.6 + 4\Delta C \approx 0.6$

$(0.1)/(\Delta C)(0.6)^4 \approx 2.1 \times 10^{13}$

$\Delta C \approx 3.7 \times 10^{-14}$

$\{(3.7 \times 10^{-14})/0.1\} \times 100 = 3.7 \times 10^{-11}\%$ (less than 5%)

$\{4 \times (3.7 \times 10^{-14})/0.6\} \times 100 = 2.47 \times 10^{-11}\%$ (less than 5%)

Assumptions $0.1 - \Delta C \approx 0.1$ and $0.6 + 4\Delta C \approx 0.6$ are valid.

Cu^{2+} ion concentration in a solution that is initially 0.10 M Cu^{2+} and 1 M NH_3

$\approx 3.7 \times 10^{-14}$ mole per liter.

13-12. Calculate the solubility of AgBr in 1M $(S_2O_3)^{2-}$.

K_{sp} (AgBr) $= 5.0 \times 10^{-13}$

$[Ag(S_2O_3)_2]^{3-}$ $K_f = 2.9 \times 10^{13}$

AgBr (s) $\rightleftarrows$ Ag^+ (aq) + Br^- (aq) $K_{sp} = 5.0 \times 10^{-13}$

Ag^+ (aq) + 2 $(S_2O_3)^{2-}$ (aq) $\rightleftarrows$ $[Ag(S_2O_3)_2]^{3-}$ (aq) $K_f = 2.9 \times 10^{13}$

AgBr (s) + 2 $(S_2O_3)^{2-}$ (aq) $\rightleftarrows$ $[Ag(S_2O_3)_2]^{3-}$ (aq) + Br^- (aq) $K = K_{sp} \times K_f = 14.5$

K>>1. So we approach the equilibrium from right to left. We create an intermediate situation in which the concentration of one of the reactants is zero.

AgBr (s) + 2 $(S_2O_3)^{2-}$ (aq) $\rightleftarrows$ $[Ag(S_2O_3)_2]^{3-}$ (aq) + Br^- (aq)

Initial concentration:	1M	0	0
Intermediate concentration:	$\{1 - (2 \times 0.5)\}$	0.5	0.5
Intermediate concentration:	0	0.5	0.5
At equilibrium:	$2\Delta C$	$0.5 - \Delta C$	$0.5 - \Delta C$

$K = (0.5 - \Delta C)(0.5 - \Delta C)/(2\Delta C)^2 = 14.5$

Assumption: $0.5 - \Delta C \approx 0.5$

$(\Delta C)^2 \approx (0.5 \times 0.5)/(4 \times 14.5) \approx 4.31 \times 10^{-3}$

$\Delta C \approx 0.06565$

$\{(0.06565/0.5) \times 100\} = 13.13\%$ (not less than 5%)

Assumption $0.5 - \Delta C \approx 0.5$ is not valid.

$\Delta C = 0.058$ by successive approximations.

Solubility of AgBr in 1 M $(S_2O_3)^{2-}$ is $(0.5 - 0.058) = 0.442$ mole per liter.

Another way to calculate the solubility:

Ag^+ (aq) + 2 $(S_2O_3^{2-})$ (aq) $\rightleftharpoons$ $[Ag (S_2O_3)_2]^{3-}$ (aq) $K_f = 2.9 \times 10^{13}$

AgBr (s) $\rightleftharpoons$ Ag^+ (aq) + Br^- (aq) $K_{sp} = 5.0 \times 10^{-13}$

If the solubility of AgBr is 'Y' then $[Ag^+] = [Br^-] = Y$ in pure water.

In pure water the solubility of AgBr = $\sqrt{(5 \times 10^{-13})} = 7.1 \times 10^{-7}$ mole per liter.

In pure water $[Ag^+] = [Br^-] = 7.1 \times 10^{-7}$ mole per liter.

When we add AgBr to 1.0 M $S_2O_3^{2-}$ solution the $[Br^-] > [Ag^+]$ due to the formation of the complex ion $[Ag (S_2O_3)_2]^{3-}$. So we have to find the concentration of Ag^+ ion in 1 M $S_2O_3^{2-}$ solution for various Br^- concentrations. When the ion product $[Ag^+] [Br^-] = K_{sp}(AgBr)$, that Br^- concentration will be the solubility of AgBr in 1 M $S_2O_3^{2-}$ solution.

$K_f \gg 1$. So we approach the equilibrium from right to left.

Case 1: Br^- concentration is 0.441 mole per liter.

	Ag^+ (aq) +	2 $(S_2O_3^{2-})$ (aq) $\rightleftharpoons$	$[Ag (S_2O_3)_2]^{3-}$ (aq)
Intermediate concentration:	0	$\{1 - (2 \times 0.441)\}$ M	0.441 M
Intermediate concentration:	0	0.118 M	0.441 M
At equilibrium	ΔC	$0.118 + 2\Delta C$	$0.441 - \Delta C$

$K_f = (0.441 - \Delta C)/\{(\Delta C) \times (0.118 + 2\Delta C)^2\} = 2.9 \times 10^{13}$

Assumptions: $0.441 - \Delta C \approx 0.441$

                 $0.118 + 2\Delta C \approx 0.118$

$(0.441)/\{(\Delta C) \times (0.118)^2\} \approx 2.9 \times 10^{13}$

$\Delta C \approx 1.092 \times 10^{-12}$

$\{(1.092 \times 10^{-12})/0.441\} \times 100 = 2.48 \times 10^{-10}\%$ (less than 5%)

$\{2 \times (1.092 \times 10^{-12})/0.118\} \times 100 = 1.85 \times 10^{-9}\%$ (less than 5%)

Assumption $0.441 - \Delta C \approx 0.441$ is valid.

Assumption $0.118 + 2\Delta C \approx 0.118$ is valid.

At equilibrium $[Ag^+] \approx 1.092 \times 10^{-12}$

[Br⁻] = 0.441

Ion product $[Ag^+] [Br^-] \approx (1.092 \times 10^{-12}) \times (0.441) \approx 4.82 \times 10^{-13}$

Ion product < K_{sp}(AgBr). So more AgBr will dissolve.

Case 2: Br⁻ concentration is 0.442 mole per liter.

$$Ag^+ (aq) + 2 (S_2O_3^{2-}) (aq) \rightleftharpoons [Ag (S_2O_3)_2]^{3-} (aq)$$

Intermediate concentration:	0	1 − (2×0.442)	0.442
Intermediate concentration:	0	0.116	0.442
At equilibrium:	ΔC	0.116 + 2ΔC	0.442 − ΔC

$K_f = (0.442 - \Delta C)/\{(\Delta C) \times (0.116 + 2\Delta C)^2\} = 2.9 \times 10^{13}$

Assumptions: $0.442 - \Delta C \approx 0.442$

$0.116 + 2\Delta C \approx 0.116$

$(0.442)/\{(\Delta C) \times (0.116)^2\} \approx 2.9 \times 10^{13}$

$\Delta C \approx 1.133 \times 10^{-12}$

$\{(1.133 \times 10^{-12})/0.442\} \times 100 = 2.56 \times 10^{-10}\%$ (less than 5%)

$\{2 \times (1.133 \times 10^{-12})/0.116\} \times 100 = 1.95 \times 10^{-9}\%$ (less than 5%)

Assumption $0.442 - \Delta C \approx 0.442$ is valid.

Assumption $0.116 + 2\Delta C \approx 0.116$ is valid.

At equilibrium $[Ag^+] \approx 1.133 \times 10^{-12}$ [Br⁻] = 0.442

Ion product $[Ag^+] [Br^-] \approx (1.133 \times 10^{-12}) \times (0.442) \approx 5.0 \times 10^{-13}$

Ion product $\approx K_{sp}$ (AgBr)

So the solubility of AgBr in 1.0 M $S_2O_3^{2-}$ solution is 0.442 mole per liter.

13-13. Calculate the solubility of AgCl in 1 M $S_2O_3^{2-}$.

K_{sp} (AgCl) = 1.77×10^{-10}

$[Ag (S_2O_3)_2]^{3-}$ $K_f = 2.9 \times 10^{13}$

$AgCl (s) \rightleftharpoons Ag^+ (aq) + Cl^- (aq)$ $K_{sp} = 1.77 \times 10^{-10}$

$Ag^+ (aq) + 2 (S_2O_3^{2-}) (aq) \rightleftharpoons [Ag (S_2O_3)_2]^{3-} (aq)$ $K_f = 2.9 \times 10^{13}$

─────────────────────────────────────

$AgCl (s) + 2 (S_2O_3^{2-}) (aq) \rightleftharpoons [Ag (S_2O_3)_2]^{3-} (aq) + Cl^- (aq)$ $K = K_{sp} \times K_f = 5133$

K>>1. So we approach the equilibrium from right to left. We create an intermediate situation in which the concentration of one of the reactants is zero.

$$AgCl \text{ (s)} + 2 \text{ } (S_2O_3)^{2-} \text{ (aq)} \rightleftharpoons [Ag \text{ } (S_2O_3)_2]^{3-} \text{ (aq)} + Cl^- \text{ (aq)}$$

	$AgCl$	$[Ag(S_2O_3)_2]^{3-}$	Cl^-
Initial concentration:	1M	0	0
Intermediate concentration:	$\{1 - (2\times0.5)\}$	0.5	0.5
Intermediate concentration:	0	0.5	0.5
At equilibrium:	$2\Delta C$	$0.5 - \Delta C$	$0.5 - \Delta C$

$K = \{(0.5 - \Delta C)\times(0.5 - \Delta C)\}/(2\Delta C)^2 = 5133$ Assumption: $0.5 - \Delta C \approx 0.5$

$(\Delta C)^2 \approx (0.5\times0.5)/(4\times5133) \approx 1.218\times10^{-5}$

$\Delta C \approx 3.490\times10^{-3}$

$\{(3.490\times10^{-3}/0.5)\times100\} = 0.70\%$ (less than 5%)   Assumption $0.5 - \Delta C \approx 0.5$ is valid.

Solubility of AgCl in 1 M $(S_2O_3)^{2-} \approx \{0.5 - (3.490\times10^{-3})\} \approx 0.4965$ mole per liter.

 Another way to calculate the solubility:

$$Ag^+ \text{ (aq)} + 2 \text{ } (S_2O_3^{2-}) \text{ (aq)} \longrightarrow [Ag \text{ } (S_2O_3)_2]^{3-} \text{ (aq)} \quad K_f = 2.9\times10^{13}$$

$$AgCl \text{ (s)} \rightleftharpoons Ag^+ \text{ (aq)} + Cl^- \text{ (aq)} \qquad\qquad K_{sp} = 1.77\times10^{-10}$$

If the solubility of AgCl is 'Y', then $[Ag^+] = [Cl^-] = Y$ in pure water.

In pure water the solubility of AgCl $= \sqrt{(1.77\times10^{-10})} = 1.33\times10^{-5}$ mole per liter.

In pure water $[Ag^+] = [Cl^-] = 1.33\times10^{-5}$ mole per liter.

When we add AgCl to 1.0 M $S_2O_3^{2-}$ solution the $[Cl^-] > [Ag^+]$ due to the formation of

the complex ion $[Ag \text{ } (S_2O_3)_2]^{3-}$. So we have to calculate the concentration of Ag^+ ion in 1.0 M

$S_2O_3^{2-}$ for various Cl^- concentrations. When the ion product $[Ag^+] [Cl^-]$ is equal to the

K_{sp} (AgCl), that Cl^- concentration will be the solubility of AgCl in 1.0 M $S_2O_3^{2-}$ solution.

$K_f \gg 1$. So we approach the equilibrium from right to left.

Case 1: Cl^- concentration is 0.4966 M.

$$Ag^+ \text{ (aq)} + \quad 2 \text{ } (S_2O_3^{2-}) \text{ (aq)} \rightleftharpoons [Ag \text{ } (S_2O_3)_2]^{3-} \text{ (aq)}$$

	Ag^+	$2(S_2O_3^{2-})$	$[Ag(S_2O_3)_2]^{3-}$
Intermediate concentration:	0	$1 - (2\times0.4966)$	0.4966 M
Intermediate concentration:	0	0.0068	0.4966
At equilibrium:	ΔC	$0.0068 + 2\Delta C$	$0.4966 - \Delta C$

$K_f = (0.4966 - \Delta C)/\{(\Delta C)\times(0.0068 + 2\Delta C)^2\} = 2.9\times10^{13}$

Assumptions: $0.4966 - \Delta C \approx 0.4966$

$0.0068 + 2\Delta C \approx 0.0068$

$(0.4966)/\{(\Delta C) \times (0.0068)^2\} \approx 2.9 \times 10^{13}$

$\Delta C \approx 3.7 \times 10^{-10}$

$\{(3.7 \times 10^{-10})/(0.4966)\} \times 100 = 7.45 \times 10^{-8}\%$ (less than 5%)

$\{2 \times (3.7 \times 10^{-10})/(0.0068)\} \times 100 = 1.09 \times 10^{-5}\%$ (less than 5%)

Assumption $0.4966 - \Delta C \approx 0.4966$ is valid.

Assumption $0.0068 + 2\Delta C \approx 0.0068$ is valid.

At equilibrium $[Ag^+] \approx 3.7 \times 10^{-10}$

$[Cl^-] = 0.4966$

Ion product $[Ag^+][Cl^-] \approx (3.7 \times 10^{-10}) \times (0.4966) \approx 1.84 \times 10^{-10}$

Ion product $> K_{sp}$ (AgCl)

So the solubility of AgCl in 1.0 M $S_2O_3^{2-}$ solution < 0.4966 mole per liter.

Case 2: Cl^- concentration is 0.4965 M.

$$Ag^+ (aq) + 2 (S_2O_3^{2-}) (aq) \rightleftharpoons [Ag(S_2O_3)_2]^{3-} (aq)$$

Intermediate concentration:	0	$1 - (2 \times 0.4965)$	0.4965 M
Intermediate concentration:	0	0.007	0.4965
At equilibrium:	ΔC	$0.007 + 2\Delta C$	$0.4965 - \Delta C$

$K_f =$ $(0.4965 - \Delta C)/\{(\Delta C) \times (0.007 + 2\Delta C)^2\} = 2.9 \times 10^{13}$

Assumptions: $0.4965 - \Delta C \approx 0.4965$

$0.007 + 2\Delta C \approx 0.007$

$(0.4965)/\{(\Delta C) \times (0.007)^2\} \approx 2.9 \times 10^{13}$

$\Delta C \approx 3.5 \times 10^{-10}$

$\{(3.5 \times 10^{-10})/(0.4965)\} \times 100 = 7.05 \times 10^{-8}\%$ (less than 5%)

$\{2 \times (3.5 \times 10^{-10})/(0.007)\} \times 100 = 1.00 \times 10^{-5}\%$ (less than 5%)

Assumption $0.4965 - \Delta C \approx 0.4965$ is valid.

Assumption $0.007 + 2\Delta C \approx 0.007$ is valid.

At equilibrium $[Ag^+] \approx 3.5 \times 10^{-10}$

$$[Cl^-] = 0.4965$$

Ion product $[Ag^+][Cl^-] \approx (3.5 \times 10^{-10}) \times (0.4965) \approx 1.74 \times 10^{-10}$

Ion product $\approx K_{sp}$ (AgCl).

Solubility of AgCl in 1 M $(S_2O_3)^{2-}$ is 0.4965 mole per liter.

13-14. Calculate the solubility of AgI in 1.0 M $S_2O_3{}^{2-}$.

K_{sp} (AgI) $= 8.52 \times 10^{-17}$

$[Ag(S_2O_3)_2]^{3-}$ $K_f = 2.9 \times 10^{13}$

AgI (s) $\rightleftharpoons$ Ag$^+$ (aq) + I$^-$ (aq)	$K_{sp} = 8.52 \times 10^{-17}$
Ag$^+$ (aq) + 2 $(S_2O_3{}^{2-})$ (aq) $\rightleftharpoons$ [Ag $(S_2O_3)_2]^{3-}$ (aq)	$K_f = 2.9 \times 10^{13}$
AgI (s) + 2 $(S_2O_3{}^{2-})$ (aq) $\rightleftharpoons$ [Ag $(S_2O_3)_2]^{3-}$ (aq) + I$^-$ (aq)	$K = K_{sp} \times K_f = 2.47 \times 10^{-3}$

K<<1. So we approach the equilibrium from left to right.

$$AgI \text{ (s)} + 2\ (S_2O_3)^{2-} \text{ (aq)} \rightleftharpoons [Ag\ (S_2O_3)_2]^{3-} \text{ (aq)} + I^- \text{ (aq)}$$

Initial concentration:	1M	0	0
At equilibrium :	$1 - 2\Delta C$	ΔC	ΔC

$K = (\Delta C)^2/(1 - 2\Delta C)^2 = 2.47 \times 10^{-3}$

Assumption : $1 - 2\Delta C \approx 1$

$(\Delta C)^2 \approx 2.47 \times 10^{-3}$

$\Delta C \approx 0.04970$

$\{2 \times (0.04970)/1\} \times 100 = 9.8\%$ (not less than 5%)

Assumption $1 - 2\Delta C \approx 1$ is not valid.

$\Delta C = 0.0452$ (by successive approximations)

Solubility of AgI in 1.0 M $S_2O_3{}^{2-}$ solution is 0.0452 mole per liter.

Another way to calculate the solubility :

Ag$^+$ (aq) + 2 $(S_2O_3{}^{2-})$ (aq) $\rightleftharpoons$ [Ag $(S_2O_3)_2]^{3-}$ (aq) $K_f = 2.9 \times 10^{13}$

AgI (s) $\rightleftharpoons$ Ag$^+$ (aq) + I$^-$ (aq) $K_{sp} = 8.52 \times 10^{-17}$

If the solubility of AgI is 'Y', then $[Ag^+] = [I^-] = Y$ in pure water.

In pure water the solubility of AgI $= \sqrt{(8.52 \times 10^{-17})} = 9.23 \times 10^{-9}$ mole per liter.

In pure water $[Ag^+] = [I^-] = 9.23 \times 10^{-9}$ mole per liter.

When we add AgI to 1.0 M $S_2O_3^{2-}$ solution the $[I^-] > [Ag^+]$ due to the formation of the complex ion $[Ag(S_2O_3)_2]^{3-}$. So we have to calculate the concentration of Ag^+ ion in 1.0 M $S_2O_3^{2-}$ for various I^- concentrations. When the ion product $[Ag^+][I^-]$ is equal to the K_{sp} (AgI), that I^- concentration will be the solubility of AgI in 1.0 M $S_2O_3^{2-}$ solution.

$K_f \gg 1$. So we approach the equilibrium from right to left.

Case 1: I^- concentration is 0.0451 M.

$$Ag^+ (aq) + 2\ (S_2O_3^{2-})\ (aq) \rightleftharpoons [Ag(S_2O_3)_2]^{3-}\ (aq)$$

Intermediate concentration:	0	$1 - (2\times0.0451)$	0.0451 M
Intermediate concentration:	0	0.9098	0.0451
At equilibrium:	ΔC	$0.9098 + 2\Delta C$	$0.0451 - \Delta C$

$K_f = (0.0451 - \Delta C)/\{(\Delta C)\times(0.9098 + 2\Delta C)^2\} = 2.9\times10^{13}$

Assumptions: $\quad 0.0451 - \Delta C \quad \approx 0.0451$

$\qquad\qquad\qquad 0.9098 + 2\Delta C \quad \approx 0.9098$

$(0.0.0451)/\{(\Delta C)\times(0.9098)^2\} \approx 2.9\times10^{13}$

$\Delta C \approx 1.879\times10^{-15}$

$\{(1.879\times10^{-15})/(0.0451)\}\times100 = 4.17\times10^{-12}\%$ (less than 5%)

$\{2\times(1.879\times10^{-15})/(0.9098)\}\times100 = 4.13\times10^{-13}\%$ (less than 5%)

Assumption $\quad 0.0451 - \Delta C \quad \approx 0.0451$ is valid.

Assumption $\quad 0.9098 + 2\Delta C \quad \approx 0.9098$ is valid.

At equilibrium $[Ag^+] \approx 1.879\times10^{-15} \qquad\qquad [I^-] = 0.0451$

Ion product $[Ag^+][I^-] \approx (1.879\times10^{-15})\times(0.0451) \approx 8.47\times10^{-17}$

Ion product $< K_{sp}$ (AgI). So more AgI will dissolve.

Case 2: I^- concentration is 0.0452 M.

$$Ag^+ (aq) + 2\ (S_2O_3^{2-})\ (aq) \rightleftharpoons [Ag(S_2O_3)_2]^{3-}\ (aq)$$

Intermediate concentration:	0	$1 - (2\times0.0452)$	0.0452 M
Intermediate concentration:	0	0.9096	0.0452
At equilibrium:	ΔC	$0.9096 + 2\Delta C$	$0.0452 - \Delta C$

$K_f = (0.0452 - \Delta C)/\{(\Delta C)\times(0.9096 + 2\Delta C)^2\} = 2.9\times10^{13}$

Assumptions: $0.0452 - \Delta C \approx 0.0452$

$0.9096 + 2\Delta C \approx 0.9096$

$(0.0452)/\{(\Delta C)\times(0.9096)^2\} \approx 2.9\times10^{13}$

$\Delta C \approx 1.884\times10^{-15}$

$\{(1.884\times10^{-15})/(0.0452)\}\times100 = 4.17\times10^{-12}\%$ (less than 5%)

$\{2\times(1.884\times10^{-15})/(0.9096)\}\times100 = 4.14\times10^{-13}\%$ (less than 5%)

Assumption $0.0452 - \Delta C \approx 0.0452$ is valid.

Assumption $0.9096 + 2\Delta C \approx 0.9096$ is valid.

At equilibrium $[Ag^+] \approx 1.884\times10^{-15}$

$[I^-] = 0.0452$

Ion product $[Ag^+][I^-] \approx (1.884\times10^{-15})\times(0.0452) \approx 8.52\times10^{-17}$

Ion product $\approx K_{sp}$ (AgI)

So the solubility of AgI in 1.0 M $S_2O_3^{2-}$ solution is 0.0452 mole per liter.

13-15. The complex ion $[Tl(Br)_6]^{3-}$ is highly stable, with log $K_f = 31.6$. What is the concentration of Tl(III) (aq) in equilibrium with a 1.12 M solution of $Na_3[Tl(Br)_6]$?

Tl^{3+} (aq) + 6Br$^-$ (aq) $\rightleftharpoons$ $[Tl(Br)_6]^{3-}$ $K_f = 3.98\times10^{31}$ (log $K_f = 31.6$)

$K_f \gg 1$. So we approach the equilibrium from right to left.

	Tl^{3+} (aq) + 6Br$^-$ (aq) $\rightleftharpoons$ $[Tl(Br)_6]^{3-}$		
Initial concentration:	0	0	1.12 M
At equilibrium:	ΔC	$6\Delta C$	$1.12 - \Delta C$

$K_f = (1.12 - \Delta C)/\{(\Delta C)(6\Delta C)^6\} = 3.98\times10^{31}$

Assumption: $(1.12 - \Delta C) \approx 1.12$

$1.12/(\Delta C)(6\Delta C)^6 \approx 3.98\times10^{31}$

$(\Delta C)^7 \approx 1.12/\{(3.98\times10^{31})\times(6)^6\} \approx 6.03\times10^{-37}$

$\Delta C \approx 6.70\times10^{-6}$

$\{(6.70\times10^{-6})/1.12\}\times100 = 5.98\times10^{-4}\%$ (less than 5%)

Assumption $(1.12 - \Delta C) \approx 1.12$ is valid.

Concentration of Tl(III)(aq) at equilibrium $\approx 6.70\times10^{-6}$ M

13-16. The ferrocyanide ion $[Fe(CN)_6]^{4-}$ is very stable with a K_f of 1×10^{35}. Calculate the concentration of cyanide ion in equilibrium with a 0.65 M solution of $K_4[Fe(CN)_6]$.

$K_f \gg 1$. So we approach the equilibrium from right to left.

$$Fe^{2+}(aq) + 6CN^- (aq) \rightleftharpoons [Fe(CN)_6]^{4-} (aq)$$

Initial: 0   0   0.65 M

At equilibrium: ΔC $6\Delta C$ $0.65 - \Delta C$

$K_f = (0.65 - \Delta C)/\{(\Delta C)(6\Delta C)^6\} = 10^{35}$

Assumption: $0.65 - \Delta C \approx 0.65$

$(\Delta C)^7 \approx (0.65\times10^{-35})/(6)^6 \approx 1.39\times10^{-40}$

$\Delta C \approx 2.02\times10^{-6}$

$\{(2.02\times10^{-6})/0.65\} \times 100 = 3.11\times10^{-4}\%$ (less than 5%)   Assumption $0.65 - \Delta C \approx 0.65$ is valid.

Concentration of CN^- at equilibrium $\approx 6\times2.02\times10^{-6} \approx 1.21\times10^{-5}$ M

13-17. Calculate the solubility of Mercury (II) iodide HgI_2 in each situation:

(a) In pure water
(b) In a 3.0 M solution of NaI assuming $[HgI_4]^{2-}$ is the only Hg-containing species present in significant amount.

$K_{sp} (HgI_2) = 2.9\times10^{-29}$

$K_f = 6.8\times10^{29}$ for $[HgI_4]^{2-}$

(a) In pure water

Let Y be the solubility of HgI_2 in pure water in moles per liter.

$$HgI_2 (s) \rightleftharpoons Hg^{2+} (aq) + 2I^- (aq)$$

 Y   Y   2Y

$K_{sp} = [Hg^{2+}][I^-]^2 = (Y)(2Y)^2$

$4Y^3 = 2.9\times10^{-29}$

$Y^3 = (2.9/4)\times 10^{-29} = 0.725\times10^{-29}$

$Y = 1.9\times10^{-10}$ M

The solubility of HgI_2 in pure water $= 1.9\times10^{-10}$ M

(b) In a 3.0 M solution of NaI

Let Z be the solubility of HgI_2 in 3.0 M NaI solution.

HgI_2 (s) $\rightleftharpoons$ Hg^{2+} (aq) + $2I^-$ (aq) $K_{sp} = 2.9 \times 10^{-29}$

Hg^{2+} (aq) + $4I^-$ (aq) $\rightleftharpoons$ $[HgI_4]^{2-}$ $K_f = 6.8 \times 10^{29}$

HgI_2 (s) + $2I^-$ (aq) $\rightleftharpoons$ $[HgI_4]^{2-}$ $K = K_{sp} \times K_f = 19.72$

At equilibrium: $(3 - 2Z)$ Z

$K = Z/(3 - 2Z)^2 = 19.72$

$Z = 1.3683$ M (by successive approximations)

$Z \approx 1.4$ M

The solubility of HgI_2 in 3.0 M NaI solution ≈ 1.4 M.

CHAPTER 14

CHEMICAL THERMODYNAMICS

INVALUABLE INFORMATION

Gas Constant (R)   : 8.31447 J/(mol K)

Gas Constant (R)   : 0.0820575 (L atm)/(mol K)

Avogadro's number   : 6.022142×10^{23}/mol

Charge on one mole of electrons (F)   : 9.6485338×10^4 C/mol

$\Delta G^0 = - nFE^0_{cell}$

$\Delta G^0 = - RT \ln K$

$E^0_{cell} = (RT/nF) \ln K = (0.0591/n) \log K$ at T = 298K

$\Delta G = \Delta G^0 + RT \ln Q$

$E_{cell} = E^0_{cell} - (0.0591/n) \log Q$ at T = 298K

14-1. The heat transferred from a system to its surroundings (or vice versa) when a chemical

reaction is run under conditions of constant pressure is equal to:

   (a) The change in the enthalpy of the system (ΔH)
   (b) The change in the energy of the system (ΔE)
   (c) The change in the entropy of the sytem (ΔS)
   (d) The change in the free energy of the system (ΔG)
   (e) The work done by the system

   $H = E + PV$

   At constant pressure $\Delta H = \Delta E + P\Delta V$

   $\Delta H = (q_p + w) + P\Delta V = (q_p - P\Delta V) + P\Delta V = q_p$ Answer(a)

14-2. The standard enthalpy of atom combination of CO_2 is $- 1608.53$ kJ/mol$_{rxn}$ and O_2

is $- 498.34$ kJ/mol$_{rxn}$ and the standard enthalpy of reaction for the following reaction is $- 283.0$

kJ/mol$_{rexn}$: CO (g) + (1/2)O_2 (g) $\rightarrow$ CO_2 (g)

What is the standard enthalpy of atom combination of CO (g) in kJ/mol?

(a) − 827.21   (b) − 1076.36   (c) − 1359.36   (d) − 1624.34

$\Delta H^0 = \sum H^0_f \text{Products} - \sum H^0_f \text{Reactants}$

$- 283.0 = - 1608.53 - \Delta H^0_{ac} CO(g) - (- 498.34/2)$

$\Delta H^0_{ac} CO(g) = - 1608.53 + (498.34/2) + 283 = - 1076.36$ kJ/mol.   Answer(b)

14-3. A reaction in which carbon monoxide spontaneously disproportionates to graphite and CO_2 causes problems in the upper part of iron blast furnaces. The enthalpy of atom combination of CO is − 1076.38 kJ/mol$_{rxn}$, graphite is − 716.6 kJ/mol$_{rxn}$, and CO_2 is −1608.53 kJ/mol$_{rxn}$. What is the enthalpy change in this disproportionation reaction at 298 K?

(a) − 613 kJ/mol$_{rxn}$ (b) − 503 kJ/mol$_{rxn}$ (c) − 283 kJ/mol$_{rxn}$

(d) − 172 kJ/mol$_{rxn}$ (e) 283 kJ/mol$_{rxn}$

$2CO (g) \rightarrow C (g) + CO_2 (g)$

$\Delta H^0 = \sum H^0_f \text{Products} - \sum H^0_f \text{Reactants} = - 716.68 - 1608.53 - 2 (-1076.38)$

$$= - 172.05 \text{ kJ/mol}$$ Answer(d)

14-4. For which of the following reactions is ΔH approximately equal to ΔE?

(a) $2H_2O_2 (g) \rightarrow 2H_2O (g) + O_2(g)$
(b) $2H_2 (g) + O_2 (g) \rightarrow 2H_2O (g)$
(c) $2NH_3 (g) \rightarrow N_2 (g) + 3H_2 (g)$
(d) $2NO (g) \rightarrow N_2 (g) + O_2 (g)$
(e) $NH_4NO_3 (s) \rightarrow N_2 (g) + (1/2)O_2 (g) + 2H_2O (g)$

At constant pressure $\Delta H = \Delta E + P\Delta V$

$\Delta V = 0$ in (d). So $\Delta H = \Delta E$.   Answer(d)

14-5. Which of the following could be described as the most disorganized at room temperature?

(a) Cl_2 (g) $S^0 = 223$ J/mol-K
(b) Diamond $S^0 = 2.43$ J/mol-K
(c) Tin $S^0 = 51.5$ J/mol-K
(d) Mercury $S^0 = 77.4$ J/mol-K
(e) N_2 (g) $S^0 = 192$ J/mol-K

S^0 is a measure of the disorder. The higher the S^0 value, the more disorganized the element will be. Cl_2 (g) has the highest S^0 value.   Answer(a)

14-6. Use your understanding of entropy to predict which of the   following processes will have the largest entropy change?

(a) $H_2O (s) \rightarrow H_2O (l)$ (b) $H_2O (s) \rightarrow H_2O (l)$ (c) $H_2O_2 (l) \rightarrow H_2O (g) + (1/2)O_2 (g)$

(d) H_2O (l) $\rightarrow$ H_2O (g)   (e) H_2O_2 (g) $\rightarrow$ H_2O (g) + $(1/2)O_2$ (g)

Entropy change will be largest when increase in disorder is largest. Solids have much more regular structures than liquids. Liquids are more disordered than solids. Gases are more disordered than the corresponding liquids. A reaction which increases the number of particles will increase the disorder. So (c) will have the largest entropy change.   Answer(c)

14-7. Which of the following will have the most positive value for ΔS^0?

(a) H_2O (l) $\rightarrow$ H_2O (s)
(b) $NaNO_3$ (s) $\rightarrow$ Na^+ (aq) + NO_3^- (aq)
(c) $2HCl$ (g) $\rightarrow$ H_2 (g) + Cl_2 (g)
(d) $2H_2$ (g) + O_2 (g) $\rightarrow$ $2H_2O$ (g)

(a) $\Delta S^0 < 0$, because solid will be more ordered.

(b) $\Delta S^0 > 0$, because ions will be disordered in solution.

(c) ΔS^0 will not be most + ve, because all are gases and the number of gas particles do not change.

(d) $\Delta S^0 < 0$, because all are gases and the number of gas particles decrease.   Answer(b)

14-8. Which of the following processes would have a negative entropy change?

(a) Decomposition of H_2O_2
(b) Sublimation of dry ice (solid CO_2)
(c) Evaporation of water
(d) Formation of Al_2O_3 from its elements.
(e) Dissolution of salt in H_2O

(a) $\Delta S > 0$, because the number of particles will increase.

    {H_2O_2 (l) $\rightarrow$ H_2O (g) + $(1/2)O_2$ (g)} or {H_2O_2 (g) $\rightarrow$ H_2O (g) + $(1/2)O_2$ (g)}

(b) $\Delta S > 0$, because gas will be more disordered.

(c) $\Delta S > 0$, because gas will be more disordered.

(d) $\Delta S < 0$, because solid will have more order than gas and solid.

    $2Al$ (s) + $(3/2)O_2$ (g) $\rightarrow$ Al_2O_3 (s)

(e) $\Delta S^0 > 0$, because ions will be disordered in solution.

    $NaCl$ (s) $\rightarrow$ Na^+ (aq) + Cl^- (aq)   Answer(d)

14-9. For which of the following highly exothermic processes would you expect ΔH^0 and ΔG^0 to be about the same.

(a) $2Al$ (s) $+ (3/2)O_2$ (g) $\rightarrow Al_2O_3$ (s)

(b) $2H_2$ (g) $+ O_2$ (g) $\rightarrow 2H_2O$ (g)

(c) $2Na$ (s) $+ 2H_2O$ (l) $\rightarrow 2NaOH$ (aq) $+ H_2$ (g)

(d) $2NO$ (g) $\rightarrow N_2O_4$ (g)

(e) $2Al$ (s) $+ Fe_2O_3$ (s) $\rightarrow 2Fe$ (s) $+ Al_2O_3$ (s)

$$\Delta G^0 = \Delta H^0 - T\Delta S^0$$

Reaction (e) involves only solids and the number particles do not change; so the change in ΔS^0 will be very low and $\Delta G^0 \approx \Delta H^0$.   Answer(e)

14-10. An exothermic chemical reaction in which 2 moles of gaseous products are formed from 3 moles of gaseous reactants is favored by:

    (a) Increasing both pressure and temperature
    (b) Increasing temperature and decreasing pressure
    (c) Decreasing both temperature and pressure
    (d) Decreasing temperature and increasing pressure
    (e) None of the above changes

In an exothermic reaction the equilibrium constant increases with decrease in temperature and forward reaction is favored. 2 moles of gaseous products are formed from 3 moles of gaseous reactants. If the pressure is increased the reaction will be shifted towards the products, because this will reduce the number of particles. So the above exothermic reaction will be favored by decreasing temperature and increasing pressure.   Answer(d)

14-11. Which of the following would have a positive value for ΔS?

    (a) $3NO$ (g) $\rightarrow NO_2$ (g) $+ N_2O$ (g)
    (b) $2CO_2$ (g) $\rightarrow 2CO$ (g) $+ O_2$ (g)
    (c) $2I$ (g) $\rightarrow I_2$ (g)
    (d) NH_3 (g) $\rightarrow NH_3$ (l)

(a) $\Delta S < 0$, because all are gases and the number of particles decrease.

(b) $\Delta S > 0$, because all are gases and the number of particles increase.

(c) $\Delta S < 0$, because all are gases and the number of particles decrease.

(d) $\Delta S < 0$, because liquid will be more ordered than gas.   Answer(b)

14-12. Which of the following would have the most positive value for ΔS?

    (a) $2H_2O$ (g) $\rightarrow 2H_2$ (g) $+ O_2$(g)
    (b) N_2 (g) $+ 3H_2$ (g) $\rightarrow 2NH_3$ (g)
    (c) CO_2 (g) $+ H_2$ (g) $\rightarrow CO$ (g) $+ H_2O$ (g)
    (d) H_2O (l) $\rightarrow H_2O$ (s)
    (e) None of these reactions would have a positive value for ΔS.

(a)$\Delta S > 0$, because all are gases and the number of particles increase.

(b) $\Delta S < 0$, because all are gases and the number of particles decrease.

(c) ΔS will not be most + ve, because all are gases and the number of particles do not change.

(d) $\Delta S < 0$, because solid will be more ordered than liquid.

Answer(a)

14-13. Which of the following reactions would have a – ve ΔS^0?

(a) $H_2 (g) \rightarrow 2H (g)$ 　(b) $H_2 (l) \rightarrow H_2 (g)$

(c) $2H_2 (g) + CO (g) \rightarrow CH_3OH (g)$ 　(d) None of the above

(a) $\Delta S^0 > 0$, because all are gases and the number of particles increase.

(b) $\Delta S^0 > 0$, because gas will be more disordered than liquid.

(c) $\Delta S^0 < 0$, because all are gases and the number of particles decrease.

Answer(c)

14-14. Use your understanding of thermodynamics to predict the signs of ΔH^0 and ΔS^0 for the following reaction:

$$2H_2 (g) + O_2 (g) \rightarrow 2H_2O (g)$$

(a) $\Delta H^0 = -$ ve and $\Delta S^0 = -$ ve 　(b) $\Delta H^0 = -$ ve and $\Delta S^0 = +$ ve

(c) $\Delta H^0 = +$ ve and $\Delta S^0 = -$ ve 　(d) $\Delta H^0 = +$ ve and $\Delta S^0 = +$ ve

(e) $\Delta H^0 = -$ ve and $\Delta S^0 = 0$

$\Delta S^0 < 0$, because all are gases and the number of particles decrease.

$\Delta H^0 < 0$, because the reaction will be exothermic due to new bond formation. 　Answer(a)

14-15. Which of the following reactions is unfavorable at low temperatures but becomes favorable as the temperature increases?

(a) $2CO (g) + O_2 (g) \rightarrow 2CO_2 (g)$ 　　$\Delta H^0 = -566$ kJ, $\Delta S^0 = -173$ J/K
(b) $2H_2O (g) \rightarrow 2H_2 (g) + O_2(g)$ 　　$\Delta H^0 = 484$ kJ, $\Delta S^0 = 90.0$ J/K
(c) $2N_2O (g) \rightarrow 2N_2 (g) + O_2 (g)$ 　　$\Delta H^0 = -164$ kJ, $\Delta S^0 = 149$ J/K
(d) $PbCl_2 (s) \rightarrow Pb^{2+} (aq) + 2Cl^- (aq)$ 　$\Delta H^0 = 23.4$ kJ, $\Delta S^0 = -12.5$ J/K

$\Delta G^0 = \Delta H^0 - T\Delta S^0$

(a) At low temperatures ΔG^0 will be – ve, so reaction will be favorable.

(b) At low temperatures ΔG^0 will be + ve, so reaction will be unfavorable. ΔG^0 will be – ve if T

　is high.

(c) At low temperatures ΔG^0 will be – ve, so reaction will be favorable.

(d) At low temperatures ΔG^0 will be + ve, so reaction will be unfavorable. ΔG^0 will be + ve at high temperatures also; so reaction will be unfavorable at high temperatures also.

Answer(b)

14-16. If the following reaction is spontaneous as written,

$$Zn \ (s) + Cu^{2+} \ (aq) \rightarrow Zn^{2+} \ (aq) + Cu \ (s)$$

Which of the following statements is true?

(a) $K_c > 1$ and $\Delta G^0 > 0$. (b) $K_c > 1$ and $\Delta G^0 < 0$. (c) $K_c < 1$ and $\Delta G^0 > 0$.

(d) $K_c < 1$ and $\Delta G^0 < 0$. (e) $K_c < 1$ and $\Delta G^0 = 0$.

$\Delta G^0 = - RT \ln K_c$ and for a spontaneous reaction ΔG^0 is – ve.

If ΔG^0 is – ve, then $\ln K_c$ must be + ve ; so K_c must be > 1.   Answer(b)

14-17. Which statement is true if the following reaction is spontaneous as written?

$Zn + Cu^{2+} \rightarrow Zn^{2+} + Cu$

(a) K_c is larger than 1, ΔG^0 is negative, and E^0 is positive
(b) K_c is larger than 1, ΔG^0 is positive, and E^0 is negative
(c) K_c is larger than 1, ΔG^0 is negative, and E^0 is negative
(d) K_c is smaller than 1, ΔG^0 is negative, and E^0 is positive
(e) K_c is smaller than 1, ΔG^0 is positive, and E^0 is negative

For a spontaneous reaction ΔG^0 is – ve

$\Delta G^0 = - nFE^0$; so E^0 has to be + ve.

At equilibrium $\Delta G^0 = - RT \ln K_c$

$\ln K_c = - (\Delta G^0 / RT)$; ΔG^0 is – ve, so $\ln K_c$ has to be + ve and K_c has to be > 1.   Answer(a)

14-18. What happens to the magnitude of the equilibrium constant for the following reaction as the temperature of the system increases?

$$2HI \ (g) \rightarrow H_2 \ (g) + I_2 \ (g) \quad \Delta H^0 = - 9.48 \ kJ \text{ and } \Delta S^0 = - 21.8 \ J/K$$

(a) K increases   (b) K decreases   (c) It is impossible to determine what happens to K

Let us consider two temperatures: $T_1 = 300K$ and $T_2 = 500K$

(i) $T_1 = 300K$ $\Delta G^0 = \Delta H^0 - (300 \times \Delta S^0) = - 9480 + (300 \times 21.8) = - 2940$

$\Delta G^0 = - RT \ln K$

$\ln K = - (\Delta G^0)/RT = - (-2940)/(8.314 \times 300) = 1.18$

$K = 3.25$

(ii) $T_2 = 500K$ $\Delta G^0 = \Delta H^0 - (500 \times \Delta S^0) = -9480 + (500 \times 21.8) = +1420$

$Ln\ K = -(\Delta G^0)/RT = -(+1420)/(8.314 \times 500) = -0.342$

$K = 0.71$

As expected for an exothermic reaction, the K (equilibrium constant) decreases as the temperature of the system increases.   Answer(b)

14-19. Which of the following always corresponds to a spontaneous reaction?

(a) $\Delta H^0 < 0$, $\Delta S^0 < 0$ (b) $\Delta H^0 > 0$, $\Delta S^0 < 0$ (c) $\Delta H^0 < 0$, $\Delta S^0 > 0$

(d) $\Delta H^0 > 0$, $\Delta S^0 > 0$ (e) $\Delta H^0 < 0$, $\Delta S^0 = 0$

$\Delta G^0 = \Delta H^0 - T\Delta S^0$

If $\Delta H^0 < 0$ and $\Delta S^0 > 0$, ΔG^0 will always be negative and the reaction will always be spontaneous.   Answer(c)

14-20. Under what conditions will a reaction that is not spontaneous at room temperature become spontaneous as the temperature increases?

(a) $\Delta H = +$ ve, and $\Delta S = +$ ve   (b) $\Delta H = +$ ve, and $\Delta S = -$ ve

(c) $\Delta H = -$ ve, and $\Delta S = +$ ve   (d) $\Delta H = -$ ve, and $\Delta S = -$ ve

(e) a reaction can not become spontaneous as the temperature increases, regardless of the signs of ΔH and ΔS.

$\Delta G = \Delta H - T\Delta S$

(a) If $\Delta H > T\Delta S$, then reaction will not be spontaneous at room temperature because ΔG will be $+$ ve. But when the temperature is such that $\Delta H < T\Delta S$ then the reaction will be spontaneous because ΔG will be $-$ ve.

(b) At room temperature ΔG will be $+$ ve and the reaction will not be spontaneous. The reaction will not be spontaneous at high temperatures also because ΔG will be $+$ ve at high temperatures also.

(c) At room temperature ΔG will be $-$ ve and the reaction will be spontaneous.

(d) At room temperature ΔG will be $+$ ve if $T\Delta S > \Delta H$ and the reaction will not be spontaneous. The reaction will not be spontaneous at high temperatures also because ΔG will be $+$ ve at high temperatures also.   Answer(a)

14-21. Consider the following reaction and data:

$$2NH_3 (g) \rightarrow N_2 (g) + 3H_2 (g)$$

Compound	S^0 (J/mol-K)
H_2 (g)	$- 98.74$
N_2 (g)	$- 114.94$
NH_3 (g)	$- 304.99$

(i)   Calculate the ΔS^0 for this reaction.

(a) $- 198.8$ J/K   (b) $- 91.26$ J/K   (c) $- 106.22$ J/K   (d) 91.26 J/K   (e) 198.8 J/K

$\Delta S^0 = \sum S^0$ Products $- \sum S^0$ Reactants

$\qquad = \{(- 114.99) + (3\times - 98.74)\} - \{2\times(- 304.99)\}$

$\qquad = 198.8$ J/K   Answer(e)

(ii)   Calculate the ΔG^0 for this reaction at 25 °C if the ΔH^0 for this reaction is 92.2 kJ/mol$_{rxn.}$

(a) $- 106.5$ kJ/mol$_{rxn}$ (b) $- 33.0$ kJ/mol$_{rxn}$ (c) 33.0 kJ/mol$_{rxn}$ (d) 106.5 kJ/mol$_{rxn}$

(e) None of the above

$\Delta G^0 = \Delta H^0 - T\Delta S^0 = 92200 - (298\times198.8) = 32958$ J $= 33.0$ kJ/mol$_{rxn}$ Answer(c)

(iii)   Calculate the equilibrium constant for this reaction.

(a) 1.7×10^{-6} (b) 1.3×10^{-3} (c) 7.8×10^2 (d) 6.1×10^5 (e) None of the above

$\Delta G^0 = - RT \ln K$

$\ln K = - (\Delta G^0/RT) = - (32958.0)/(8.314\times298) = - 13.30$

$K = 1.7\times10^{-6}$ Answer(a)

14-22. Calculate the ΔS^0 for the following reaction:

$$Sn (s) + 2Cl_2 (g) \rightarrow SnCl_4 (l)$$

	S^0 (J/mol-K)
Sn (s)	$- 124.35$
Cl_2 (g)	$- 107.33$
$SnCl_4$ (l)	$- 570.77$

(a) $- 339.0$ (b) $- 231.7$ (c) 231.7 (d) 339.0 (e) None of the above

$\Delta S^0 = \sum S^0$ Products $- \sum S^0$ Reactants

$= -570.77 - \{(-124.35) + (-107.33 \times 2)\} = -570.77 + 339.01$

$= -231.7 \text{ J/K}$ Answer(b)

14-23. Consider the following reaction and data:

$$CO(g) + 2H_2(g) \rightarrow CH_3OH(g)$$

Compound	ΔH^0_{ac} (kJ/mol)	S^0 (J/mol-K)
CO (g)	− 1076.38	− 121.48
H$_2$ (g)	− 435.3	− 98.78
CH$_3$OH (g)	− 2037.11	− 538.19

(i) What is the value of ΔS^0 per mole of CH$_3$OH produced in this reaction?

$\Delta S^0 = \sum S^0 \text{ Products} - \sum S^0 \text{ Reactants}$

$= -538.19 - \{(-121.48) + (-98.78 \times 2)\} = -219.2 \text{ J/K}$

(ii) What is the value of ΔG^0 per mole of CH$_3$OH produced in this reaction?

$\Delta H^0 = \sum H^0_f \text{ Products} - \sum H^0_f \text{ Reactants}$

$= -2037.11 - \{(-1076.38) + (-435.3 \times 2)\} = -90.13 \text{ kJ/mol} = -90130 \text{ J/mol}$

$\Delta G^0 = \Delta H^0 - T\Delta S^0 = -90130 - (-219.2 \times 298) = -24808.4 \text{ J} = -24.8 \text{ kJ}$

(iii) Which of the following statements about this reaction is true?

(a) The reaction can never become favorable
(b) The reaction can never become unfavorable
(c) The reaction will become more favorable as the temperature is increased
(d) The reaction will become less favorable as the temperature is increased
(e) Temperature has no effect on whether this reaction is favorable

At 298K $\quad \Delta G^0 = \Delta H^0 - T\Delta S^0 = -24.8 \text{ kJ}$

At 500K $\quad \Delta G^0 = \Delta H^0 - T\Delta S^0 = -90130 - (-219.2 \times 500) = +19.47 \text{ kJ}$

The reaction becomes unfavorable as the temperature is increased. Answer(d)

(iv) Calculate the equilibrium constant for this reaction at 25 °C.

$\Delta G^0 = -24808 \text{ J at } 25 \text{ °C.}$

$\Delta G^0 = -RT \ln K$

$\ln K = -(\Delta G^0/RT) = -(-24808)/(8.314 \times 298) = 10.0$

$K = 2.2 \times 10^4$

14-24. Consider the following reaction and data:

$$2Fe^{3+} (aq) + Cu (s) \rightarrow 2Fe^{2+} (aq) + Cu^{2+} (aq)$$

Compound	ΔH^0_{ac} (kJ/mol)	S^0 (J/mol-K)
Cu (s)	− 338.32	− 133.23
Cu^{2+} (aq)	− 273.55	− 266.00
Fe^{2+} (aq)	− 505.4	− 318.20
Fe^{3+} (aq)	− 464.8	− 496.40

(i)   Calculate the ΔS^0 for this reaction.

$$\Delta S^0 = \sum S^0 \text{ Products} - \sum S^0 \text{ Reactants}$$

$$= \{(-318.2 \times 2) + (-266.0)\} - \{(-496.4 \times 2) + (-133.23)\} = 223.63 \text{ J/K}$$

(ii)   Calculate the equilibrium constant for this reaction.

$$\Delta H^0 = \sum H^0_f \text{ Products} - \sum H^0_f \text{ Reactants}$$

$$= \{(-505.4 \times 2) + (-273.55)\} - \{(-464.8 \times 2) + (-338.32)\} = -16430 \text{ J/mol}$$

$$\Delta G^0 = \Delta H^0 - T\Delta S^0 = -16430 - (298 \times 223.63) = -66642 \text{ J.}$$

$$\Delta G^0 = -RT \ln K$$

$$\text{Ln } K = -(\Delta G^0/RT) = -(-66642)/(8.314 \times 298) = +26.90 \qquad K = 4.8 \times 10^{11}$$

(iii)   If $\Delta G^0 = -nFE^0$, then

(a) E^0 is between − 1.00 V and − 0.5 V and the reaction is not spontaneous.
(b) E^0 is between − 0.5 V and 0.0 V and the reaction is not spontaneous.
(c) E^0 is between 0.0 V and 0.5 V and the reaction is spontaneous.
(d) E^0 is between 0.5 V and 1.0 V and the reaction is spontaneous.
(e) None of the above.

Oxidation: $Cu \rightarrow Cu^{2+} + 2e^-$

Reduction: $\underline{2\{Fe^{3+} + e^- \rightarrow Fe^{2+}\}}$

$$2Fe^{3+} + Cu \rightarrow Cu^{2+} + 2Fe^{2+} \qquad n = 2$$

$$\Delta G^0 = -nFE^0$$

ΔG^0 is negative. So the reaction will be spontaneous.

$$-66642 = -2 \times 96485 \times E^0$$

$E^0 = 0.34$ V   Answer(c)

14-25. Which statement correctly describes the following reaction?

$$PbO\ (s) + C\ (s) \rightarrow Pb\ (s) + CO\ (s) \quad \Delta S^0 = 188\ J/K, \quad \Delta H^0 = 106\ kJ/mol_{rxn}$$

(a) The reaction is spontaneous at all temperatures
(b) The reaction is spontaneous only below 200 °C
(c) The reaction becomes spontaneous at about 300 °C
(d) The reaction becomes spontaneous at about 560 °C
(e) The reaction is spontaneous only above 2000 °C

$$\Delta G^0 = \Delta H^0 - T\Delta S^0$$

T	ΔG^0
563K (290 °C)	+ 156 J
564K (291 °C)	− 32 J
573K (300 °C)	− 1724 J

The reaction becomes spontaneous at 291 °C.   Answer(c)

14-26. The entropy change in the process

$$2H\ (g) \rightarrow H_2\ (g) \text{ is } - 98.74\ J/mol\text{-}K$$

and ΔH^0_{ac} for H_2 (g) $= - 435$ kJ/mol. Estimate the temperature above which the formation of H_2

molecules from H atoms will no longer be a spontaneous process.

(a) 500K   (b) 1250K   (c) 2300K   (d) 4400K   (e) 6500K

$$\Delta G^0 = \Delta H^0 - T\Delta S^0$$

T	ΔG^0
2300K	− 207898 J
4405K	− 50.30 J
4406K	+ 48.44 J

Above 4405K ($\approx$4400K) the formation will not be spontaneous.   Answer(d)

14-27. Which of the following statements is true?

(a) $\Delta G = \Delta G^0$ when the reaction is at equilibrium
(b) ΔG is a measure of how far the reaction is from equilibrium
(c) ΔG^0 is positive for reactions that have too much reactants in the standard state.
(d) The value of ΔG does not depend on the temperature of the reaction
(e) None of these statements is true

$\Delta G^0 = \Delta H^0 - T\Delta S^0$ and $\Delta G = \Delta G^0 + RT \ln Q$

$\Delta G = \Delta G^0$ under standard-state conditions.

$\Delta G^0 > 0$ when $\Delta H^0 > 0$ and $\Delta S^0 < 0$

When the reaction is far from equilibrium the ΔG value will be high. A – ve ΔG value means that the reaction has to shift to right to reach equilibrium and a + ve ΔG value means that the reaction has to shift to left to reach equilibrium. The magnitude of ΔG will get smaller as the reaction approaches equilibrium and at equilibrium $\Delta G = 0$. Answer(b)

14-28. Which of the following statements is false?

(a) ΔG is equal to ΔG^0 when the system is at the standard-state
(b) ΔG is zero when the system is at equilibrium
(c) ΔG measures how far the reaction is from equilibrium
(d) When ΔG is + ve, the reaction should proceed forward to form more product
(e) All of the above statements is true

$\Delta G = \Delta G^0 + RT \ln Q$

$\Delta G = \Delta G^0$ under standard-state conditions.

$\Delta G = 0$ at equilibrium and $K_p = Q_p$

When the reaction is far from equilibrium the value of ΔG will be high. The magnitude of ΔG will get smaller as the reaction approaches equilibrium. If ΔG is – ve, the reaction must shift to right to reach equilibrium and if ΔG is + ve, the reaction must shift to left to reach equilibrium.

Answer(d)

14-29. When the reaction is at equilibrium, which of the following is true?

(a) $\Delta G = 0$ (b) $\Delta G = \Delta G^0$ (c) $\Delta G^0 = 0$ (d) $Q = 0$ (e) $\ln K = 0$

$\Delta G = \Delta G^0 + RT \ln Q$

When the reaction is at equilibrium, there is no driving force behind the reaction and $\Delta G = 0$.

Answer (a)

14-30. $\Delta G^0 = - 53.6$ kJ/mol$_{rxn}$ for the following reaction at 25 ^{0}C:

$$Zn^{2+} + 4NH_3 \rightarrow Zn (NH_3)_4^{2+}$$

What is the value of the complex formation equilibrium constant for the reaction?

$\Delta G^0 = - RT \ln K_f$

$\ln K_f = - (\Delta G^0)/(RT) = - (- 53600)/(8.314 \times 298) = 21.63$

$K_f = 2.5 \times 10^9$

14-31. What is the equilibrium constant for the following reaction if ΔG^0 at 298K is

$- 16.5$ kJ/mol$_{rxn}$?

$$N_2 \text{ (g)} + 3H_2 \text{ (g)} \rightarrow 2NH_3 \text{ (g)}$$

$\Delta G^0 = - RT \ln K$

$\ln K = - (\Delta G^0)/(RT) = - (- 16500)/(8.314 \times 298) = 6.66$ $\qquad$ $K = 780$

14-32. $E^0 = 0.62$ V for the following reaction:

$$Sn^{2+} \text{ (aq)} + 2Fe^{3+} \text{ (aq)} \rightarrow Sn^{4+} \text{ (aq)} + 2Fe^{2+} \text{ (aq)}$$

What is the ΔG^0 for this reaction at 25 °C?

$\Delta G^0 = - nFE^0 = - 2 \times 96485 \times 0.62 = - 119.6$ kJ

14-33. K_a for benzoic acid is 3.0×10^{-5}. What is the free energy change for the dissociation of this acid at 25 °C?

$\Delta G^0 = - RT \ln K = - 8.314 \times 298 \times \ln (3.0 \times 10^{-5})$

$= 25.8$ kJ

14-34. The following reaction is used to produce a mixture of CO and H_2 known as synthetic gas:

$$C \text{ (s)} + H_2O \text{ (g)} \rightarrow CO \text{ (g)} + H_2 \text{ (g)}$$

If $\Delta H^0 = 131.3$ kJ/mol$_{rxn}$ and $\Delta S^0 = 134.4$ J/mol-K, at what temperature must the reaction be carried out so that the equilibrium constant is larger than 1?

$\qquad$ $\Delta G^0 = \Delta H^0 - T\Delta S^0$ $\qquad\qquad$ $\Delta G^0 = - RT \ln K$

T	ΔG^0 J	K
976K	125.6	0.98
977K	$- 8.8$	1.001

The reaction must be carried out at or above 977K if the equilibrium constant has to be > 1.

14-35. For the reaction,

$$2SO_3 \text{ (g)} \rightarrow 2SO_2 \text{ (g)} + O_2 \text{ (g)}$$

the following thermodynamic data are available:

Compound	ΔH^0_{ac} (kJ/mol)	S^0 (J/mol-K)
SO_3 (g)	− 1422.04	− 394.23
SO_2 (g)	− 1073.95	− 241.21
O_2 (g)	− 498.34	−116.97

What is the equilibrium constant for this reaction at 25 °C?

$\Delta S^0 = \sum S^0$ Products $- \sum S^0$ Reactants $= \{(- 241.21 \times 2) + (- 116.97)\} - (- 394.23 \times 2)$

$= 189.07$ J/K

$\Delta H^0 = \sum H_f^0$ Products $- \sum H_f^0$ Reactants $= \{(- 1073.95 \times 2) + (- 498.34)\} - (-1422.04 \times 2)$

$= 197.84$ kJ

$\Delta G^0 = \Delta H^0 - T\Delta S^0 = 197840 - (298 \times 189.07) = 141.5$ kJ

$\Delta G^0 = - RT \ln K$

$\ln K = - (\Delta G^0)/RT = - (141500)/(8.314 \times 298) = - 57.1$

$K = 1.6 \times 10^{-25}$

14-36. The standard-state cell potential for the following reaction is 0.23 V:

$$2Fe^{3+} \text{ (aq)} + 2I^- \text{ (aq)} \rightarrow 2Fe^{2+} \text{ (aq)} + I_2 \text{ (aq)}$$

If $\Delta G^0 = - nFE^0$ and F = 96485 C, what is the equilibrium constant for this reaction at 25 °C?

Oxidation: $2I^- \rightarrow I_2 + 2e^-$

Reduction: $2Fe^{3+} + 2e^- \rightarrow 2Fe^{2+}$

$2Fe^{3+} + 2I^- \rightarrow 2Fe^{2+} + I_2$ n = 2

$\Delta G^0 = - 2 \times 96485 \times 0.23 = - 44383$ J

$\ln K = - (\Delta G^0)/RT = - (- 44383)/(8.314 \times 298) = 17.91$

$K = 6 \times 10^7$

14-37. An excess of H_2 is allowed to react with equal amounts of F_2, Cl_2, Br_2, and I_2 in a sealed

flask. Which of the following products is present in the largest concentration at equilibrium?

(a) HF (b) HCl (c) HBr (d) HI

(e) all four products are present in equal concentrations

H_2 (g) + F_2 (g) $\rightarrow$ 2HF $\Delta G^0 = - 190$ kJ

H_2 (g) + Cl_2 (g) $\rightarrow$ 2HCl $\Delta G^0 = - 546$ kJ

H_2 (g) + Br_2 (g) → 2HBr $\Delta G^0 = -106$ kJ

H_2 (g) + I_2 (g) → 2HI $\Delta G^0 = 4$ kJ

Let us consider the temperature 298K

Ln K = $- (\Delta G^0)/RT$

(a) HF: ln K = $-(-190000)/(8.314 \times 298) = 76.7$ K= 2×10^{33}

(b) HCl: ln K = $-(-546000)/(8.314 \times 298) = 220.4$ K= 5.1×10^{95}

(c) HBr: ln K = $-(-106000)/(8.314 \times 298) = 42.78$ K= 3.8×10^{18}

(d) HI: ln K = $-(4000)/(8.314 \times 298) = -1.61$ K= 0.2

The K for H_2 (g) + Cl_2 (g) → 2HCl (g) is the largest. So theoretically HCl will be present in the largest concentration at equilibrium.

CHAPTER 15

ELECTROCHEMISTRY

INVALUABLE INFORMATION

Gas Constant (R)	: 8.31447 J/(mol K)
Gas Constant (R)	: 0.0820575 (L atm)/(mol K)
Avogadro's number	: 6.022142×10^{23}/mol
Charge on one mole of electrons (F)	: 9.6485338×10^4 C/mol

$\Delta G^0 = - nFE^0_{cell}$

$\Delta G^0 = - RT \ln K$

$E^0_{cell} = (RT/nF) \ln K = (0.0591/n) \log K$ at T = 298K

$\Delta G = \Delta G^0 + RT \ln Q$

$E_{cell} = E^0_{cell} - (0.0591/n) \log Q$ at T = 298K

15-1. What is the E^0_{cell} for

$$2Al (s) + 3Sn^{2+} \rightarrow 2Al^{3+} + 3Sn (s)?$$

$$Al^{3+} + 3e^- \rightarrow Al (s) \quad E^0_{red} = - 1.66 \text{ V}$$

$$Sn^{2+} + 2e^- \rightarrow Sn (s) \quad E^0_{red} = - 0.137 \text{ V}$$

Standard-state reduction potential is an intrinsic property meaning that if the stoichiometric coefficients are multiplied by 'x', the standard state potential will not change.

$2 \{ Al (s) \rightarrow Al^{3+} + 3e^- \}$ $E^0_{oxi} = - E^0_{red} = + 1.66 \text{ V}$ $n_1 = 6$ $\Delta G^0_{oxi} = - 6FE^0_{oxi}$

$3 \{Sn^{2+} + 2e^- \rightarrow Sn (s)\}$ $E^0_{red} = - 0.137 \text{ V}$ $n_2 = 6$ $\Delta G^0_{red} = - 6FE^0_{red}$
_____

$2Al (s) + 3Sn^{2+} \rightarrow 2Al^{3+} + 3Sn (s)$ $E^0_{cell} = ?$ $n_3 = 6$

ΔG^0 is a state function. So $\Delta G^0_{cell} = \Delta G^0_{oxi} + \Delta G^0_{red}$

$- 6FE^0_{cell} = - 6FE^0_{oxi} - 6FE^0_{red}$

$E^0_{cell} = E^0_{oxi} + E^0_{red} = 1.66 - 0.137 = 1.539$ V

So in redox reactions where $n_1 = n_2 = n_3$ and where we are eliminating electrons we can use the equation $E^0_{cell} = E^0_{oxi} + E^0_{red}$

15-2. What is the E^0_{red} for half-reaction $Fe^{3+} + 3e^- \rightarrow Fe$ (s)?

$$Fe^{3+} + e^- \rightarrow Fe^{2+} \qquad E^0_{red} = 0.771 \text{ V}$$

$$Fe^{2+} + 2e^- \rightarrow Fe \text{ (s)} \quad E^0_{red} = -0.44 \text{ V}$$

When we add two half-reactions to obtain a third half-reaction where $n_1, n_2,$ and n_3 are not same and where we are not eliminating electrons we cannot use the equation $E^0_3 = E^0_1 + E^0_2$ (E^0 is not a state function). We have to use the equation $\Delta G^0_3 = \Delta G^0_1 + \Delta G^0_2$.

$\Delta G^0 = - nFE^0$

$Fe^{3+} + e^- \rightarrow Fe^{2+} \qquad E^0_{red1} = 0.771 \text{ V} \qquad n_1 = 1 \qquad \Delta G^0 = -1FE^0_{red1}$

$Fe^{2+} + 2e^- \rightarrow Fe \text{ (s)} \quad E^0_{red2} = -0.44 \text{ V} \quad n_2 = 2 \quad \Delta G^0 = -2FE^0_{red2}$

$Fe^{3+} + 3e^- \rightarrow Fe \text{ (s)} \quad E^0_{red3} = ? \qquad\qquad n_3 = 3$

$\Delta G^0_3 = \Delta G^0_1 + \Delta G^0_2$

$- 3FE^0_{red3} = - 1FE^0_{red1} - 2FE^0_{red2}$

$3E^0_{red3} = (1 \times 0.771) + \{2 \times (-0.44)\} = - 0.109$

$E^0_{red3} = (- 0.109)/3 = - 0.036$ V

15-3. Determine whether the following reaction is likely to occur spontaneously under standard-state conditions:

$4MnO_2$ (s) $+ 3O_2$ (g) $+ 4OH^-$ (aq) $\rightarrow 4MnO_4^-$ (aq) $+ 2H_2O$

$4\{MnO_2 + 4OH^- \rightarrow MnO_4^- + 2H_2O + 3e^-\} \qquad E^0_{oxi} = - 0.59$ V

$3\{O_2 + 2H_2O + 4e^- \rightarrow 4OH^-\} \qquad\qquad E^0_{red} = 0.401$ V

$4MnO_2 + 3O_2 + 4OH^- \rightarrow 4MnO_4^- + 2H_2O \quad E^0_{cell} = E^0_{oxi} + E^0_{red} = - 0.59 + 0.401 = - 0.189$ V

$\Delta G^0 = - nFE^0$

$\Delta G^0 > 0$. The reaction will not occur spontaneously.

15-4. Calculate the ΔG^0 for the reduction of ferric ion by iodide:

$2Fe^{3+}$ (aq) $+ 2I^-$ (aq) $\rightarrow 2Fe^{2+}$ (aq) $+ I_2$ (s)

Will the reaction occur spontaneously?

$2I^- (aq) \rightarrow I_2 (s) + 2e^-$ $E^0_{oxi} = -0.5355$ V

$\underline{2\{Fe^{3+} (aq) + e^- \rightarrow Fe^{2+} (aq)\}}$ $E^0_{red} = 0.770$ V

$2Fe^{3+} (aq) + 2I^- (aq) \rightarrow 2Fe^{2+} (aq) + I_2 (s)$ $E^0_{cell} = ?$ $n = 2$

$E^0_{cell} = E^0_{oxi} + E^0_{red} = -0.5355 + 0.770 = 0.2345$ V

$\Delta G^0 = -nFE^0 = -2 \times 96485 \times 0.2345 = -4.5 \times 10^4$ J/mol $= -45$ kJ/mol

$\Delta G^0 < 0$. The reaction will occur spontaneously.

15-5. Calculate the equilibrium constant for the reaction of Sn^{2+} with O_2 to produce Sn^{4+} (aq) and water under standard-state conditions.

$2Sn^{2+} (aq) + O_2 (g) + 4H^+ (aq) \rightleftharpoons 2Sn^{4+} (aq) + 2H_2O (l)$

$2\{Sn^{2+} \rightarrow Sn^{4+} + 2e^-\}$ $E^0_{oxi} = -0.15$ V

$\underline{O_2 (g) + 4H^+ + 4e^- \rightarrow 2H_2O (l)}$ $E^0_{red} = 1.229$ V

$2Sn^{2+} + O_2 + 4H^+ \rightarrow 2Sn^{4+} + 2H_2O$ $E^0_{cell} = -0.15 + 1.229 = 1.079$ V $n = 4$

$E^0_{cell} = (RT/nF) \ln K = (0.0591/n) \log K$ at $T = 298K$

$E^0_{cell} = (0.0591/4) \log K = 1.079$

$\log K = (4 \times 1.079)/(0.0591) = 73.03$

$K = 1.07 \times 10^{73}$

15-6. $4MnO_2 (s) + 3O_2 (g) + 4OH^- (aq) \rightarrow 4MnO_4^- (aq) + 2H_2O (l)$ $E^0_{cell} = -0.189$ V

Calculate E_{cell} for the above reaction under the following non-standard conditions and decide whether the reaction will occur spontaneously:

pH = 10, $P_{Oxygen} = 0.2$ atm. and $(MnO_4^-) = 1.0 \times 10^{-4}$M.

$4\{MnO_2 + 4OH^- \rightarrow MnO_4^- + 2H_2O + 3e^-\}$ $E^0_{oxi} = -0.59$ V

$\underline{3\{O_2 + 2H_2O + 4e^- \rightarrow 4OH^-\}}$ $E^0_{red} = 0.401$ V

$4MnO_2 + 3O_2 + 4OH^- \rightarrow 4MnO_4^- + 2H_2O$ $E^0_{cell} = -0.59 + 0.401 = -0.189$ V $n = 12$

$n = 12$, pH = 10; $(OH^-) = 10^{-4}$ M

$P_{Oxygen} = 0.2$, $(MnO_4^-) = 1.0 \times 10^{-4}$ M

$E_{cell} = (E^0_{cell}) - (0.0591/12) \log [(MnO_4^-)^4/\{(P_{Oxygen})^3 \times (OH^-)^4\}]$

$\quad = (E^0_{cell}) - (0.0591/12) \log [(10^{-4})^4/\{(0.2)^3 \times (10^{-4})^4\}]$

$\quad = -0.189 - (0.0591/12) \log 125$

$$= -0.189 - 0.01 = -0.199 \text{ V}$$

$\Delta G = -nFE$ $\Delta G > 0$. The reaction will not be spontaneous.

15-7. A concentration cell contains 1.0 M solution of Lanthanum nitrate $La(NO_3)_3$ in one compartment and a 1.0 M solution of sodium fluoride saturated with LaF_3 in another. Metallic La strip is inserted into each compartment and the circuit is closed. The measured potential is 0.32 V. What is the K_{sp} of LaF_3?

Cathode: La^{3+} (aq) $+ 3e^- \rightleftarrows$ La (s)   Anode: La (s) $\rightleftarrows$ $La^{3+} + 3e^-$ $n = 3$

K_{sp} of $LaF_3 = [La^{3+}] [F^-]^3 = [La^{3+}] [1]^3 = [La^{3+}]$

Overall reaction : La^{3+} (aq,con) $\rightarrow La^{3+}$ (aq, dil)

Concentration cell: $E^0{}_{cell} = E^0{}_{oxi} + E^0{}_{red} = 0$

E_{cell} $= 0 - (0.0591/3) \log \{(La^{3+} \text{ dil})/(La^{3+} \text{ con})\}$

0.32 $= - (0.0591/3) \log (K_{sp}/1)$

$\text{Log } K_{sp} = - (3 \times 0.32)/(0.0591) = -16.24$ $K_{sp} = 5.75 \times 10^{-17}$

15-8. Suppose you work for an environmental lab and want to use an electrochemical method to measure the concentration of Pb^{2+} in ground water. You construct a galvanic cell using a standard oxygen electrode in one compartment ($E^0{}_{cathode} = 1.23$ V). The other compartment contains a strip of lead in a sample of ground water to which you have added sufficient acetic acid, a weak organic acid, to ensure electrical conductivity. The cell diagram is as follows:

Pb (s) | Pb^{2+} (aq, ? M) || H^+ (aq, 1.0 M) | O_2 (g, 1atm) | Pt (s)

When the circuit is closed ,the cell has a measured potential of 1.62 V. What is the concentration of Pb^{2+} in ground water?

$2\{Pb \text{ (s)} \rightarrow Pb^{2+} + 2e^-\}$ $E^0{}_{oxi} = 0.126$ V

$O_2 + 4H^+ + 4e^- \rightarrow 2H_2O$ $E^0{}_{red} = 1.23$ V

$2Pb \text{ (s)} + O_2 + 4H^+ \rightarrow 2Pb^{2+} + 2H_2O$ $E^0{}_{cell} = 1.23 + 0.126 = 1.356$ V $n = 4$

$E_{cell} = (E^0{}_{cell}) - (0.0591/4) \log \{[Pb^{2+}]^2/([H^+]^4 \times [P_{Oxygen}])\}$

$E_{cell} - E^0{}_{cell} = - (0.0591/4) \log [Pb^{2+}]^2$

$1.62 - 1.356 = - (0.0591/4) \log [Pb^{2+}]^2$

$\log [Pb^{2+}]^2 = - (4 \times 0.264)/(0.0591) = -17.87$

$2 \log [Pb^{2+}] = -17.87$ $\log [Pb^{2+}] = -8.935$ $[Pb^{2+}] = 1.2 \times 10^{-9}$ M

15-9. Calculate the potential in the following cell when 99.99% of the Cu^{2+} ions have been consumed:

$$Zn \mid Zn^{2+} \ (1.00 \ M) \parallel Cu^{2+} \ (1.00 \ M) \mid Cu$$

$Zn \ (s) \rightarrow Zn^{2+} \ (aq) + 2e^-$ $E^0_{oxi} = 0.7628$ V

$Cu^{2+} \ (aq) + 2e^- \rightarrow Cu \ (s)$ $E^0_{red} = 0.3402$ V

$\overline{Zn \ (s) + Cu^{2+} \rightarrow Zn^{2+} + Cu \ (s)}$ $E^0_{cell} = 0.7628 + 0.3402 = 1.1030$ V $n = 2$

$E_{cell} = (E^0_{cell}) - (0.0591/2) \log\{(Zn^{2+})/(Cu^{2+})\}$

	Zn (s)	+	Cu^{2+}	$\rightarrow$	Zn^{2+}	+ Cu (s)

Initial   1.0 M   1.0 M $E_1 = E^0_{cell} = 1.1030$ V

At time 't'(99.99% of Cu^{2+}

have been consumed) $(1 - 0.9999)$ M $(1 + 0.9999)$ M $E_2 = ?$

$E_2 = (E^0_{cell}) - (0.0591/2) \log \{(1 + 0.9999)/(1 - 0.9999)\}$

   $= 1.1030 - \{(0.0591/2) \times (4.3)\}$

   $= 1.1030 - 0.127 = 0.976$ V

15-10. The standard state potential for the following half-reaction

$$C_6H_{12}O_6 + H_2O \rightarrow C_6H_{12}O_7 + 2H^+ + 2e^- \ \text{is} \ E^0_{oxi} = -0.05 \ V.$$

Calculate the potential for the above half-reaction if the reaction is run is at pH = 11.

$E = (E^0_{oxi}) - (0.0591/2) \log (H^+)^2$;   pH= 11, so $(H^+) = 10^{-11}$

$E = (E^0_{oxi}) - (0.0591/2) \log (10^{-11})^2$

$E = (-0.05) - (0.0591) \log (10^{-11})$

   $= (-0.05) + (0.650) = 0.600$ V

15-11. Which of the following isn't an example of an oxidation reduction reaction?

   (a) Ca_3P_2 (s) + $6H_2O$ (l) $\rightarrow$ $3Ca^{2+}$ (aq) + $6OH^-$ (aq) + $2PH_3$ (g)
   (b) PH_3 (g) + $2O_2$ (g) $\rightarrow$ H_3PO_4 (s)
   (c) P_4 (s) + $5O_2$ (g) $\rightarrow$ P_4O_{10} (s)
   (d) $6Ca$ (s) + P_4 (s) $\rightarrow$ $2Ca_3P_2$ (s)
   (e) All of the above

(a) $Ca_3 P_2$ (s) + $6H_2 O$ (l) $\rightarrow$ $3Ca^{2+}$ (aq) + 6 O H$^-$ (aq) + 2 P H$_3$

   +2 −3   +1 −2   +2   −2 +1   −3 +1   oxidation number

No change in the oxidation number of the atoms in this reaction. So (a) isn't an example of an oxidation-reduction reaction. Answer(a)

15-12. Identify the oxidizing agents in the following reaction:

$3P_4$ (s) + 20HNO$_3$ (aq) + 8H$_2$O (l) → 12H$_3$PO$_4$ (aq) + 20NO (g)

(a) P_4 and H_3PO_4 (b) P_4 and NO   (c) P_4 and NH_3 (d) HNO_3 and H_3PO_4

(e) HNO_3 and NO

$$3P_4 \text{ (s)} + 20H \quad N \quad O_3 + \quad 8H_2O \quad \rightarrow 12H_3 \quad P \quad O_4 + 20N \quad O$$

Oxidation number :	0	+1 +5 −2	+1 −2		+1 +5 −2	+2 −2
	Reducing	Oxidising			Oxidising	Reducing
	agent	agent			agent	agent

The oxidation number of P increases from 0 to +5. So it is a reducing agent.

So HNO_3 and H_3PO_4 are oxidizing agents. Answer(d)

15-13. For the reaction

$3Sn^{2+}$ (aq) + $Cr_2O_7^{2-}$ (aq) + 14H$^+$ (aq) → 3Sn^{4+} (aq) + 2Cr^{3+} (aq) + 7H$_2$O (l)

Which of the following statements is true?

  (a) Both Sn^{2+} and H$^+$ are oxidizing agents
  (b) $Cr_2O_7^{2-}$ is the oxidizing agent
  (c) Sn^{2+} is reduced
  (d) The acid is not important to the reaction
  (e) None of the above is true

$$3Sn^{2+} + Cr_2O_7^{2-} + 14H^+ \rightarrow 3Sn^{4+} + 2Cr^{3+} + 7H_2O$$

Oxidation number: +2   +6 −2   +1   +4   +3   +1 − 2

Sn^{2+} is oxidized to Sn^{4+}, Sn^{2+} is a reducing agent.

The oxidation number of Cr decreases from +6 to +3. So $Cr_2O_7^{2-}$ is the oxidizing agent.

Answer(b)

15-14. Write a balanced equation for the following reaction:

MnO_2 (s) + PbO$_2$ (s) + H$^+$ (aq) → MnO_4^- (aq) + Pb^{2+} (aq)

Rule: Balance the excess 'O' atom with H_2O and balance H using H$^+$ in acidic medium.

Oxidation : $MnO_2 \rightarrow MnO_4^-$

$2\{MnO_2 + 2H_2O \rightarrow MnO_4^- + 4H^+ + 3e^-\}$

Reduction: $PbO_2 \rightarrow Pb^{2+}$

$3\{4H^+ + PbO_2 + 2e^- \rightarrow Pb^{2+} + 2H_2O\}$
$\overline{\phantom{3\{4H^+ + PbO_2 + 2e^- \rightarrow Pb^{2+} + 2H_2O\}}}$

$3PbO_2 + 2MnO_2 + 4H^+ \rightarrow 3Pb^{2+} + 2MnO_4^- + 2H_2O$

$3PbO_2 \text{ (s)} + 2MnO_2 \text{ (s)} + 4H^+ \text{ (aq)} \rightarrow 3Pb^{2+} \text{ (aq)} + 2MnO_4^- \text{ (aq)} + 2H_2O\text{(l)}$

15-15. Write a balanced equation for the following reaction which can be used to standardize aqu. permanganate ion solutions:

$H_2C_2O_4 \text{ (aq)} + MnO_4^- \text{ (aq)} + H^+ \text{ (aq)} \rightarrow CO_2 \text{ (g)} + Mn^{2+} \text{ (aq)}$

Oxidation: $H_2C_2O_4 \rightarrow CO_2$

$5\{H_2C_2O_4 \rightarrow 2CO_2 + 2H^+ + 2e^-\}$

Reduction: $MnO_4^- \rightarrow Mn^{2+}$

$2\{MnO_4^- + 8H^+ + 5e^- \rightarrow Mn^{2+} + 4H_2O\}$
$\overline{\phantom{2\{MnO_4^- + 8H^+ + 5e^- \rightarrow Mn^{2+} + 4H_2O\}}}$

$5H_2C_2O_4 + 2MnO_4^- + 6H^+ \rightarrow 10CO_2 + 2Mn^{2+} + 8H_2O$

$5H_2C_2O_4 \text{ (aq)} + 2MnO_4^- \text{ (aq)} + 6H^+ \text{ (aq)} \rightarrow 10CO_2 \text{ (g)} + 2Mn^{2+} \text{ (aq)} + 8H_2O \text{ (l)}$

15-16. Balance the following redox reaction. Clearly show the two balanced half-reactions and indicate which represents oxidation and which reduction?

$H_2O_2 \text{ (aq)} + NO \text{ (g)} + H^+ \text{ (aq)} \rightarrow NO_3^- \text{ (aq)}$

Oxidation: $NO \rightarrow NO_3^-$

$2\{NO + 2H_2O \rightarrow NO_3^- + 4H^+ + 3e^-\}$

Reduction: $H_2O_2 \rightarrow 2H_2O$

$3\{H_2O_2 + 2H^+ + 2e^- \rightarrow 2H_2O\}$
$\overline{\phantom{3\{H_2O_2 + 2H^+ + 2e^- \rightarrow 2H_2O\}}}$

$3H_2O_2 + 2NO \rightarrow 2NO_3^- + 2H^+ + 2H_2O$

$3H_2O_2 \text{ (aq)} + 2NO \text{ (g)} \rightarrow 2NO_3^- \text{ (aq)} + 2H^+ \text{ (aq)} + 2H_2O\text{(l)}$

15-17. What is the coefficient of sulfur in the following redox reaction?

$S \text{ (s)} + KClO_3 \text{ (s)} \rightarrow SO_2 \text{ (g)} + KCl \text{ (s)}$

(a) 1 (b) 2 (c) 3 (d) 4 (e) 5

Oxidation: $S \rightarrow SO_2$

$3\{2H_2O + S \rightarrow SO_2 + 4H^+ + 4e^-\}$

Reduction: $KClO_3 \rightarrow KCl$

$2\{6H^+ + KClO_3 + 6e^- \rightarrow KCl + 3H_2O\}$

$\overline{\phantom{2\{6H^+ + KClO_3 + 6e^- \rightarrow KCl + 3H_2O\}}}$

$3S + 2KClO_3 \rightarrow 3SO_2 + 2KCl$

$3S\ (s) + 2KClO_3\ (s) \rightarrow 3SO_2\ (g) + 2KCl\ (s)$

The coefficient of S = 3   Answer(c)

15-18. Balance the following reaction:

$$Cr_2O_7^{2-}\ (aq) + I^-\ (aq) + H^+\ (aq) \rightarrow Cr^{3+}\ (aq) + I_2\ (aq)$$

Oxidation: $I^- \rightarrow I_2$

$3\{2I^- \rightarrow I_2 + 2e^-\}$

Reduction: $Cr_2O_7^{2-} \rightarrow Cr^{3+}$

$Cr_2O_7^{2-} + 14H^+ + 6e^- \rightarrow 2Cr^{3+} + 7H_2O$

$\overline{\phantom{Cr_2O_7^{2-} + 14H^+ + 6e^- \rightarrow 2Cr^{3+} + 7H_2O}}$

$Cr_2O_7^{2-} + 14H^+ + 6I^- \rightarrow 3I_2 + 2Cr^{3+} + 7H_2O$

$Cr_2O_7^{2-}\ (aq) + 14H^+\ (aq) + 6I^-\ (aq) \rightarrow 3I_2\ (aq) + 2Cr^{3+}\ (aq) + 7H_2O(l)$

15-19. Consider the following reaction:

$$OCl^-\ (aq) \xrightarrow{\quad OH^- \quad} ClO_2\ (aq) + Cl_2\ (aq)$$

(i)   How many OCl^- ions are consumed in the balanced equation for this reaction?

      (a) 2   (b) 3   (c) 4 (d) 5 (e) 6

Oxidation: $OCl^- \rightarrow ClO_2^-$

$OCl^- + 2OH^- \rightarrow ClO_2^- + H_2O + 2e^-$

Reduction: $OCl^- \rightarrow Cl_2$

$2H_2O + 2OCl^- + 2e^- \rightarrow Cl_2 + 4OH^-$

$\overline{}$

$H_2O + 3OCl^- \rightarrow Cl_2 + ClO_2^- + 2OH^-$

$H_2O(l) + 3OCl^-\ (aq) \rightarrow Cl_2\ (aq) + ClO_2^-\ (aq) + 2OH^-\ (aq)$

3 OCl^- ions are consumed.   Answer(b)

(ii)   How many OH^- ions are involved in the balanced equation for this reaction?

(a) 0 (b) 1 (c) 2 (d) 3 (e) 4

$$H_2O(l) + 3OCl^- (aq) \rightarrow Cl_2 (aq) + ClO_2^- (aq) + 2OH^- (aq)$$

2 OH^- ions are involved.   Answer(c)

(iii)   What is the net effect on the OH^- ion concentration in this reaction?
(a) The OH^- ion concentration increases
(b) The OH^- ion concentration decreases
(c) The OH^- ion concentration remains the same
(d) There is no way to predict what happens to the OH^- ion concentration in this reaction

$$H_2O + 3OCl^- + 2OH^- \rightarrow Cl_2 + ClO_2^- + 4OH^-$$

The OH^- ion concentration increases.   Answer(a)

15-20. Consider the following reaction:

$$CrO_4^{2-} + PH_3 \xrightarrow{OH^-} Cr(OH)_4^- + P_4$$

(i)   Which of the following are oxidizing agents in this reaction?

(a) CrO_4^{2-} and PH_3 (b) CrO_4^{2-} and P_4 (c) $Cr(OH)_4^-$ and PH_3

(d)$Cr(OH)_4^-$ and P_4 (e) CrO_4^{2-} and OH^-

$$CrO_4^{2-} + P\ H_3 \xrightarrow{OH^-} Cr(OH)_4^- + P_4$$

Oxidation number: +6 −2   −3 +1   +3 −2 +1   0

The oxidation number of Cr decreases from + 6 to +3; CrO_4^{2-} is an oxidizing agent.

The oxidation number of P increases from − 3 to 0; PH_3 is a reducing agent.

$$CrO_4^{2-} + \quad PH_3 \xrightarrow{OH^-} Cr(OH)_4^- + \quad P_4$$

Oxidizing   Reducing   Reducing   Oxidizing

agent   agent   agent   agent   Answer(b)

(ii)   How many electrons are transferred in the half reaction which involves CrO_4^{2-}?

(a) 1 (b) 2 (c) 3 (d) 4 (e) 5

Reduction: $CrO_4^{2-} \rightarrow Cr(OH)_4^-$

$$4H_2O + CrO_4^{2-} + 3e^- \rightarrow Cr(OH)_4^- + 4OH^-$$

3 electrons are transferred. Answer(c)

    (iii)   How many OH^- ions are produced in the half-reaction which involves CrO_4^{2-}?

        (a) 1 (b) 2 (c) 3 (d) 4 (e) 5

$$4H_2O + CrO_4^{2-} + 3e^- \rightarrow Cr(OH)_4^- + 4OH^-$$

4 OH^- ions are produced Answer(d)

    (iv)   In the simplest balanced chemical equation, is OH^- produced or consumed?

Oxidation: $PH_3 \rightarrow P_4$

$$4PH_3 + 12OH^- \rightarrow P_4 + 12H_2O + 12e^-$$

Reduction: $CrO_4^{2-} \rightarrow Cr(OH)_4^-$

$$4\{4H_2O + CrO_4^{2-} + 3e^- \rightarrow Cr(OH)_4^- + 4OH^-\}$$
$$\overline{4CrO_4^{2-} + 4H_2O + 4PH_3 \rightarrow 4Cr(OH)_4^- + P_4 + 4OH^-}$$

OH^- is produced.

    (v)   How many CrO_4^{2-} ions are consumed in the balanced chemical equation?

        (a) 1 (b) 2 (c) 3 (d) 4 (e) 6

$4CrO_4^{2-} + 4H_2O + 4PH_3 \rightarrow 4Cr(OH)_4^- + P_4 + 4OH^-$ 4 CrO_4^{2-} ions are consumed.

Answer(d)

15-21. Consider the following reaction in acid solution:

H^+

$CuS\ (s) + NO_3^-\ (aq) \longrightarrow Cu^{2+}\ (aq) + SO_4^{2-}\ (aq) + NO\ (g)$

    (i)   In the complete balanced overall reaction, what is the coefficient of the NO_3^- ion?

        (a) 3 (b) 4 (c) 5 (d) 6 (e) None of the above

Oxidation: $CuS \rightarrow Cu^{2+} + SO_4^{2-}$

$$3\{CuS + 4H_2O \rightarrow Cu^{2+} + SO_4^{2-} + 8H^+ + 8e^-\}$$

Reduction: $NO_3^- \rightarrow NO$

$$8\{NO_3^- + 4H^+ + 3e^- \rightarrow NO + 2H_2O\}$$
$$\overline{8NO_3^- + 8H^+ + 3CuS \rightarrow 8NO + 4H_2O + 3Cu^{2+} + 3SO_4^{2-}}$$

The coefficient of NO_3^- ion is 8. Answer(e)

    (ii)   In the balanced half-reaction for the NO_3^- ion, how many electrons are involved?

(a) 1 (b) 2 (c) 3 (d) 4 (e) None of these

Reduction: $NO_3^- \rightarrow NO$

$NO_3^- + 4H^+ + 3e^- \rightarrow NO + 2H_2O$

3 electrons are involved.   Answer(c)

(iii)   Which of the following isn't true?
         (a) The reducing agent is CuS
         (b) CuS is oxidized
         (c) The NO_3^- is the oxidizing agent
         (d) The NO_3^- ion is reduced
         (e) The oxidation number of the copper changes from 0 to +2

$$8NO_3^- + 8H^+ + 3CuS \rightarrow 8NO + 4H_2O + 3Cu^{2+} + 3SO_4^{2-}$$

Oxidation number: +5 −2   +1   +2 −2   +2 −2   +1 −2   +2   +6 −2

In the above reaction CuS is oxidized and NO_3^- ion is reduced. The oxidation number of copper

does not change.   Answer(e)

15-22. How many moles of Fe^{2+} can be oxidized to Fe^{3+} by 0.5 moles of sodium permanganate

$(NaMnO_4)$?

    (a) 1 (b) 2 (c) 2.5 (d) 3 (e) None of the above

Oxidation: $Fe^{2+} \rightarrow Fe^{3+}$

$5\{Fe^{2+} \rightarrow Fe^{3+} + e^-\}$

Reduction: $MnO_4^- \rightarrow Mn^{2+}$

$8H^+ + MnO_4^- + 5e^- \rightarrow Mn^{2+} + 4H_2O$
_____
$5Fe^{2+} + 8H^+ + MnO_4^- \rightarrow 5Fe^{3+} + Mn^{2+} + 4H_2O$

1 mole of sodium permanganate oxidizes 5 moles of Fe^{2+}

0.5 mole of sodium permanganate will oxidize 2.5 moles of Fe^{2+} Answer(c)

15-23. How many moles of iodide will be oxidized by one mole of permanganate ion in the

following reaction?

    MnO_4^- (aq) + I^- (aq) + H^+ (aq) $\rightarrow$ Mn^{2+} (aq) + I_2 (aq) + H_2O (l)

    (a) 1 (b) 2 (c) 3 (d) 4 (e) 5

$$\text{Oxidation: } I^- \rightarrow I_2$$

$$5\{2I^- \rightarrow I_2 + 2e^-\}$$

$$\text{Reduction: } MnO_4^- \rightarrow Mn^{2+}$$

$$\underline{2\{8H^+ + MnO_4^- + 5e^- \rightarrow Mn^{2+} + 4H_2O\}}$$

$$16H^+ + 2MnO_4^- + 10I^- \rightarrow 2Mn^{2+} + 8H_2O + 5I_2$$

Two moles of permanganate ion oxidize 10 moles of Iodide.

One mole of permanganate ion will oxidize 5 moles of Iodide.   Answer(e)

15-24. If permanganate ion reacts with oxalic acid to form CO_2 and Mn^{2+} according to the following unbalanced equation,

$$H^+$$

$$MnO_4^- \text{ (aq)} + H_2C_2O_4 \text{ (aq)} \longrightarrow CO_2 \text{ (g)} + Mn^{2+} \text{ (aq)}$$

and 50.1 ml of MnO_4^- is needed to titrate 115 ml of 0.0250 M $H_2C_2O_4$, what is the concentration of the MnO_4^- ion solution?

   (a) 0.0115 M (b) 0.0230 M (c) 0.0574 M (d) 0.115 M (e) 0.143 M

$$\text{Oxidation: } H_2C_2O_4 \rightarrow CO_2$$

$$5\{H_2C_2O_4 \rightarrow 2CO_2 + 2H^+ + 2e^-\}$$

$$\text{Reduction: } MnO_4^- \rightarrow Mn^{2+}$$

$$\underline{2\{8H^+ + MnO_4^- + 5e^- \rightarrow Mn^{2+} + 4H_2O\}}$$

$$5H_2C_2O_4 + 2MnO_4^- + 6H^+ \rightarrow 10CO_2 + 2Mn^{2+} + 8H_2O$$

5 moles of $H_2C_2O_4$ require 2 moles of MnO_4^-

(115×0.025) millimoles of $H_2C_2O_4$ will require (2/5)×2.875 millimoles of MnO_4^-

Molarity×50.1 = (2/5)×2.875

Molarity   = {(2/5)×(2.875)/(50.1)} = 0.023 M   Answer(b)

15-25. A sample of impure zinc metal of mass 2.45g is analyzed by titration with a $KBrO_3$ solution. The sample required 40.4 ml of 0.247 M $KBrO_3$ solution. Find the percentage of zinc metal in the sample, assuming that the sample does not contain any other reducing agent (AW: Zn= 65.38 amu)

The unbalanced equation for this reaction is:

$$Zn\,(s) + BrO_3^-\,(aq) \rightarrow Zn^{2+} + Br_2\,(aq)$$

Oxidation: $Zn \rightarrow Zn^{2+}$

$$5\{Zn \rightarrow Zn^{2+} + 2e^-\}$$

Reduction: $BrO_3^- \rightarrow Br_2$

$$2BrO_3^- + 12\,H^+ + 10e^- \rightarrow Br_2 + 6H_2O$$

$$2BrO_3^- + 12\,H^+ + 5Zn \rightarrow Br_2 + 5Zn^{2+} + 6H_2O$$

2 moles of BrO_3^- will oxidize 5 moles of Zn.

2 millimoles of BrO_3^- will oxidize 5 millimoles of Zn.

(40.4×0.247) millimoles of BrO_3^- will oxidize $\{(5/2)\times(40.4\times0.247)\}$ millimoles of Zn.

9.98 millimoles of BrO_3^- will oxidize 24.95 millimoles of Zn.

9.98 millimoles of BrO_3^- will oxidize 24.95×65.38 mg of Zn.

% of Zinc = $\{(24.95\times65.38)/(2.45\times1000)\}\times100 = 66.6\%$

15-26. If a 10.0 ml sample of a $K_2C_2O_4$ solution is diluted to a total volume of 50.00 ml and then requires 15.00 ml of a 0.25 M $KMnO_4$ solution to titrate in acidic medium, what was the molarity of the original $K_2C_2O_4$ solution?

$$5H_2C_2O_4 + 2MnO_4^- + 6H^+ \rightarrow 10CO_2 + 2Mn^{2+} + 8H_2O$$

5 moles of $H_2C_2O_4$ will consume 2 moles of MnO_4^-.

5 millimoles of $H_2C_2O_4$ will consume 2 moles of MnO_4^-.

(15×0.25) millimoles of MnO_4^- will be required by $\{(5/2)\times(15\times0.25)\}$ millimoles of $H_2C_2O_4$.

3.75 millimoles of MnO_4^- will be required by 9.375 millimoles of $H_2C_2O_4$.

10 ml of original $K_2C_2O_4$ solution was used for titration.

9.375 millimoles of $K_2C_2O_4$ in 10 ml. 0.94 moles of $K_2C_2O_4$/liter   Molarity = 0.94 M

15-27. What weight of ferrous chloride can be oxidized by 3.2 ml of 3M $KMnO_4$? (AW: Fe = 55.85 amu; Cl = 35.45 amu)

$$5Fe^{2+} + 8H^+ + MnO_4^- \rightarrow 5Fe^{3+} + Mn^{2+} + 4H_2O$$

1 millimole of MnO_4^- will oxidize 5 millimoles of $Fe^{2+.}$

(3.2×3) millimoles of MnO_4^- will oxidize $(5\times3.2\times3)$ millimoles of Fe^{2+}.

9.6 millimoles of MnO_4^- will oxidize 48 millimoles of Fe^{2+}.

Molecular weight of $FeCl_2 = 55.85 + 35.45 + 35.45 = 126.75$ g/mol

3.2 ml of 3 M $KMnO_4^-$ will oxidize $\{(48 \times 126.75)/(1000)\} = 6.084$ g of $FeCl_2$.

15-28. Cobalt (III)oxide reacts with hydrogen gas to form cobalt metal and water :

$$Co_2O_3 \text{ (s)} + 3H_2 \text{ (g)} \rightarrow 2Co \text{ (s)} + 3H_2O$$

What does this reaction tell you about the relative strength of the oxidizing and reducing agents

in this reaction?

    (a) The Co^{3+} ion is a stronger reducing agent than cobalt metal
    (b) Cobalt metal is a better reducing agent than water
    (c) Cobalt is a weaker reducing agent than hydrogen
    (d) The Co^{3+} ion is a weaker oxidizing agent than water
    (e) Statements (a) through (d) above are all false

	Co_2O_3	+	$3H_2$	$\rightarrow$	$2Co$	+	$3H_2O$
Oxidation number:	+3 −2		0		0		+1 −2
	Stronger oxidizing agent		Stronger reducing agent		Weaker reducing agent		Weaker oxidizing agent

Cobalt is a weaker reducing agent than H_2.   Answer(c)

15-29. Which is the strongest oxidizing agent among the following?

    (a) Cl_2 gas at 1 atm pressure
    (b) O_2 gas at 1 atm pressure
    (c) F_2 gas at 1 atm pressure
    (d) Ag metal
    (e) H_2O_2 in acid solution

$Cl_2 \text{ (g)} + 2e^- \rightleftharpoons 2Cl^- \text{ (aq)}$ $E^0_{red} = 1.3583$

$O_2 \text{ (g)} + 4H^+ + 4e^- \rightleftharpoons 2H_2O \text{ (l)}$ $E^0_{red} = 1.229$

$F_2 \text{ (g)} + 2e^- \rightleftharpoons 2F^- \text{ (aq)}$ $E^0_{red} = 2.870$

$Ag^+ \text{ (aq)} + e^- \rightleftharpoons Ag \text{ (s)}$ $E^0_{red} = 0.7996$

$H_2O_2 + 2H^+ + 2e^- \rightleftharpoons 2H_2O \text{ (l)}$ $E^0_{red} = 1.776$

F_2 (with the highest + ve E^0_{red} value) is the strongest oxidizing agent.   Answer(c)

15-30. Which of the following is the strongest reducing agent?

    (a) H^+ (b) H_2 (c) H^- (d) H_2O (e) O_2

O_2 is an oxidizing agent.

$$H_2 + 2e^- \rightarrow 2H^- \qquad\qquad E^0_{red} = -2.230$$

$$2H^+ + 2e^- \rightarrow H_2 \qquad\qquad E^0_{red} = 0.000$$

$$O_2 + 4H^+ + 4e^- \rightarrow 2H_2O \qquad E^0_{red} = 1.229$$

H_2 has the highest $-$ ve E^0_{red} value. So H^- is the strongest reducing agent.   Answer(c)

15-31. Which of the following statements correctly describes the voltaic cell represented by the following notation?

   Zn (s)|Zn^{2+} (1.0 M) || H^+ (1.0 M)|H_2(g)|Pt

   (a) H^+ ions flow towards the cathode
   (b) The Zn^{2+} ion concentration increases with time
   (c) H_2 gas is given off at the cathode
   (d) Electrons flow from the Zn electrode to the Pt electrode
   (e) Statements (a) through (d) are all true

Anode: $Zn (s) \rightarrow Zn^{2+} + 2e^-$ $E^0_{oxi} = 0.76$

Cathode: $\underline{2H^+ + 2e^- \rightarrow H_2}$ $E^0_{red} = 0.00$

            $Zn (s) + 2H^+ \rightarrow Zn^{2+} + H_2$ $E^0_{cell} = 0.76 + 0.00 = 0.76$

$\Delta G^0 = -nFE^0$

$\Delta G^0 < 0$. The reaction will be spontaneous.

Electrons flow from the Zn electrode to the Pt electrode where they combine with H^+

(which flow towards the cathode) to form H atoms which immediately combine to form H_2.

The Zn^{2+} concentration will increase with time since the cell reaction is spontaneous.

                                       Answer(e)

15-32. Use line notation to describe the following galvanic cell: The anode is the standard

hydrogen electrode and the cathode is a fluorine gas electrode. The gas is in the standard state

but the solution in which the gas is in contact has $(F^-) = 0.10$ M. Use Pt wire for both

electrodes.

Rules for line notation for voltaic cells:

- A single vertical line indicates a change in state or phase.
- Within a half-cell the reactants are listed before the products.
- Concentrations of aq. solutions are written in parenthesis after the symbol for the ion or molecule.

- A double vertical line is used to indicate the junction between the half-cells.
- The line notation for the anode (oxidation) is written before the line notation for the cathode (reduction).

Anode: Oxidation $\qquad H_2 \rightarrow 2H^+ + 2e^-$

Cathode: Reduction $\qquad F_2 + 2e^- \rightarrow 2F^-$

Line notation: $\qquad$ Pt|H$_2$ (g)|H$^+$ (aq, 1M) || F$_2$ (g)|F$^-$ (aq, 0.10M)|Pt

15-33. Consider the following generalized half-reactions:

$$A^+ + e^- \rightleftharpoons A$$

$$B^+ + e^- \rightleftharpoons B$$

If the E^0_{red} for the half-reaction involving A is positive and if E^0_{red} for the half-reaction involving B is negative, then at standard conditions:

(a) A will reduce B^+ $\qquad\qquad$ (b) B will reduce A^+

(c) No reaction will occur

The E^0_{red} for half-reaction involving B is negative and the E^0_{red} for half-reaction involving A is positive, meaning that B is a more powerful reducing agent than A. So B will reduce A^+.

$\qquad\qquad\qquad\qquad\qquad\qquad\qquad\qquad\qquad\qquad\qquad\qquad\qquad\qquad$ Answer(b)

15-34. The following electrochemical cell has a voltage of 0.48 V:

$$Pt|PtCl_4^{2-}, Cl^- \ || \ MnO_2|H^+, Mn^{2+}|Pt$$

For the spontaneous reaction,

(a) Mn^{2+} will be oxidized to MnO_2
(b) $PtCl_4^{2-}$ will be reduced to Pt
(c) MnO_2 will be reduced to Mn^{2+}
(d) MnO_2 will be oxidized to Mn^{2+}
(e) It is not possible to identify the species reduced or oxidized from the information given

The line notation for the anode (oxidation) is written before the line notation for the cathode (reduction). So,

Oxidation: $\qquad\qquad Pt + 4Cl^- \rightarrow PtCl_4^{2-} + 2e^-$

Reduction: $\qquad\qquad MnO_2 + 4H^+ + 2e^- \rightarrow Mn^{2+} + 2H_2O$

MnO_2 will be reduced to Mn^{2+}. $\qquad\qquad\qquad\qquad\qquad\qquad\qquad\qquad\qquad$ Answer(c)

15-35. Determine the E^0 for the following cell:

$Pt|Fe^{2+}, Fe^{3+} ||Cl^-|AgCl,Ag$

Oxidation:	$Fe^{2+} \rightarrow Fe^{3+} + e^-$	$E^0_{oxi} = -0.770$ V
Reduction:	$AgCl + e^- \rightarrow Ag + Cl^-$	$E^0_{red} = 0.222$ V
	$Fe^{2+} + AgCl \rightarrow Fe^{3+} + Ag + Cl^-$	$E^0_{cell} = -0.770 + 0.222 = -0.548$ V

15-36. Which of the following cells has a potential larger than the standard-state potential of the cell?

(a) $Zn|Zn^{2+}$ (1 M)$||Cu^{2+}$ (1 M)$|$ Cu   (b) $Zn|Zn^{2+}$ (2 M) $|| Cu^{2+}$ (1 M)$|$Cu

(c)$Zn|Zn^{2+}$ (1 M) $||Cu^{2+}$ (2 M)$|$Cu   (d) $Zn|Zn^{2+}$ (1 M) $|| Cu^{2+}$ (0.1 M)$|$Cu

(e) $Zn|Zn^{2+}$ (0.1 M)$||Cu^{2+}$ (0.1 M)$|$Cu

$Zn + Cu^{2+} \rightarrow Zn^{2+} + Cu$

$E_{cell} = E^0_{cell} - (0.0591/2) \log \{(Zn^{2+})/(Cu^{2+})\}$

(a) $Zn^{2+} = 1M$, $Cu^{2+} = 1M$ $E_{cell} = E^0_{cell}$

(b) $Zn^{2+} = 2M$, $Cu^{2+} = 1M$ $E_{cell} = E^0_{cell} - (0.0591/2) \log 2 = E^0_{cell} - 0.009$

(c) $Zn^{2+} = 1M$, $Cu^{2+} = 2M$ $E_{cell} = E^0_{cell} - (0.0591/2) \log (1/2) = E^0_{cell} + 0.009$

(d) $Zn^{2+} = 1M$, $Cu^{2+} = 0.1M$ $E_{cell} = E^0_{cell} - (0.0591/2) \log 10 = E^0_{cell} - 0.02955$

(e) $Zn^{2+} = 0.1M$, $Cu^{2+} = 0.1M$ $E_{cell} = E^0_{cell} - (0.0591/2) \log 1 = E^0_{cell}$ Answer(c)

15-37. Calculate the standard-state potential for the following reaction:

$$4Cr^{2+} (aq) + O_2 (g) + 4H^+ (aq) \rightleftharpoons 4Cr^{3+} (aq) + 2H_2O (l)$$

What will be the cell potential if the reaction is run at pH =7?

Oxidation:	$4\{Cr^{2+} \rightarrow Cr^{3+} + e^-\}$	$E^0_{oxi} = 0.410$ V
Reduction:	$O_2 + 4H^+ + 4e^- \rightarrow 2H_2O$	$E^0_{red} = 1.229$ V
	$4Cr^{2+} + O_2 + 4H^+ \rightarrow 4Cr^{3+} + 2H_2O$	$E^0 = 0.410 + 1.229 = 1.639$ V $n = 4$

Reaction at pH = 7

$H^+ = 10^{-7}$

$E = E^0 - (0.0591/4) \log \{(1)/ (10^{-7})^4\}$

$= E^0 - (0.0591/4) \log 10^{28}$

$= 1.639 - \{(0.0591/4) \times 28\} = 1.225$ V

15-38. Calculate the E^0 for the following cell.

$$Zn \text{ (s) } |Zn^{2+} \parallel H^+|O_2 \text{ (g)}|H_2O|Pt \text{ (s)}$$

Oxidation: $2\{Zn \rightarrow Zn^{2+} + 2e^-\}$ $E^0_{oxi} = 0.763$ V

Reduction: $O_2 + 4H^+ + 4e^- \rightarrow 2H_2O$ $E^0_{red} = 1.229$ V

$2Zn + O_2 + 4H^+ \rightarrow 2Zn^{2+} + 2H_2O$ $E^0_{cell} = 0.763 + 1.229 = 1.992$ V

15-39. Use a table of standard potentials to decide which of the following reactions doesn't occur spontaneously.

(a) $Zn \text{ (s) } + 2H^+ \text{ (aq) } \rightleftharpoons Zn^{2+} \text{ (aq) } + H_2 \text{ (g)}$
(b) $I_2 \text{ (s) } + 2Fe^{2+} \text{ (aq)} \rightleftharpoons 2I^- \text{ (aq) } + 2Fe^{3+} \text{ (aq)}$
(c) $2Al \text{ (s) } + (3/2) O_2 \text{ (g) } + 6H^+ \text{ (aq)} \rightleftharpoons 2Al^{3+} \text{ (aq) } + 3H_2O \text{ (l)}$
(d) $Mg \text{ (s) } + Cl_2 \text{ (g)} \rightleftharpoons MgCl_2 \text{ (s)}$
(e) All of these are spontaneous

(a) Oxidation: $Zn \rightarrow Zn^{2+} + 2e^-$ $E^0_{oxi} = 0.763$ V

Reduction: $2H^+ + 2e^- \rightarrow H_2$ $E^0_{red} = 0.000$ V

$Zn + 2H^+ \rightarrow Zn^{2+} + H_2$ $E^0 = 0.763 + 0.000 = 0.763$ V

$\Delta G^0 = -nFE^0$ $\Delta G^0 < 0$. The reaction will be spontaneous.

(b) Oxidation: $2\{Fe^{2+} \rightarrow Fe^{3+} + e^-\}$ $E^0_{oxi} = -0.770$ V

Reduction: $I_2 + 2e^- \rightarrow 2I^-$ $E^0_{red} = 0.5355$ V

$2Fe^{2+} + I_2 \rightarrow Fe^{3+} + 2I^-$ $E^0 = -0.770 + 0.5355 = -0.2345$ V

$\Delta G^0 = -nFE^0$ $\Delta G^0 > 0$. The reaction will not be spontaneous.

(c) Oxidation: $4\{Al \rightarrow Al^{3+} + 3e^-\}$ $E^0_{oxi} = 1.706$ V

Reduction: $3\{O_2 + 4H^+ + 4e^- \rightarrow 2H_2O\}$ $E^0_{red} = 1.229$ V

$4Al + 3O_2 + 12H^+ \rightarrow 4Al^{3+} + 6H_2O$ $E^0 = 1.706 + 1.229 = 2.935$ V

$\Delta G^0 = -nFE^0$ $\Delta G^0 < 0$. The reaction will be spontaneous.

(d) Oxidation: $Mg \rightarrow Mg^{2+} + 2e^-$ $E^0_{oxi} = 2.375$ V

Reduction: $Cl_2 + 2e^- \rightarrow 2Cl^-$ $E^0_{red} = 1.358$ V

$Mg + Cl_2 \rightarrow MgCl_2$ $E^0 = 2.375 + 1.358 = 3.733$ V

$\Delta G^0 = -nFE^0$ $\Delta G^0 < 0$. The reaction will be spontaneous. Answer(b)

15-40. What is the potential for a voltaic cell from the following half-reactions, if all ions are present at 1.0 M concentrations?

Oxidation: $Al \rightleftharpoons Al^{3+} + 3e^-$

Reduction: $Zn^{2+} + 2e^- \rightleftharpoons Zn$

(a) -3.59 V (b) -0.943 V (c) 0.943 V (d) 1.12 V (e) 3.59 V

Oxidation: $2\{Al \rightleftharpoons Al^{3+} + 3e^-\}$ $\qquad E^0_{oxi} = 1.706$ V

Reduction: $\dfrac{3\{Zn^{2+} + 2e^- \rightleftharpoons Zn\}}{2Al + 3Zn^{2+} \rightarrow 2Al^{3+} + 3Zn}$ $\qquad \begin{matrix} E^0_{red} = -0.763 \text{ V} \\ E^0_{cell} = 1.706 - 0.763 = 0.943 \text{ V} \end{matrix}$

All ions are present at 1.0 M concentrations. So standard-state potential must be calculated.

$E^0_{cell} = 0.943$ V $\hfill$ Answer(c)

15-41. Consider a voltaic cell containing Cu^{2+}/Cu and Zn^{2+}/Zn half-reactions.

(i) What is the cell potential if $(Zn^{2+}) = 0.500$ M and $(Cu^{2+}) = 0.01$ M?

(a) 1.00 V (b) 1.05 V (c) 1.10 V (d) 1.20 V

Anode: Oxidation $\qquad Zn \rightleftharpoons Zn^{2+} + 2e^-$ $\quad E^0_{oxi} = 0.763$ V

Cathode: Reduction: $\qquad \dfrac{Cu^{2+} + 2e^- \rightleftharpoons Cu}{Zn + Cu^{2+} \rightarrow Zn^{2+} + Cu}$ $\quad \begin{matrix} E^0_{red} = 0.340 \text{ V} \\ E^0_{cell} = 0.763 + 0.340 = 1.103 \text{ V} \quad n = 2 \end{matrix}$

$E_{cell} = E^0_{cell} - (0.0591/2) \log \{(Zn^{2+})/(Cu^{2+})\} = 1.103 - (0.0591/2) \log (0.5/0.01)$

$= 1.103 - 0.05$ V $= 1.053$ V $\hfill$ Answer(b)

(ii) What happens to the concentration of the Zn^{2+} ion in the above voltaic cell?

(a) The concentration of Zn^{2+} increases
(b) The concentration of Zn^{2+} decreases
(c) The concentration of Zn^{2+} remains the same

$Zn + Cu^{2+} \rightarrow Zn^{2+} + Cu$ $\qquad\qquad E_{cell} = 1.053$ V

$\Delta G = -nFE$ $\Delta G < 0$. The reaction will be spontaneous. The reaction is spontaneous in the

forward direction. So Zn^{2+} concentration will increase. $\hfill$ Answer(a)

(iii) Which of the following statements is true in the above voltaic cell?

(a) Cu metal is produced at the cathode
(b) Cu metal is produced at the anode
(c) Zn metal is produced at the cathode
(d) Zn metal is produced at the anode
(e) No metal is produced either at the cathode or at the anode

The reaction is spontaneous in the forward direction. So Cu metal is produced at the cathode.

$\hfill$ Answer(a)

(iv)   In the above voltaic cell electrons flow

(a) From the Zn electrode to the Cu electrode
(b) From the Cu electrode to the Zn electrode

Oxidation at the anode (Zn electrode). So electrons will flow from Zn electrode to the Cu

electrode.   Answer(a)

15-42. What is the potential of the following cell?

$$Zn|Zn^{2+} (1\ M)||\ Sn^+ (1M)|Sn$$

(a) $- 0.90$ V (b) $- 0.63$ V (c) 0.63 V (d) 0.90 V

Oxidation:	$Zn \rightarrow Zn^{2+} + 2e^-$	$E^0_{oxi} = 0.763$ V
Reduction:	$Sn^{2+} + 2e^- \rightarrow Sn$	$E^0_{red} = - 0.1364$ V
	$Zn + Sn^{2+} \rightarrow Zn^{2+} + Sn$	$E^0_{cell} = 0.763 - 0.1364 = 0.627$ V

All ions are present at 1 M concentrations. So standard-state potential must be calculated.

$E = E^0_{cell} = 0.627$ V   Answer(c)

15-43. What is the potential for the following cell?

$$Pt|Sn^{2+}(1\ M),Sn^{4+}(1\ M)||I_2\ (1\ M),\ I^-\ (1\ M)|Pt$$

(a) $- 0.69$ V (b) $- 0.39$ V (c) 0.39 V (d) 0.69 V

Oxidation:	$Sn^{2+} \rightarrow Sn^{4+} + 2e^-$	$E^0_{oxi} = - 0.15$ V
Reduction:	$I_2 + 2e^- \rightarrow 2I^-$	$E^0_{red} = 0.5355$ V
	$Sn^{2+} + I_2 \rightarrow Sn^{4+} + 2I^-$	$E^0_{cell} = - 0.15 + 0.5355 = 0.39$ V n = 2

$E_{cell} = (E^0_{cell}) - (0.0591/2) \log 1 = E^0_{cell} = 0.39$ V   Answer(c)

15-44. What is the potential for the following cell?

$$Mn|Mn^{2+}(0.00010\ M)||Zn^{2+} (1.5\ M)|Zn$$

(a) 0.15 V (b) 0.28 V (c) 0.40 V (d) 1.80 V (e) None of the above

Oxidation:	$Mn \rightarrow Mn^{2+} + 2e^-$	$E^0_{oxi} = 1.04$ V
Reduction:	$Zn^{2+} + 2e^- \rightarrow Zn$	$E^0_{red} = - 0.763$ V
	$Mn + Zn^{2+} \rightarrow Mn^{2+} + Zn$	$E^0_{cell} = 1.04 - 0.763 = 0.277$ V n = 2

$E_{cell} = (E^0_{cell}) - (0.0591/2) \log \{(Mn^{2+})/(Zn^{2+})\} = 0.277 - \{(0.0591/2) \log (0.0001/1.5)\}$

$= 0.277 - \{(0.0591/2) \times (- 4.17)\}$

$= 0.277 + 0.123 = 0.40 \text{ V}$ 　　　　　　　　　　　　　　　　　　　Answer(c)

15-45. What is the potential for the following cell?

$$Ag|Ag^+ (0.0010 \text{ M})|| Ag^+ (0.1 \text{ M})|Ag$$

(a) -0.12 V　(b) 0.12 V　(c) 0.68 V　(d) 0.92 V

Oxidation: 　　　　$Ag \rightarrow Ag^+ + e^-$ 　　　　　$E^0_{oxi} = -0.7996 \text{ V}$

Reduction: 　　　　$Ag^+ + e^- \rightarrow Ag$ 　　　　　$E^0_{red} = 0.7996 \text{ V}$

　　　　　　　　　　　　　　　　　　　　$E^0_{cell} = 0.0 \text{ V}$ 　　　　$n = 1$

$E_{cell} = E^0_{cell} - \{(0.0591/1) \log (0.001/0.1)\} = 0.0 - \{(0.0591) \times (-2)\} = 0.1182 \text{ V}$ 　　　Answer(b)

15-46. What is the standard-state potential for the following redox reaction?

$$2Fe^{3+} (aq) + 2I^- (aq) \rightarrow 2Fe^{2+} (aq) + I_2 (aq)$$

(a) -1.3 V　(b) -0.24 V　(c) 0.234 V　(d) 1.30 V　(e) None of the above

Oxidation: 　　　　$2I^- \rightarrow I_2 + 2e^-$ 　　　　　$E^0_{oxi} = -0.536 \text{ V}$

Reduction: 　　　$\underline{2\{Fe^{3+} + e^- \rightarrow Fe^{2+}\}}$ 　　　$E^0_{red} = 0.770 \text{ V}$

　　　　　　　$2Fe^{3+} + 2I^- \rightarrow 2Fe^{2+} + I_2$ 　　　$E^0 = -0.536 + 0.770 = 0.234$　Answer(c)

15-47. What is the magnitude of the standard-state cell potential for the following redox

reaction?

$$2Al (s) + 3 Pb^{2+} (aq) \rightarrow 2 Al^{3+} (aq) + 3Pb (s)$$

(a) 1.58 V　(b) 1.83 V　(c) 3.03 V　(d) 3.790 V　(e) 4.866 V

Oxidation: 　　　　$2\{Al \rightarrow Al^{3+} + 3e^-\}$ 　　　$E^0_{oxi} = 1.706 \text{ V}$

Reduction: 　　　$\underline{3\{Pb^{2+} + 2e^- \rightarrow Pb\}}$ 　　　$E^0_{red} = -0.1263 \text{ V}$

　　　　　　　$2Al + 3Pb^{2+} \rightarrow 2Al^{3+} + 3Pb$ 　　　$E^0_{cell} = 1.706 - 0.1263 = 1.58 \text{ V}$　Answer(a)

15-48. The equilibrium constant for the following reaction is 1.2×10^5 :

$$H_2 (g) + Sn^{4+} (aq) \rightarrow 2H^+ (aq) + Sn^{2+} (aq)$$

What is the value of E^0 for the reaction?

(a) -0.15 V　(b) 0.15 V　(c) 0.30 V　(d) 0.35 V

Oxidation: 　　　　$H_2 \rightarrow 2H^+ + 2e^-$

Reduction: 　　　　$Sn^{4+} + 2e^- \rightarrow Sn^{2+}$

　　　　　　　$\underline{Sn^{4+} + H_2 \rightarrow Sn^{2+} + 2H^+}$ 　　　　$n = 2$

$E^0 = (0.0591/n) \log K$ at $T = 298$

$E^0 = (0.0591/2) \log K = (0.0591/2) \log (1.2 \times 10^5) = (0.0591/2) \times (5.08) = 0.15$

$\quad = 0.15$ V $\hspace{7cm}$ Answer(b)

15-49. What is the equilibrium constant at 25 °C for the following reaction,

$$Zn \ (s) + Cu^{2+} \ (aq) \rightarrow Zn^{2+} \ (aq) + Cu$$

if the value of E^0 is 1.10 V?

$\quad$ (a) 1.9 $\quad$ (b) 37 $\quad$ (c) 1.7×10^{37} $\quad$ (d) 1.7×10^{-37} $\quad$ (e) -37

Oxidation: $\hspace{2cm} Zn \rightarrow Zn^{2+} + 2e^-$

Reduction: $\hspace{2cm} \underline{Cu^{2+} + 2e^- \rightarrow Cu}$

$\hspace{3cm} Zn + Cu^{2+} \rightarrow Zn^{2+} + Cu \hspace{1.5cm} E^0 = 1.10$ V $\hspace{1.5cm} n = 2$

$E^0 = (0.0591/n) \log K$ at $T = 298$

$E^0 = (0.0591/2) \log K$

$\text{Log } K = (2) \times (1.1)/(0.0591) = 37.225 \hspace{2cm} K = 1.68 \times 10^{37} \hspace{2cm}$ Answer(c)

15-50. Calculate the value of K_{sp} for the CdS from the following data.

$$CdS + 2e^- \rightleftharpoons Cd + S^{2-} \hspace{2cm} E^0_{red} = -1.21 \text{ V}$$

$$Cd^{2+} + 2e^- \rightleftharpoons Cd \hspace{2cm} E^0_{red} = -0.40 \text{ V}$$

$\quad$ (a) 3×10^{-55} $\quad$ (b) 4×10^{-28} $\quad$ (c) 2×10^{-14} $\quad$ (d) 3×10^{27}

$CdS + 2e^- \rightleftharpoons Cd + S^{2-} \hspace{2cm} E^0_{red} = -1.21$ V

$\underline{Cd \rightleftharpoons Cd^{2+} + 2e^- \hspace{2.5cm} E^0_{oxi} = 0.40 \text{ V}}$

$CdS \rightleftharpoons Cd^{2+} + S^{2-} \hspace{2cm} E^0 = 0.40 - 1.21 = -0.81$ V $\hspace{1cm} n = 2$

$E = E^0 - (0.0591/2) \log \{(Cd^{2+})(S^{2-})\}$

At equilibrium $E = 0$.

$0 = E^0 - (0.0591/2) \log K_{sp}$

$0 = -0.81 - (0.0591/2) \log K_{sp}$

$\text{Log } K_{sp} = -(2 \times 0.81)/(0.0591) = -27.41$

K_{sp} for CdS $= 3.9 \times 10^{-28} \hspace{6cm}$ Answer(b)

15-51. There are two half-reactions for the reduction of O_2 to H_2O.

$$O_2 + 4H^+ + 4e^- \rightleftharpoons 2H_2O \hspace{2cm} E^0_{red} = 1.229 \text{ V}$$

$$O_2 + 2H_2O + 4e^- \rightleftharpoons 4\,OH^- \qquad E^0_{red} = 0.401\ V$$

Use these half-reactions to calculate the ionization constant for water, K_w.

$$H_2O\ (l) \rightleftharpoons H^+\ (aq) + OH^-\ (aq) \qquad K_w = ?$$

Oxidation: $(1/4)\{2H_2O \rightleftharpoons O_2 + 4H^+ + 4e^-\} \qquad E^0_{oxi} = -1.229\ V$

Reduction: $(1/4)\{O_2 + 2H_2O + 4e^- \rightleftharpoons 4\,OH^-\} \qquad E^0_{red} = 0.401\ V$

$$H_2O \rightleftharpoons H^+ + OH^- \qquad E^0 = 0.401 - 1.229 = -0.828\ V \quad n = 1$$

$E = (E^0) - (0.0591/1)\ \log\ \{(H^+)\ (OH^-)\}$

At equilibrium $E = 0$.

$0 = (E^0) - (0.0591)\ \log K_w$

$0 = -0.828 - (0.0591)\ \log K_w$

$\text{Log}\ K_w = -(0.828)/(0.0591) = -14 \qquad K_w = 1 \times 10^{-14}$

15-52. The half-reaction reduction potentials for the mercury (I) and silver (I) ions are given below.

$$Hg_2^{2+} + 2e^- \rightleftharpoons 2\ Hg \qquad E^0_{red} = 0.7961\ V$$

$$Ag^+ + e^- \rightleftharpoons Ag \qquad E^0_{red} = 0.7996\ V$$

Calculate the equilibrium constant for the following reaction under standard- state conditions:

$$Hg_2^{2+}\ (aq) + 2Ag\ (s) \rightleftharpoons 2\ Hg\ (l) + 2Ag^+\ (aq)$$

Oxidation: $\qquad 2\{Ag \rightarrow Ag^+ + e^-\} \qquad E^0_{oxi} = -0.7996\ V$

Reduction: $\qquad Hg_2^{2+} + 2e^- \rightarrow 2Hg \qquad E^0_{red} = 0.7961\ V$

$$Hg_2^{2+} + 2Ag \rightarrow 2Hg + 2Ag^+ \qquad E^0 = 0.7961 - 0.7996 = -0.0035\ V \quad n = 2$$

$E = E^0 - (0.0591/2)\ \log\ \{\ (Ag^+)^2/(Hg_2^{2+})\}$

At equilibrium $E = 0$.

$0 = E^0 - (0.0591/2)\ \log K\ = -0.0035 - (0.0591/2)\ \log K$

$\text{Log}\ K = -(2 \times 0.0035)/(0.0591) = -0.12 \qquad\qquad K = 0.759$

15-53. Calculate the complex dissociation constant K_d for the $Cd(NH_3)_4^{2+}$ complex ion from the following data.

$$Cd(NH_3)_4^{2+} + 2e^- \rightleftharpoons Cd + 4NH_3 \qquad E^0_{red} = -0.59\ V$$

$$Cd^{2+} + 2e^- \rightleftharpoons Cd \qquad E^0_{red} = -0.403\ V$$

(a) 2.7×10^{-7} (b) 0.19 (c) 6.6 (d) 3.6×10^6 (e) None of the above

Oxidation: $\quad\quad\quad\quad Cd \rightarrow Cd^{2+} + 2e^- \quad\quad\quad\quad E^0_{oxi} = 0.403$ V

Reduction: $\quad\quad\quad Cd(NH_3)_4^{2+} + 2e^- \rightarrow Cd + 4NH_3 \quad E^0_{red} = -0.597$ V

$\quad\quad\quad\quad\quad\quad Cd(NH_3)_4^{2+} \rightarrow Cd^{2+} + 4NH_3 \quad\quad E^0 = 0.403 - 0.590 = -0.194$ V $\quad n = 2$

$E = (E^0) - (0.0591/2) \log [\{(Cd^{2+})(NH_3)^4\}/\{Cd(NH_3)_4^{2+}\}]$

At equilibrium E = 0.

$0 = -0.194 - (0.0591/2) \log K_d$

$\log K_d = -(2) \times (0.194) /(0.0591) = -6.57$

K_d for the $Cd(NH_3)_4^{2+}$ complex ion $= 2.7 \times 10^{-7}$ $\quad\quad\quad\quad\quad\quad\quad\quad$ Answer(a)

15-54. Which of the following describes what happens when an aqueous solution of $MgCl_2$ is

electrolyzed?

(a) Mg metal forms at the cathode
(b) Mg^{2+} ions are oxidized at the cathode
(c) Cl^- ions are oxidized to Cl_2 gas at the cathode
(d) Cl^- ions flow toward the cathode
(e) H_2 gas is formed at the cathode

$Mg^{2+} + 2e^- \rightarrow Mg \quad\quad\quad\quad E^0_{red} = -2.37$ V

$2H_2O + 2e^- \rightarrow H_2 + 2OH^- \quad\quad E^0_{red} = -0.83$ V

Because it is easier to reduce water than Mg^{2+} ions, the product formed at the cathode is H_2 gas.

Cahode (−) : $\quad\quad 2H_2O$ (l) $+ 2e^- \rightarrow H_2$ (g) $+ 2OH^-$ (aq) $\quad\quad\quad\quad$ Answer(e)

15-55. Which of the following is the correct half-cell reaction for the anode process in the

electrolysis of an aqueous solution of potassium sulfate?

(a) $SO_4^{2-} \rightarrow SO_2$ (g) $+ O_2$ (g) $+ 2e^-$
(b) $H_2O \rightarrow (1/2) O_2$ (g) $+ 2H^+ + 2e^-$
(c) $H_2O + (1/2) O_2 + 2e^- \rightarrow 2OH^-$
(d) $SO_4^{2-} + 4H^+ + 2e^- \rightarrow SO_2 + 2H_2O$
(e) $3H_2O + e^- \rightarrow (1/2) H_2$ (g) $+ 3OH^- + 2H^+$

$2SO_4^{2-} \rightarrow S_2O_8^{2-} + 2e^- \quad\quad\quad E^0_{oxi} = -2.05$ V

$2H_2O \rightarrow O_2 + 4H^+ + 4e^- \quad\quad E^0_{oxi} = -1.229$ V

Because it is easier to oxidize water than SO_4^{2-} ions, the correct half-cell reaction for the anode

process is: Anode (+) : $H_2O \rightarrow (1/2) O_2$ (g) $+ 2H^+ + 2e^-$ $\quad\quad\quad\quad\quad\quad$ Answer(b)

15-56. Which of the following statements about the electrolysis of an aqueous solution of NaCl is true?

    (a) O_2 is liberated at the cathode
    (b) H_2 is liberated at the cathode
    (c) Na is liberated at the cathode
    (d) Cl_2 is liberated at the cathode
    (e) The decomposition potential must not be exceeded

$Na^+ + e^- \rightarrow Na$ $E^0_{red} = -2.71$ V

$2H_2O + 2e^- \rightarrow H_2 + 2OH^-$ $E^0_{red} = -0.83$ V

Because it is easier to reduce water than Na^+, the product formed at the cathode is hydrogen

gas.

Cathode $(-)$: $2H_2O + 2e^- \rightarrow H_2$ (g) $+ 2OH^-$ Answer(b)

15-57. In the electrolysis of NaBr (aq), which of the following products is least likely to be

found?

    (a) H_2 (g) (b) O_2 (g) (c) Br_2 (g) (d) Na (s) (e) NaOH (aq)

$2 Br^- \rightarrow Br_2 + 2e^-$ $E^0_{oxi} = -1.087$ V

$2H_2O \rightarrow O_2 + 4H^+ + 4e^-$ $E^0_{oxi} = -1.229$ V

It is easier to oxidize Br^- than H_2O.

$Na^+ + e^- \rightarrow Na$ $E^0_{red} = -2.7109$ V

$2H_2O + 2e^- \rightarrow H_2 + 2(OH)^-$ $E^0_{red} = -0.83$ V

Because it is easier to reduce water than Na^+ ions, the product formed at the cathode is

hydrogen gas. So Na is least likely to be found.   Answer(d)

15-58. Which of the following compounds would give the largest weight of metal if a 10.00

amp current is passed through molten samples of these salts for 2 hours?

    (a) KCl (AW: K= 39.08 amu; Cl = 35.45 amu)
    (b) $CaCl_2$ (AW: Ca = 40.08 amu; Cl = 35.45 amu)
    (c) $ScCl_3$ (AW: Sc = 44.96 amu; Cl = 35.45 amu)
    (d) $TiCl_4$ (AW: Ti = 47.90 amu; Cl = 35.45 amu)
    (e) The same weight of metal would be obtained from the electrolysis of each of these salts

Amount of electrical charge that passes through the molten sample: $10 \times 2 \times 60 \times 60 = 72000$ C

Number of mole of electrons that carry this charge: $(72{,}000)/(96485) = 0.746$ mol e^-

At cathode: $K^+ + e^- \rightarrow K$; $Ca^{2+} + 2e^- \rightarrow Ca$; $Sc^{3+} + 3e^- \rightarrow Sc$; $Ti^{4+} + 4e^- \rightarrow Ti$

(a) 1 mole of electrons produce 1 mole of K. 0.746 mole of electrons will produce 0.746 mole of K. So 0.746×39.08 = 29.15 g of K will be produced.

(b) 2 moles of electrons produce 1 mole of Ca. 0.746 mole of electrons will produce

(0.746/2) = le of Ca. So 0.373×40.08 = 14.95 g of Ca will be produced.

(c) 3 moles of electrons produce 1 mole of Sc. 0.746 mole of electrons will produce

(0.746/3) = 0.249 mole of Sc. So 0.249×44.96 = 11.20 g of Sc will be produced.

(d) 4 moles of electrons will produce 1 mole of Ti. 0.746 mole of electrons will produce

(0.746/4) = 0.1865 mole of Ti. So 0.1865×47.90 = 8.93 g of Ti will be produced.   Answer(a)

15-59. Calculate the weight of sodium metal that would be produced by the electrolysis of

molten sodium chloride for 1 hour with a 10.0 amp. current. (AW: Na = 22.99 amu)

Amount of electrical charge that passes through the molten sample: 10×1×60×60 =36000 C

Number of mole of electrons that carry this charge: (36000)/(96485) = 0.373 mol e^-

Cathode: $Na^+ + e^- \rightarrow Na$ 1 mole of electrons will produce 1 mole of Na.

0.373 mole of electrons will produce 0.373 mole of Na. So 0.373×22.99 = 8.58 g of Na will be

produced.

15-60. What is the ratio of the weight of Cl_2 produced at the anode to the weight of Al

produced at the cathode when 5 Faradays of electric charge are passed through a molten

sample of $AlCl_3$? (AW: Al = 27.0 amu; Cl = 35.45 amu)

   (a) About 2   (b) About 4 (c) About 8 (d) About 16 (e) None of the above

Amount of electrical charge that passes through the molten sample: 5×96485 C

Number of moles of electrons that carry this charge: (5×96485)/(96485) = 5 mol e^-

Cathode: $Al^{3+} + 3e^- \rightarrow Al$ 3 moles of electrons produce 1 mole of Al.

5 moles of electrons will produce (5/3) moles of Al.   So (5/3)×27 = 45 g of Al will be

produced.

Anode: $2Cl^- \rightarrow Cl_2 + 2e^-$ 2 moles of electrons are involved in the production of 1 mole of Cl_2.

 5 moles of electrons will be involved in the production of (5/2) moles of Cl_2.

  So (5/2)×2×35.45 = 177.25 g of Cl_2 will be produced.

(Weight of Cl_2 produced at the anode)/(Weight of Al produced at the cathode) = 177.25/45

= 3.9   Answer(b)

15-61. Calculate the amount of Al produced in 1.00 hr by the electrolysis of molten $AlCl_3$ if the current is 10.0 amperes. (AW: Al = 27.0 amu)

Amount of electrical charge that passes through the molten sample: $10 \times 1 \times 60 \times 60 = 36000$ C

Number of mole of electrons that carry this charge: $(36000)/(96485) = 0.373$ mol e^-

Cathode: $Al^{3+} + 3e^- \rightarrow Al$ 3 moles of electrons produce 1 mole of Al.

0.373 mole of electrons will produce $(0.373/3)$ mole of Al. So $(0.373/3) \times 27 = 3.357$ g of Al will be produced.

15-62. Suppose a beer can weighs 40.0 g. For how long would a current of 100.0 amp. need to be passed through a molten AlF_3 electrolysis cell to produce enough Al to replace a discarded beer can? (AW: Al = 27.0 g/mol)

First let us calculate the amount of Al produced in 1 hour.

Amount of electrical charge that passes through the molten sample: $100 \times 1 \times 60 \times 60 = 360000$ C

Number of moles of electrons that carry this charge: $(360000)/(96485) = 3.73$ mol e^-

Cathode: $Al^{3+} + 3e^- \rightarrow Al$ 3 moles of electrons produce 1 mole of Al.

3.73 moles of electrons will produce $(3.73/3)$ moles of Al.

So $(3.73/3) \times 27 = 33.57$ g of Al will be produced in 1 hour.

Number of hours required to produce 40.0 g of Al $= (1/33.57) \times 40.0 = 1.2$ hours

15-63. Which of the following electrolysis processes will produce the largest volume of Cl_2 at STP?

(a) Passing 35 amp. for 20 hours through an aqueous Na_2SO_4 solution
(b) Passing 10 amp. for 2.69 hours through 1M NaCl solution
(c) Passing 2 Faradays through a 1 M $CaCl_2$ solution
(d) Passing 1.5 Faradays through a 5 M $CaCl_2$ solution
(e) Passing 1.4 Faradays through a 5 M $AlCl_3$ solution

Anode (+): $2Cl^- \rightarrow Cl_2 + 2e^-$

2 moles of electrons are involved in the production of 1 mole of Cl_2.

(a) No Cl_2 will be produced

(b) Amount of electrical charge that passes though the solution: $10 \times 2.69 \times 60 \times 60 = 96840$ C

Number of mole of electrons that carry this charge: $(96840)/(96485) = 1$ mol e^-

1 mole of electrons will be involved in the production of 0.5 mole of Cl_2.

$PV = nRT$; $V = nRT/P = 0.5 \times 0.08206 \times 273.15/1 = 11.21$ L

(c) Amount of electrical charge that passes through the solution: 2×96485 C

Number of moles of electrons that carry this charge: $(2 \times 96485)/(96485) = 2$ mol e^-

2 moles of electrons will be involved in the production of 1 mole of Cl_2.

$V = 1 \times 0.08206 \times 273.15/1 = 22.41$ L

(d) Amount of electrical charge that passes through the solution: 1.5×96485 C

Number of moles of electrons that carry this charge: $1.5 \times 96485/96485 = 1.5$ mol e^-

1.5 moles of electrons will be involved in the production of $(1.5/2) = 0.75$ mole of Cl_2.

$V = 0.75 \times 0.08206 \times 273.15/1 = 16.81$ L

(e) Amount of electrical charge that passes through the solution: 1.4×96485 C

Number of moles of electrons that carry this charge $= 1.4 \times 96485/96485 = 1.4$ mol e^-

1.4 moles of electrons will be involved in the production of $(1.4/2) = 0.7$ mole of Cl_2.

$V = 0.7 \times 0.08206 \times 273.15/1 = 15.69$ L   Answer(c)

15-64. A molten sample of $TiCl_4$ was electrolyzed for 10 hours at 12 amp. What is the ratio of the weight of Cl_2 produced compared to that of Ti? (AW: Ti = 47.90 amu; Cl= 35.45 amu)

   (a) 0.34 g Cl_2/ 1.00 g Ti   (b) 0.68 g Cl_2/ 1.00 g Ti   (c) 0.74 g Cl_2/ 1.00 g Ti

   (d) 1.48 g Cl_2/ 1.00 g Ti   (e) 2.96 g Cl_2/ 1.00 g Ti

Cathode (−) : $Ti^{4+} + 4e^- \rightarrow$ Ti   4 moles of electrons produce 1 mole of Ti

 Anode (+):$2Cl^- \rightarrow Cl_2 + 2e^-$

2 moles of electrons are involved in the production of 1 mole of Cl_2.

Amount of electrical charge that passes through the molten sample: $12 \times 10 \times 60 \times 60 = 432000$ C

Number of moles of electrons that carry this charge: $432000/96485 = 4.48$ mol e^-

4.48 moles of electrons will produce $(4.48/4) = 1.12$ moles of Ti.

So $1.12 \times 47.90 = 53.65$ g of Ti will be produced.

4.48 moles of electrons will be involved in the production of $(4.48/2) = 2.24$ moles of Cl_2.

So $2.24 \times 2 \times 35.45 = 158.82$ g of Cl_2 will be produced.

(Weight of Cl_2 produced)/(Weight of Ti produced) $= 158.82/53.65 = 2.96$ Answer(e)

15.65. A current of 1.5 amp. is applied to a 1.00L solution of 0.100 M hydrochloric acid for 1 hour. What is the pH of the solution after electrolysis is complete?

Cathode (−): $2H^+ + 2e^- \rightarrow H_2$ $\qquad\qquad E^0_{red} = 0.00$ V

$\qquad\qquad 2H_2O + 2e^- \rightarrow H_2 + 2OH^-$ $\qquad E^0_{red} = -0.83$ V

2 moles of electrons consume 2 moles of H^+.

Amount of electrical charge that passes through the solution: $1.5 \times 1 \times 60 \times 60 = 5400$ C

Number of mole of electrons that carry this charge: $5400/96485 = 0.06$ mol e^-

0.06 mole of electrons will consume 0.06 mole of H^+.

Initial H^+ concentration = 0.1 M

H^+ concentration after electrolysis is complete = 0.1 − 0.06 = 0.04 M

pH = − log 0.04 = 1.4

15-66. An electric current is passed through a solution of $CuSO_4$ (aq) producing Cu (s) at the

cathode and O_2 at the anode. If 3.48 L of O_2 (g) measured at STP is produced at the anode how

many grams of Cu must have been deposited on the cathode? (AW: Cu = 63.55 amu;

S = 32.06 amu ; O= 16.00 amu ; 1F= 96485 C; R = 0.08206 L-atm/mol-K)

Number of mole of O_2 produced = PV/RT = $1 \times 3.48/(0.08206 \times 273.15) = 0.155$ mole

Anode (+): $2H_2O \rightarrow O_2 + 4H^+ + 4e^-$ Production of 1 mole of O_2 involves 4 moles of

electrons. Production of 0.155 mole of O_2 will involve $4 \times 0.155 = 0.62$ mole of electrons.

Cathode (−): $Cu^{2+} + 2e^- \rightarrow Cu$ $\qquad$ 2 moles of electrons produce 1 mole of Cu.

0.62 mole of electrons will produce 0.62/2 = 0.31 mole of Cu.

So $0.31 \times 63.55 = 19.7$ g of Cu will be deposited.

15-67. What is the ratio by weight of Br_2 to Cr if a molten sample of $CrBr_2$ is electrolyzed for

4 hrs at 10.0 amp.? (AW: Cr =52.0 amu ; Br = 79.9 amu)

$\quad$ (a) 1.05 g Br_2 / 1 g Cr (b) 1.54 g Br_2 / 1 g Cr (c)2.05 g Br_2 / 1 g Cr

$\quad$ (d) 3.07 g Br_2 / 1 g Cr (e) 4.61 g Br_2 / 1 g Cr

Amount of electrical charge that passes through the molten sample: $10 \times 4 \times 60 \times 60 = 144000$ C

Number of moles of electrons that carry this charge: $144000/96485 = 1.492$ mol e^-

Cathode (−) : $Cr^{2+} + 2e^- \rightarrow Cr$ $\qquad$ 2 moles of electrons produce 1 mole of Cr.

1.492 moles of electrons will produce 1.492/2 = 0.746 mole of Cr.

So $0.746 \times 52.0 = 38.8$ g of Cr will be produced.

Anode (+): $2Br^- \rightarrow Br_2 + 2e^-$

2 moles of electrons are involved in the production of 1 mole of Br_2.

1.492 moles of electrons will be involved in the production of $1.492/2 = 0.746$ mole of Br_2.

So $0.746 \times 2 \times 79.9 = 119.2$ g of Br_2 will be produced.

(Weight of Br_2 produced)/(Weight of Cr produced) = $119.2/38.8 = 3.07$ Answer(d)

15-68. How much Cl_2 gas would be collected when a 2.0 M NaCl (aq) solution is electrolyzed for 2 hours with a current of 15 amp.? (AW: Cl = 35.45 amu)

     (a) 0.0220 g   (b) 39.7 g   (c) 79.4 g   (d) 159 g   (e) None of the above

Amount of electrical charge that passes through the solution: $15 \times 2 \times 60 \times 60 = 108000$ C

Number of moles of electrons that carry this charge: $108000/96485 = 1.12$ mol e^-

Anode(+): $2Cl^- \rightarrow Cl_2 + 2e^-$

2 moles of electrons are involved in the collection of 1 mole of Cl_2.

1.12 moles of electrons will be involved in the production of $1.12/2 = 0.56$ mole of Cl_2.

So $0.56 \times 2 \times 35.45 = 39.7$ g of Cl_2 will be collected.   Answer(b)

15-69. A current of 10.0 amp. over a period of 3 hours is passed through a molten sample of KCl. What weight of K metal is produced? (AW: K = 39.1 amu)

     (a) 0.365 g   (b) 0.729 g   (c) 21.9 g   (d) 43.8 g   (e) 87.5 g

Amount of electrical charge that passes through the molten sample: $10 \times 3 \times 60 \times 60 = 108000$ C

Number of moles of electrons that carry this charge: $108000/96485 = 1.12$ mol e^-

Cathode (−): $K^+ + e^- \rightarrow K$ 1 mole of electrons produce 1 mole of K

1.12 moles of electrons will produce 1.12 moles of K.

So $1.12 \times 39.1 = 43.8$ g of K will be produced.   Answer(d)

15-70. A current of 2.0 amp. passed through a molten sample of MCl_2 for a period of 30 minutes produces 1.32 g of Cl_2 gas and 0.453 g of M. Which of the following metals is present in this compound?

     (a) Mg (AW: 24.3 amu)   (b) Ca (AW: 40.1 amu)   (c) Zn (AW: 65.4 amu)

     (d) Sr (AW: 87.6 amu)   (e) Hg (AW: 200.6 amu)

Amount of electrical charge that passes through the molten sample: $2 \times 30 \times 60 = 3600$ C

Number of moles of electrons that carry this charge: $3600/96485 = 0.0373$ mol e^-

Cathode $(-)$: $M^{2+} + 2e^- \rightarrow M$ 2 moles of electrons produce 1 mole of M

0.0373 mole of electrons will produce $0.0373/2 = 0.01865$ mole of M

Atomic weight of M $=(0.453)/(0.01865) = 24.3$ amu   Answer(a)

15-71. What is the oxidation state of Osmium atom in an unknown salt if 26.7 g of Osmium

plate out when a current of 15 amp. is passed through a solution of this salt for 1 hour?(AW:

Os = 190.2 amu)

     (a) Os^+ (b) Os^{2+} (c) Os^{3+} (d) Os^{4+} (e)Os^{5+}

Cathode $(-)$: $Os^{x+} + xe^- \rightarrow Os$ x moles of electrons are required to produce 1 mole of Os.

Amount of electrical charge that passes through the solution: $15 \times 1 \times 60 \times 60 = 54000$ C

Number of moles of electrons that carry this charge: $54000/96485 = 0.56$ mol e^-

Number of moles of Os produced = $26.7/190.2 = 0.140$ mole

Number of moles of electrons required to produce 0.140 mole of Os = 0.56 mol e^-

Number of moles of electrons required to produce 1 mole of Os = $0.56/0.14 = 4$

So the oxidation state of Os = 4+   Cathode $(-)$: $Os^{4+} + 4e^- \rightarrow Os$ Answer(d)

15-72. What is the oxidation state of Cerium atom in Ce_xZ_y if a current of 1.2 amp. for 3 hours

deposits 4.70 g of Ce at the cathode? (AW: Ce = 140.1 amu)

        (a) 0   (b) 1+   (c) 2+   (d) 3+   (e) 4+

Cathode $(-)$: $Ce^{x+} + xe^- \rightarrow Ce$ x moles of electrons are required to produce 1 mole of Ce.

Amount of electrical charge that passes through the solution: $1.2 \times 3 \times 60 \times 60 = 12960$ C

Number of moles of electrons that carry this charge: $12960/96485 = 0.134$ mol e^-

Number of moles of Ce produced = $4.7/140.1 = 0.0335$ mole

Number of moles of electrons required to produce 0.0335 mole of Ce= 0.134 mol e^-

Number of moles of electrons required to produce 1 mole of Ce = $0.134/0.0335 = 4$

So the oxidation state of Ce = 4+   Cathode $(-)$: $Ce^{4+} + 4e^- \rightarrow Ce$ Answer(e)

15-73. What is the oxidation state of tin atom in an unknown salt if 14.8 g of tin are plated out

when a current of 40 amp. is passed through a molten solution of this salt for 10 minutes?

(AW: Sn = 118.7 amu)

(a) 1+ (b) 2+ (c) 3+ (d) 4+ (e) 6+

Cathode (−): $Sn^{x+} + xe^- \rightarrow Sn$ x moles of electrons are required to produce 1 mole of Sn.

Amount of electrical charge that passes through the molten solution: $40 \times 10 \times 60 = 24000$ C

Number of moles of electrons that carry this charge: $24000/96485 = 0.25$ mol e^-

Number of moles of Sn produced $= 14.8/118.7 = 0.125$ mole

Number of moles of electrons required to produce 0.125 mole of Sn$= 0.25$ mol e^-

Number of moles of electrons required to produce 1 mole of Sn $= 0.25/0.125 = 2$

So the oxidation state of Sn $= 2+$ Cathode (−): $Sn^{2+} + 2e^- \rightarrow Sn$ Answer(b)

15-74. Which of the following is the strongest oxidizing agent ?

(a) H_2O_2 in base (b) H_2O_2 in acid (c) O_2 in acid (d) CrO_4^{2-} in acid (e) Br_2

Oxidizing	$H_2O_2 + 2e^- \rightleftharpoons 2OH^-$	$E^0_{red} = 0.880$ V
power	$Br_2 (aq) + 2e^- \rightleftharpoons 2Br^-$	$E^0_{red} = 1.087$ V
increases	$CrO_4^{2-} + 8H^+ + 3e^- \rightleftharpoons Cr^{3+} + 4H_2O$	$E^0_{red} = 1.195$ V
	$O_2 + 4H^+ + 4e^- \rightleftharpoons 2H_2O$	$E^0_{red} = 1.229$ V
	$H_2O_2 + 2e^- + 2H^+ \rightleftharpoons 2H_2O$	$E^0_{red} = 1.776$ V

H_2O_2 in acid is the strongest oxidizing agent (most positive E^0_{red})   Answer(b)

15-75. Which of the following is the strongest reducing agent?

(a) $KMnO_4$ (b) Zn (c) Cu (d) O_2 (e) Na

O_2 is an oxidising agent.

$Na^+ + e^- \rightleftharpoons Na$	$E^0_{red} = -2.7109$ V	
$Zn^{2+} + 2e^- \rightleftharpoons Zn$	$E^0_{red} = -0.763$ V	Reducing
$Cu^{2+} + 2e^- \rightleftharpoons Cu$	$E^0_{red} = +0.3402$ V	power
$MnO_4^{2-} + 2H_2O + 3e^- \rightleftharpoons MnO_2 + 4(OH)^-$	$E^0_{red} = +0.588$ V	increases

Na is the strongest reducing agent (Na^+ has the most negative E^0_{red}) .   Answer(e)

15-76. Which is the strongest oxidizing agent?

(a) Ce^{4+} (b) Ce^{3+} (c) H^+ (d) Cr^{2+} (e) Mg

Oxidizing	$Mg^{2+} + 2e^- \rightleftharpoons Mg$	$E^0_{red} = -2.37$ V
power	$Cr^{2+} + 2e^- \rightleftharpoons Cr$	$E^0_{red} = -0.91$ V
increases	$2H^+ + 2e^- \rightleftharpoons H_2$	$E^0_{red} = -0.00$ V
	$Ce^{4+} + e^- \rightleftharpoons Ce^{3+}$	$E^0_{red} = 1.61$ V

Ce^{4+} (most positive E^0_{red}) is the strongest oxidizing agent. **Answer(a)**

15-77. Which of the following reagents should react with H^+ to produce H_2?

(a) Both Mg^{2+} and Cr^{2+} (b) Both Pd and Cr^{2+} (c) Only Pd^{2+}

(d)Both Mg and Cr (e) Mg,Cr, and Pd

Oxidizing	$Mg^{2+} + 2e^- \rightleftharpoons Mg$	$E^0_{red} = -2.37$ V	
power	$Cr^{2+} + 2e^- \rightleftharpoons Cr$	$E^0_{red} = -0.91$ V	
increases	$2H^+ + 2e^- \rightleftharpoons H_2$	$E^0_{red} = 0.000$ V	Reducing
	$Pd^{2+} + 2e^- \rightleftharpoons Pd$	$E^0_{red} = 0.987$ V	power
	$Ce^{4+} + e^- \rightleftharpoons Ce^{3+}$	$E^0_{red} = 1.61$ V	increases

Mg is a stronger reducing agent than H_2. It will react with H^+ to form H_2.

Cr is a stronger reducing agent than H_2. It will react with H^+ to form H_2.

H_2 is a stronger reducing agent than Pd. It will reduce Pd^{2+} to Pd.

H_2 is a stronger reducing agent than Ce^{3+}. It will reduce Ce^{4+} to Ce^{3+}. **Answer(d)**

15-78. Use the table of electrode potentials to determine which of the following reactions is not

spontaneous.

(a) $Zn\ (s) + 2H^+\ (aq) \rightarrow Zn^{2+}\ (aq) + H_2\ (g)$
(b) $2Ag\ (s) + Zn^{2+}\ (aq) \rightarrow 2Ag^+\ (aq) + Zn\ (s)$
(c) $Cl_2\ (aq) + 2Fe^{2+}\ (aq) \rightarrow 2Cl^-\ (aq) + 2Fe^{3+}\ (aq)$
(d) $2Al\ (s) + (3/2)O_2\ (g) + 6H^+\ (aq) \rightarrow 2Al^{3+}\ (aq) + 3H_2O\ (l)$
(e) $Mg\ (s) + Cl_2\ (g) \rightarrow MgCl_2\ (s)$

(a) $Zn \rightarrow Zn^{2+} + 2e^-$ $E^0_{oxi} = 0.763$ V

$2H^+ + 2e^- \rightarrow H_2$ $E^0_{red} = 0.000$ V

$\overline{Zn + 2H^+ \rightarrow Zn^{2+} + H_2}$ $E^0 = 0.763 + 0.000 = 0.763$ V

$\Delta G^0 = -nFE^0$

$\Delta G^0 < 0$. The reaction will occur spontaneously.

(b) $2\{Ag \rightarrow Ag^+ + e^-\}$ $E^0_{oxi} = -0.7996$ V

        $Zn^{2+} + 2e^- \rightarrow Zn$ $E^0_{red} = -0.763$ V

        $\overline{Zn^{2+} + 2Ag \rightarrow Zn + 2Ag^+}$ $E^0 = -0.7996 - 0.763 = -1.5626$ V

        $\Delta G^0 = -nFE^0$

        $\Delta G^0 > 0$. The reaction will not occur spontaneously.

(c) $2\{Fe^{2+} \rightarrow Fe^{3+} + e^-\}$ $E^0_{oxi} = -0.770$ V

        $Cl_2 + 2e^- \rightarrow 2Cl^-$ $E^0_{red} = +1.36$ V

        $\overline{2Fe^{2+} + Cl_2 \rightarrow 2Fe^{3+} + 2Cl^-}$ $E^0 = 1.36 - 0.770 = 0.59$ V

        $\Delta G^0 = -nFE^0$

        $\Delta G^0 < 0$. The reaction will occur spontaneously.

(d) $2\{Al \rightarrow Al^{3+} + 3e^-\}$ $E^0_{oxi} = 1.706$ V

        $(3/2)\{O_2 + 4H^+ + 4e^- \rightarrow 3H_2O\}$ $E^0_{red} = 1.229$ V

        $\overline{2Al + (3/2)\, O_2 + 6H^+ \rightarrow 2Al^{3+} + 3H_2O}$ $E^0 = 1.229 + 1.706 = 2.935$

        $\Delta G^0 = -nFE^0$

        $\Delta G^0 < 0$. The reaction will occur spontaneously.

(e) $Mg \rightarrow Mg^{2+} + 2e^-$ $E^0_{oxi} = 2.375$ V

        $Cl_2 + 2e^- \rightarrow 2Cl^-$ $E^0_{red} = 1.36$ V

        $\overline{Mg + Cl_2 \rightarrow MgCl_2 \text{ (s)}}$ $E^0 = 2.375 + 1.36 = 3.735$ V

        $\Delta G^0 = -nFE^0$

        $\Delta G^0 < 0$. The reaction will occur spontaneously.   Answer(b)

15-79. Which of the following pairs of ions can't coexist in aqu. solution under standard-state conditions because a spontaneous reaction should take place?

   (a) Sn^{4+} and Fe^{3+} (b) Sn^{4+} and Fe^{2+} (c) Sn^{2+} and Fe^{3+}

   (d) Sn^{2+} and Fe^{2+} (e) None of these can coexist

        $Sn^{4+} + 2e^- \rightleftharpoons Sn^{2+}$ $E^0_{red} = 0.154$ V

        $Fe^{3+} + e^- \rightleftharpoons Fe^{2+}$ $E^0_{red} = 0.770$ V

Sn^{2+} is a stronger reducing agent than Fe^{2+}. So it will reduce Fe^{3+} to Fe^{2+}.

$$Sn^{2+} \rightleftharpoons Sn^{4+} + 2e^- \qquad\qquad E^0_{oxi} = -0.154 \text{ V}$$

$$\underline{2\{Fe^{3+} + e^- \rightleftharpoons Fe^{2+}\} \qquad\qquad E^0_{red} = 0.770 \text{ V}}$$

$$Sn^{2+} + Fe^{3+} \rightleftharpoons Sn^{4+} + Fe^{2+} \qquad\qquad E^0 = 0.770 - 0.154 = 0.616 \text{ V}$$

$$\Delta G^0 = -nFE^0$$

$\Delta G^0 < 0$. The reaction will occur spontaneously.   So Sn^{2+} and Fe^{3+} can't coexist.   Answer(c)

15-80. Calculate the value of K_{sp} of AgBr. The silver-silver bromide has a standard potential of

0.07133 V.

$$Ag\ (s) \rightleftharpoons Ag^+\ (aq) + e^- \qquad E^0_{oxi} = -0.7996 \text{ V}$$

$$\underline{AgBr\ (s) + e^- \rightleftharpoons Ag\ (s) + Br^-\ (aq) \quad E^0_{red} = 0.07133 \text{ V}}$$

$$AgBr\ (s) \rightleftharpoons Ag^+\ (aq) + Br^-\ (aq) \qquad E^0 = 0.07133 - 0.7996 = -0.7283 \text{ V}$$

$$E = E^0 - \{(0.0591/1) \log\ [Ag^+][Br^-]\}$$

We have to assume that the reaction is at equilibrium.

At equilibrium $E = 0$

$$0 = E^0 - \{(0.0591/1) \log\ [Ag^+][Br^-]\}$$

$$0 = (-0.7283) - \{(0.0591/1) \log\ K_{sp}\ (AgBr)\}$$

$$\log K_{sp}\ (AgBr) = -12.323$$

$$K_{sp}\ (AgBr) = 4.75 \times 10^{-13}$$

15-81. Explain why the sum of the potentials for the half reactions

$$Sn^{4+}\ (aq) + 2e^- \rightleftharpoons Sn^{2+}\ (aq) \quad \text{and} \quad Sn^{2+}\ (aq) + 2e^- \rightleftharpoons Sn\ (s)$$

does not equal the potential for the reaction $Sn^{4+}\ (aq) + 4e^- \rightleftharpoons Sn\ (s)$

What is the net cell potential? Compare the values of ΔG^0 for the sum of the potentials and the

actual net cell potential.

$$Sn^{4+}\ (aq) + 2e^- \rightleftharpoons Sn^{2+}\ (aq) \quad E^0_{red} = 0.1500 \text{ V}$$

$$Sn^{2+}\ (aq) + 2e^- \rightleftharpoons Sn\ (s) \quad E^0_{red} = -0.1364 \text{ V}$$

$$Sn^{4+}\ (aq) + 4e^- \rightleftharpoons Sn\ (s) \quad E^0_{red} = ?$$

 When we add two half-reactions to obtain a third half-reaction where $n_1, n_2,$ and n_3 are not

same and where we are not eliminating electrons we cannot use the equation $E^0_3 = E^0_1 + E^0_2$

(E^0 is not a state function). We have to use the equation $\Delta G^0_3 = \Delta G^0_1 + \Delta G^0_2$.

$\Delta G^0 = -nFE^0$

$Sn^{4+}(aq) + 2e^- \rightleftharpoons Sn^{2+}(aq)$ $E^0_{red1} = 0.1500$ V $n_1 = 2$ $\Delta G^0_1 = -2FE^0_{red1}$

$Sn^{2+}(aq) + 2e^- \rightleftharpoons Sn(s)$ $E^0_{red2} = -0.1364$ V $n_2 = 2$ $\Delta G^0_2 = -2FE^0_{red2}$

$Sn^{4+}(aq) + 4e^- \rightleftharpoons Sn(s)$ $E^0_{red3} = ?$ $n_3 = 4$ $\Delta G^0_3 = -4FE^0_{red3}$

$\Delta G^0_3 = \Delta G^0_1 + \Delta G^0_2$

$-4FE^0_{red3} = -2FE^0_{red1} - 2FE^0_{red2}$

$2E^0_{red3} = E^0_{red1} + E^0_{red2} = -0.1364 + 0.15 = 0.0136$ V

$E^0_{red3} = 0.0136/2 = 0.0068$ V

Actual net potential $= 0.0068$ V

$\Delta G^0_3 = -4 \times 96485 \times 0.0068 = -2624$ J/mol $= -2.624$ kJ/mol

Sum of the potentials $= E^0_{red1} + E^0_{red2} = -0.1364 + 0.15 = 0.0136$ V

$\Delta G^0 = -4 \times 96485 \times 0.0136 = -5249$ J/mol $= -5.249$ kJ/mol

15-82. For the cell represented as

Al (s) | Al^{3+} (aq) || Sn^{4+} (aq), Sn^{2+} (aq) | Pt (s) , how many electrons are transferred in the redox

reaction? What is the standard cell potential? What is the ΔG^0 ? Is this a spontaneous process?

$2 (Al \rightleftharpoons Al^{3+} + 3e^-)$ $E^0_{oxi} = 1.706$ V

$3 (Sn^{4+} + 2e^- \rightleftharpoons Sn^{2+})$ $E^0_{red} = 0.150$ V

$2Al + 3Sn^{4+} \rightleftharpoons 2Al^{3+} + 3Sn^{2+}$ $E^0_{cell} = 1.706 + 0.150 = 1.856$ V

6 electrons are transferred in this reaction.

$\Delta G^0 = -(6 \times 96485 \times 1.856)/2$ J/mol Al

$= -537$ kJ/mol Al

ΔG^0 is negative. So this is a spontaneous process.

15-83. If $E^0_{red} = 0.158$ V for $Cu^{2+} + e^- \rightleftharpoons Cu^+$

and $E^0_{red} = 0.522$ V for $Cu^+ + e^- \rightleftharpoons Cu$

do you agree that E^0_{red} for $Cu^{2+} + 2e^- \rightleftharpoons Cu$ will be equal to 0.68 V?

When we add two half-reactions to obtain a third half-reaction where $n_1, n_2,$ and n_3 are not

same and where we are not eliminating electrons we cannot use the equation

$E^0_3 = E^0_1 + E^0_2$ (E^0 is not a state function). We have to use the equation $\Delta G^0_3 = \Delta G^0_1 + \Delta G^0_2$.

$$\Delta G^0 = -nFE^0$$

$Cu^{2+} + e^- \rightleftharpoons Cu^+$ $E^0_{red1} = 0.158$ V $n_1 = 1$ $\Delta G^0_1 = -1FE^0_{red1}$

$Cu^+ + e^- \rightleftharpoons Cu$ $E^0_{red2} = 0.522$ V $n_2 = 1$ $\Delta G^0_2 = -1FE^0_{red2}$

$Cu^{2+} + 2e^- \rightleftharpoons Cu$ $E^0_{red3} = ?$ $n_3 = 2$ $\Delta G^0_3 = -2FE^0_{red3}$

$$\Delta G^0_3 = \Delta G^0_1 + \Delta G^0_2$$

$$-2FE^0_{red3} = -FE^0_{red1} - FE^0_{red2}$$

$$2E^0_{red3} = E^0_{red1} + E^0_{red2} = 0.158 + 0.522 = 0.68 \text{ V}$$

$$E^0_{red3} = 0.68/2 = 0.34 \text{ V}$$

E^0_{red} for $Cu^{2+} + 2e^- \rightleftharpoons Cu$ will be 0.34 V (not 0.68 V).

15-84. Draw the cell diagram for a galvanic cell with a SHE and a zinc electrode that carries out this overall reaction:

$$Zn\,(s) + 2H^+\,(aq) \rightarrow Zn^{2+}\,(aq) + H_2\,(g)$$

$$Zn\,(s) \,|\, Zn^{2+}\,(aq) \,\|\, H^+\,(aq, 1M) \,|\, H_2\,(g, 1atm) \,|\, Pt\,(s)$$

15-85. For the reduction of O_2 to water, $E^0_{red} = 1.23$ V. What is the potential for this half reaction at pH 7.00? What is the potential in a 0.85 M solution of NaOH?

   (a) pH = 7

$O_2 + 4H^+ + 4e^- \rightleftharpoons 2H_2O$ $E^0_{red} = 1.23$ V

$$[H]^+ = 10^{-7} \text{ M}$$

$$E = E^0 - (0.0591/4) \log\{1/[H^+]^4\}$$

$$= E^0 - (0.0591/4) \log\{[H^+]^{-4}\}$$

$$= E^0 - (0.0591/4)\{-4 \log[H^+]\}$$

$$= E^0 - 0.0591\{-\log[H^+]\}$$

$$= E^0 - 0.0591\{-\log[10^{-7}]\}$$

$$= 1.23 - (0.0591 \times 7)$$

$$= 0.8163 \text{ V}$$

Potential at pH 7 = 0.8163 V

   (b) In a 0.85 M solution of NaOH

$O_2 + 4H^+ + 4e^- \rightleftharpoons 2H_2O$ $E^0_{red} = 1.23$ V

$[OH^-] = 0.85$

$[H^+][OH^-] = 10^{-14}$

$[H^+](0.85) = 10^{-14}$

$[H^+] = (10^{-14})/0.85 = 1.18 \times 10^{-14}$

$E = E^0 - 0.0591\{ - \log [H^+]\}$

$\quad = E^0 - 0.0591\{ - \log (1.18 \times 10^{-14})\}$

$\quad = E^0 - (0.0591 \times 13.93)$

$\quad = 1.23 - 0.823$

$\quad = 0.407 \text{ V}$

Another way:

$O_2 + 2H_2O + 4e^- \rightleftharpoons 4OH^- \quad E^0_{red} = 0.401 \text{ V}$

$OH^- = 0.85 \text{ M}$

$E = E^0 - (0.0591/4) \log \{[OH^-]^4\}$

$\quad = E^0 - (0.0591/4) \times 4 \log [OH^-]$

$\quad = E^0 - 0.0591 \log [OH^-]$

$\quad = E^0 - 0.0591 \log [0.85]$

$\quad = E^0 - \{(0.0591) \times (- 0.07)\}$

$\quad = 0.401 + 0.004 = 0.405 \text{ V}$

15-86. If $E^0_{red} = - 2.336$ V for $Ce^{3+} + 3e^- \rightleftharpoons Ce \text{ (s)}$

and $E^0_{red} = 1.44$ V for $Ce^{4+} + e^- \rightleftharpoons Ce^{3+}$

What is the E^0_{red} for $Ce^{4+} + 4e^- \rightleftharpoons Ce \text{ (s)}$?

When we add two half-reactions to obtain a third half-reaction where $n_1, n_2,$ and n_3 are not same and where we are not eliminating electrons we cannot use the equation $E^0_3 = E^0_1 + E^0_2$ (E^0 is not a state function). We have to use the equation $\Delta G^0_3 = \Delta G^0_1 + \Delta G^0_2$.

$\Delta G^0 = - nFE^0$

$$Ce^{4+} + e^- \rightleftharpoons Ce^{3+} \qquad E^0_{red1} = 1.44 \text{ V} \qquad n_1 = 1 \qquad \Delta G^0_1 = -1FE^0_{red1}$$

$$Ce^{3+} + 3e^- \rightleftharpoons Ce \text{ (s)} \qquad E^0_{red2} = -2.336 \text{ V} \qquad n_2 = 3 \qquad \Delta G^0_2 = -3FE^0_{red2}$$

$$\overline{Ce^{4+} + 4e^- \rightleftharpoons Ce \text{ (s)} \qquad E^0_{red3} = ? \qquad n_3 = 4 \qquad \Delta G^0_3 = -4FE^0_{red3}}$$

$$\Delta G^0_3 = \Delta G^0_1 + \Delta G^0_2$$

$$-4FE^0_{red3} = -FE^0_{red1} - 3FE^0_{red2}$$

$$4E^0_{red3} = E^0_{red1} + 3E^0_{red2} = 1.44 + (-2.336 \times 3) = -5.568 \text{ V}$$

$$E^0_{red3} = -(5.568/4) = -1.392 \text{ V}$$

15-87. The standard electrode potential for the half-reaction

Ni^{2+} (aq) $+ 2e^- \rightleftharpoons Ni$ (s) is -0.257 V. What pH is needed for this reaction to take place

in the presence of 1 atm H_2 (g) as the reductant if $[Ni^{2+}]$ is 1.00 M?

$$Ni^{2+} + 2e^- \rightleftharpoons Ni \text{ (s)} \qquad E^0_{red} = -0.257 \text{ V}$$

$$\underline{H_2 \rightleftharpoons 2H^+ + 2e^- \qquad\qquad E^0_{oxi} = 0.000 \text{ V}}$$

$$Ni^{2+} + H_2 \rightleftharpoons Ni \text{ (s)} + 2H^+ \qquad E^0_{cell} = -0.257 \text{ V}$$

$$E = E^0 - (0.0591/2) \log [H^+]^2 = E^0 - 0.0591 \log [H^+] = E^0 + (0.0591 \times pH)$$

$$E = E^0 + (0.0591 \times pH)$$

$$E = -0.257 + (0.0591 \times pH)$$

E will be positive if $(0.0591 pH) > 0.257$.   E will be positive if pH $> (0.257/0.0591)$.

E will be positive if pH > 4.349 \qquad\qquad For the reaction to take place the pH > 4.349

15-88. Hydrogen gas reduces Ni^{2+} according to the following reaction:

Ni^{2+} (aq) $+ H_2$ (g) $\rightarrow Ni$ (s) $+ 2H^+$ (aq) $E^0_{cell} = -0.257$ V, $\Delta H = 54$ kJ/mol

- (i)   What is the K for this reaction?
- (ii)   Is this reaction likely to occur?
- (iii)   What conditions can be changed to increase the likelihood that the reaction will occur as written?

(i)

$$Ni^{2+} + 2e^- \rightleftharpoons Ni \text{ (s)} \qquad E^0_{red} = -0.257 \text{ V}$$

$$\underline{H_2 \rightleftharpoons 2H^+ + 2e^- \qquad\qquad E^0_{oxi} = 0.000 \text{ V}}$$

$$Ni^{2+} + H_2 \rightleftharpoons Ni \text{ (s)} + 2H^+ \qquad E^0_{cell} = -0.257 \text{ V}$$

$$E = E^0 - (0.0591/2) \log K$$

At equilibrium E = 0.

$E^0 = (0.0591/2) \log K = -0.257$

$\log K = -(2 \times 257)/0.0591 = -8.7$ $\quad\quad\quad K = 2 \times 10^{-9}$

(ii)

ΔG^0 is positive. So the reaction will not be spontaneous under standard state conditions.

(iii)

$Ni^{2+} + H_2 \rightleftharpoons Ni\ (s) + 2H^+$ $\quad\quad E^0_{cell} = -0.257\ V$

$E = E^0 - (0.0591/2) \log [H^+]^2$

If E is positive, then ΔG will be negative and the reaction will be spontaneous.

$E = E^0 - (0.0591/2) \log [H^+]^2 = E^0 - 0.0591 \log [H^+] = -0.257 + (0.0591 \times pH)$

E will be positive if $(0.0591 \times pH) > 0.257$ $\quad\quad$ E will be positive if $pH > (0.257/0.0591)$

E will be positive if $pH > 4.349$

CHAPTER 16

CHEMICAL KINETICS

16-1. The rate of a zero-order reaction:

    (a) Increases as reactant is consumed
    (b) Depends on the concentration of products
    (c) Decreases as reactant is consumed
    (d) Is independent of temperature
    (e) Is independent of the concentration of reactants and products

For a zero-order reaction,   Rate $= k (A)^0 = k$

The rate of a zero-order reaction is independent of the concentration of reactants

and products.   Answer(e)

16-2. Which of the following choices is a correct expression for the rate of the forward

reaction for the following reaction?

$$2NO\ (g) + O_2\ (g) \rightarrow 2NO_2\ (g)$$

(a) $d(NO)^2(O_2)/dt$ (b) $- d(NO)^2(O_2)/dt$ (c) $- d(NO)^2/dt$

(d) $- d(O_2)/dt$ (e) $d(O_2)/dt$

The correct expression is $- d(O_2)/dt$.

The rate of consumption of NO : $- d(NO)/dt = 2\{- d(O_2)/dt\}$ Answer(d)

16-3. What is the correct rate law for the reaction?

$$2NO\ (g) + O_2\ (g) \rightarrow 2\ NO_2\ (g)$$

(a) Rate $= k(NO)^2$ (b) Rate $= k(NO)^2(O_2)$ (c) Rate $= k(NO)(O_2)$

(d) Rate $= k(NO)$ (e) The rate is impossible to determine from the

Information given

It is impossible to determine the rate law without the mechanism for the reaction.   Answer(e)

16-4. For a one-step second-order reaction, the forward reaction is found to have a rate

constant, $k_f = 5 \times 10^{13}\ M^{-1}s^{-1}$. For this process the equilibrium constant has a value of 1×10^{25}.

What is the rate constant in $M^{-1}s^{-1}$ for the reverse process?

$K_c = (k_f)/(k_r)$ $k_r = (k_f)/(K_c) = (5\times10^{13})/(1\times10^{25}) = 5\times10^{-12}\,M^{-1}s^{-1}$

16-5. The following is a one step reaction for which the forward rate constant is 0.52×10^{-6}

$M^{-1}s^{-1}$: $CH_3Cl + I^- \rightarrow CH_3I + Cl^-$

The equilibrium constant for this reaction is 3.5×10^4. What is the rate constant for the reverse

reaction, in units of $M^{-1}s^{-1}$?

$K_c = (k_f)/(k_r)$ $k_r = (k_f)/(K_c) = (0.52\times10^{-6})/(3.5\times10^4) = 1.5\times10^{-11}\,M^{-1}s^{-1}$

16-6. For the reaction, $CO\ (g) + Cl_2\ (g) \rightarrow COCl_2\ (g)$,

K_f is $1\times10^7\,M^{-1}s^{-1}$, and $k_r = 2\times10^2\,s^{-1}$. What is the equilibrium constant for this reaction?

$K_c = (k_f)/(k_r) = (1\times10^7)/(2\times10^2) = 5\times10^4$

16-7. For the reaction,

$\quad$ $CH_3I\ (aq) + OH^-\ (aq) \rightarrow CH_3OH\ (aq) + I^-\ (aq)$

The following data were obtained:

	Initial (CH_3I)	Initial (OH^-)	Initial instantaneous rate of reaction
Trial 1	1.35 M	0.10 M	8.8×10^{-6} M/s
Trial 2	1.00 M	0.10 M	6.5×10^{-6} M/s
Trial 3	0.85 M	0.10 M	5.5×10^{-6} M/s
Trial 4	0.85 M	0.15 M	8.3×10^{-6} M/s
Trial 5	0.85 M	0.25 M	1.4×10^{-5} M/s

(i)   The rate law for this reaction is :

$\quad$ (a) Zero-order in CH_3I (b) Half-order in CH_3I (c) First order in CH_3I

$\quad$ (d) Second-order in CH_3I (e) None of the above.

(Rate for trial 1)/(Rate for trial 3) $= (8.8\times10^{-6})/(5.5\times10^{-6}) = 1.6$

In trial 3 the initial concentration of CH_3I was 0.85M

In trial 1 the initial concentration of CH_3I was 1.35M

In both trials the initial concentration of OH^- was 0.10M

When only the concentration of CH_3I increases $1.35/0.85 = 1.6$ times, the initial instantaneous

rate increases 1.6 times. So the rate of this reaction is directly proportional to the concentration of CH_3I and the rate law for this reaction is first-order in CH_3I.   Answer(c)

    (ii)   The rate law for this reaction is:

        (a) Zero-order in OH^- (b) Half-order in OH^- (c) First-order in OH^-

        (d) Second order in OH^- (e) None of the above

(Rate for trial 5)/(Rate for trial 3) $= (1.4 \times 10^{-5})/(5.5 \times 10^{-6}) = 2.5$

(Initial concentration of OH^- in trial 5)/(Initial concentration of OH^- in trial 3) $= 0.25/0.10 = 2.5$

(Initialconcentration of CH_3I in trial 5)/(Initial concentration of CH_3I in trial 3)$=0.85/0.85 = 1.0$

When only the concentration of OH^- increases 2.5 times, the initial instantaneous rate of

reaction increases 2.5 times. So the rate of this reaction is directly proportional to the

concentration of OH^- and the rate law for this reaction is first-order in OH^-.   Answer(c)

    (iii)   Calculate the rate constant for this reaction .

Rate $= k\,(CH_3I)\,(OH^-)$

Rate for Trial 2 $= 6.5 \times 10^{-6} = k\,(1)\,(0.1)$

k (rate constant) $= (6.5 \times 10^{-6})/\{(1) \times (0.1)\} = 6.5 \times 10^{-5}$

16-8. Use the following data on the initial rate of reaction to determine the overall order of the

reaction:

BrO_3^- (aq) $+ 5Br^-$ (aq) $+ 6H^+$ (aq) $\rightarrow 3Br_2$ (aq) $+ 3H_2O$ (l)

	Initial (BrO_3^-)	Initial (Br^-)	Initial (H^+)	Initial rate of reaction
Trial 1	0.10 M	0.10 M	0.10 M	9.3×10^{-4} M/s
Trial 2	0.10 M	0.15 M	0.10 M	1.4×10^{-3} M/s
Trial 3	0.25 M	0.15 M	0.10 M	3.5×10^{-3} M/s
Trial 4	0.25 M	0.15 M	0.25 M	8.8×10^{-3} M/s

    (a) First-order overall   (b) Second-order overall (c) Third-order overall

    (d) Fourth-order overall (e) None of the above

In trials 3 and 2, the initial concentration of Br^- was 0.15 M.

In trials 3 and 2, the initial concentration of H^+ was 0.10 M.

(Rate of trial 3)/(Rate of trial 2) $= (3.5 \times 10^{-3})/(1.4 \times 10^{-3)} = 2.5$

{Initial concentration of BrO_3^- in trial 3}/{Initial concentration of BrO_3^- in trial 2} = 0.25/0.10

= 2.5

So the rate of this reaction is directly proportional to the concentration of BrO_3^- and the rate law for this reaction is first-order in BrO_3^-.

In trials 2 and 1, the initial concentration of H^+ was 0.10 M.

In trials 2 and 1, the initial concentration of BrO_3^- was 0.10 M.

(Rate of trial 2)/(Rate of trial 1) = $(1.4 \times 10^{-3})/(9.3 \times 10^{-4})$ = 1.5

{Initial concentration of Br^- in trial 2}/{Initial concentration of Br^- in trial 1} = 0.15/0.10 = 1.5

So the rate of this reaction is directly proportional to the concentration of Br^- and the rate law for this reaction is first-order in Br^-.

In trials 4 and 3, the initial concentration of BrO_3^- was 0.25 M.

In trials 4 and 3, the initial concentration of Br^- was 0.15 M.

(Rate of trial 4)/(Rate of trial 3) = $(8.8 \times 10^{-3})/(3.5 \times 10^{-3})$ = 2.5

(Initial concentration of H^+ in trial 4)/(Initial concentration of H^+ in trial 3) = 0.25/0.10 = 2.5

So the rate of this reaction is directly proportional to the concentration of H^+ and the rate law for this reaction is first-order in H^+.

Rate = k (BrO_3^-) (Br^-) (H^+)

The reaction is first-order in BrO_3^- , first-order in Br^- , first-order in OH^- and third-order overall.   Answer(c)

16-9. The rate law for the following reaction

BrO_3^- (aq) + $5Br^-$ (aq) + $6H^+$ (aq) $\rightarrow$ $3Br_2$ + $3H_2O$ (l)

is Rate = k (BrO_3^-) (Br^-) (H^+) . What are the units for k?

Unit for rate = M/s

Unit for Br_3^- = M

Unit for Br^- = M

Unit for H^+ = M

Units for k = $(M/s)/(M^3)$ = $M^{-2}s^{-1}$

16-10.   NO (g) + O_3 (g) $\rightarrow$ NO_2 + O_2 (g)

For the above reaction the following data were obtained.

	Initial (NO)	Initial (O_3)	Initial rate of reaction
Trial 1	2.1×10^{-6} M	2.1×10^{-6} M	1.6×10^{-5} M/s
Trial 2	4.2×10^{-6} M	2.1×10^{-6} M	3.2×10^{-5} M/s
Trial 3	6.3×10^{-6} M	2.1×10^{-6} M	4.8×10^{-5} M/s
Trial 4	6.3×10^{-6} M	4.2×10^{-6} M	9.6×10^{-5} M/s
Trial 5	6.3×10^{-6} M	6.3×10^{-6} M	14.4×10^{-5} M/s

(i)   The rate law for the reaction would be :

(a) Zero-order in NO   (b) First-order in NO   (c) Second-order in NO

(d) Third-order in NO   (e) None of the above

In trials 3 and 1, the initial concentration of O_3 was 2.1×10^{-6} M.

(Rate for trial 3)/(Rate for trial 1) = $(4.8 \times 10^{-5})/(1.6 \times 10^{-5})$ = 3

(Initial concentration of NO in trial 3)/(Initial concentration of NO in trial 1)

= $(6.3 \times 10^{-6})/(2.1 \times 10^{-6})$ = 3

So the rate of this reaction is directly proportional to the concentration of NO and the rate law

for this reaction is first-order in NO.   Answer(b)

(ii)   The rate law for the reaction would be :

(a) Zero-order in O_3 (b) First-order in O_3 (c) Second-order in O_3

(d) Third-order in O_3 (e) None of the above.

In trials 3 and 5, the initial concentration of NO was 6.3×10^{-6} M.

(Rate for trial 5)/(Rate for trial 3) = $(14.4 \times 10^{-5})/(4.8 \times 10^{-5})$ = 3

 (Initial concentration of O_3 in trial 5)/(Initial concentration of O_3 in trial 3)

 = $(6.3 \times 10^{-6})/(2.1 \times 10^{-6})$ = 3

So the rate of this reaction is directly proportional to the concentration of O_3 and the rate law

for this reaction is first-order in O_3.   Answer(b)

(iii)   If these were obtained by watching the rate at which NO disappears, what would be
        the initial instantaneous rate of disappearance of O_3 in trial 3?

For this reaction $-d(NO)/dt = -d(O_3)/dt$

So for trial 3 the initial instantaneous rate of disappearance of $O_3 = 4.8 \times 10^{-5}$ M/s

    (iv)   If a trial was run in which the initial (NO) was 3.15×10^{-6} and the initial (O_3) was

        3.15×10^{-6}, what would be the initial rate of reaction?

Rate = k (NO) (O_3)

Trial 1   Rate = k $(2.1 \times 10^{-6}) \times (2.1 \times 10^{-6}) = 1.6 \times 10^{-5}$

            $k = (1.6 \times 10^{-5})/\{(2.1 \times 10^{-6}) \times (2.1 \times 10^{-6})\} = 0.3628 \times 10^7$

If the initial (NO) was 3.15×10^{-6} M and the initial (O_3) was 3.15×10^{-6} M, the initial rate

of reaction would be = $(0.3628 \times 10^7) \times (3.15 \times 10^{-6}) \times (3.15 \times 10^{-6}) = 3.6 \times 10^{-5}$ M/s

16-11. For reaction, $2NO\ (g) + Br_2\ (g) \rightarrow 2NOBr\ (g)$

The following data were obtained:

	Initial (NO)	Initial (Br_2)	Initial rate of reaction
Trial 1	0.025 M	0.040 M	1.2 M/s
Trial 2	0.025 M	0.080 M	2.4 M/s
Trial 3	0.025 M	0.120 M	3.6 M/s
Trial 4	0.050 M	0.040 M	4.8 M/s
Trial 5	0.075 M	0.040 M	10.8 M/s

    (i)   The rate law for this reaction would be

        (a) Zero-order in NO   (b) Half-order in NO   (c) First-order in NO

        (d) Second-order in NO (e) None of the above

In trials 5 and 1, the initial concentration of Br_2 was 0.040 M.

(Initial concentration of NO in trial 5)/(Initial concentration of NO in trial 1) = 0.075/0.025 = 3

(Rate for trial 5)/(Rate for trial 1) = (10.8)/(1.2) = 9

When only the concentration of NO increases 3 times the rate increases 9 times. So the rate

law for this reaction is second-order in NO.   Answer(d)

    (ii)   The rate law for this reaction would be:

        (a) Zero-order in Br_2 (b) Half-order in Br_2 (c) First-order in Br_2

        (d) Second-order in Br_2 (e) None of the above

In trials 3 and 1, the initial concentration of NO was 0.025 M.

(Initial concentration of Br_2 in trial 3)/(Initial concentration of Br_2 in trial 1) = 0.120/0.040 = 3

(Rate for Trial 3)/(Rate for trial 1) = 3.6/1.2 = 3

So the rate of this reaction is directly proportional to the concentration of Br_2 and the rate law

for this reaction is first-order in Br_2.   Answer(c)

    (iii)   If these data were obtained by watching the rate at which NO disappears, what would be the initial instantaneous rate of disappearance of Br_2 in trial 1?

According to the stoichiometry of this reaction two NO moles are consumed for every mole of

Br_2 consumed. This means that the rate of disappearance of NO is twice as fast as the rate of

disappearance of Br_2.

   $-d(NO)/dt = 2\{-d(Br_2)/dt\}$

Trial 1: $-d(Br_2)/dt = (1/2)\times\{-d(NO)/dt\} = (1/2)\times(1.2)$ M/s = 0.6 M/s

    (iv)   The magnitude of rate constant k, for this reaction is:

       (a) Less than 1×10^{-4} (b) Between 10^{-4} and 10^{-2} (c) Between 0.01 and 100

       (d) Between 100 and 10,000 (e) More than 10,000

Rate = $k (NO)^2 (Br_2)$

Trial 1 :   1.2 M/s = $k (0.025)^2 (0.040)$

          $k = (1.2)/\{(0.025)^2\times(0.040)\} = 48000$ $M^{-2}s^{-1}$ Answer(e)

    (v)   A plot of (NO) versus time would most closely resemble

       (a) A straight line with a positive slope
       (b) A straight line with a negative slope
       (c) A straight line with a slope of zero
       (d) A curve in which the NO concentration increases rapidly at first and then slows down until eventually a maximum concentration is achieved
       (e) A curve in which the NO concentration decreases rapidly at first and then slows down until eventually a minimum concentration is achieved

The rate law for this reaction is second-order in NO. So a plot of (NO) versus time will not be

a straight line. Only a plot of 1/(NO) versus time will be a straight line. NO is consumed in the

reaction. So the NO concentration will decrease. Since a plot of 1/(NO) versus time will be a

straight line with a positive slope, a plot of (NO) versus time will provide a curve in which the

(NO) will decrease rapidly at first and then slow down until a minimum concentration is achieved.   Answer(e)

16-12. For the reaction,

$$Cr_2O_7^{2-} (aq) + 6I^- (aq) + 14H^+ (aq) \rightarrow 2Cr^{3+} (aq) + 3I_2 (aq) + 7H_2O (l)$$

the following data were obtained:

	Initial ($Cr_2O_7^{2-}$)	Initial (I^-)	Initial (H^+)	Initial rate of reaction
Trial 1	0.0040 M	0.01 M	0.02 M	5×10^{-4} M/s
Trial 2	0.0040 M	0.02 M	0.02 M	20×10^{-4} M/s
Trial 3	0.0040 M	0.03 M	0.02 M	46×10^{-4} M/s
Trial 4	0.0040 M	0.03 M	0.04 M	179×10^{-4} M/s
Trial 5	0.0080 M	0.03 M	0.04 M	355×10^{-4} M/s

(i)   If the initial rate data are for the rate at which I^- disappears $-d(I^-)/dt$, what would be the rate of disappearance of $Cr_2O_7^{2-}$ in trial 1?

According to the stoichiometry of the reaction 6 moles (I^-) are consumed for every mole of $Cr_2O_7^{2-}$ consumed. This means that the rate of disappearance of I^- is 6 times as fast as the rate of disappearance $Cr_2O_7^{2-}$.

$$-d(I^-)/dt = 6 \{-d(Cr_2O_7^{2-})/dt\}$$

Trial 1 : $-d(Cr_2O_7^{2-})/dt = (1/6) \{-d(I^-)/dt\} = (1/6)\times(5\times10^{-4}) = 0.83\times10^{-4}$ M/s

(ii)   The rate law for this reaction would be:

(a) Zero-order in $Cr_2O_7^{2-}$ (b) Half-order in $Cr_2O_7^{2-}$ (c) First-order in $Cr_2O_7^{2-}$

(d) Second-order in $Cr_2O_7^{2-}$ (e) None of the above

In trials 4 and 5, the initial concentration of I^- was 0.030 M.

In trials 4 and 5, the initial concentration of H^+ was 0.040 M.

(Initial concentration of $Cr_2O_7^{2-}$ in trial 5)/(Initial concentration of $Cr_2O_7^{2-}$ in trial 4)

$= 0.008/0.004 = 2$

(Rate for Trial 5)/(Rate for trial 4) = $(355\times10^{-4})/(179\times10^{-4}) = 1.98 \approx 2.0$

The rate of this reaction is directly proportional to the concentration of $Cr_2O_7^{2-}$ and the rate law for this reaction is first-order in $Cr_2O_7^{2-}$.   Answer(c)

(iii)   The rate law for this reaction would be:

(a) First-order in both I^- and H^+ (b) First-order in I^- and second-order in H^+

(c) First-order in H^+ and second-order in I^- (d) Second-order in both H^+ and I^-

(e) None of the above

In trials 1 and 2, the initial concentration of $Cr_2O_7^{2-}$ was 0.0040 M.

In trials 1 and 2, the initial concentration of H^+ was 0.020 M.

(Initial concentration of I^- in trial 2)/(Initial concentration of I^- in trial 1) = 0.02/0.01 = 2

(Rate for trial 2)/(Rate for trial 1) = $(20\times10^{-4})/(5\times10^{-4})$ = 4

When only the concentration of I^- increases 2 times, the rate increases 4 times. So the rate law

for this reaction is second-order in I^-.

In trials 4 and 3, the initial concentration of $Cr_2O_7^{2-}$ was 0.0040 M.

In trials 4 and 3, the initial concentration of I^- was 0.030 M.

(Initial concentration of H^+ in trial 4)/(Initial concentration of H^+ in trial 3) = 0.04/0.02 = 2

(Rate for trial 4)/(Rate for trial 3) = $(179\times10^{-4})/(46\times10^{-4})$ = 3.9 ≈ 4

When only the concentration of H^+ increases 2 times , the rate increases 4 times. So the rate

law for this reaction is second-order in H^+.

The rate law for this reaction is second-order in both H^+ and I^-.   Answer(d)

   (iv)   Calculate the rate constant for this reaction.

Rate = k $(Cr_2O_7^{2-})$ $(I^-)^2(H^+)^2$

Trial 1 : 5×10^{-4} M/s = k $(0.004)\times (0.01)^2\times(0.02)^2$

                  k = $(5\times10^{-4})/\{(0.004)\times(0.01)^2\times(0.02)^2\}$ = 3.125×10^6 $M^{-4}s^{-1}$

   (v)   A graph of $(Cr_2O_7^{2-})$ versus time would most closely resemble:

      (a) A straight line with a negative slope
      (b) A straight line with a slope of zero
      (c) A straight line with appositive slope
      (d) A curve in which the $Cr_2O_7^{2-}$ concentration decreases rapidly at first, and then the rate at
           which $Cr_2O_7^{2-}$ disappears slows down with time
      (e) A curve in which the $Cr_2O_7^{2-}$ concentration increases rapidly at first, and then
           the rate at which $Cr_2O_7^{2-}$ appears slows down with time.

The rate law for this reaction is first-order in $Cr_2O_7^{2-}$. So a graph of $(Cr_2O_7^{2-})$ versus time will

not be a straight line; only a plot of ln $(Cr_2O_7^{2-})$ versus time will be a straight line. $Cr_2O_7^{2-}$ is

consumed in the reaction; so $Cr_2O_7^{2-}$ concentration will not increase. Since a plot of ln

$(Cr_2O_7^{2-})$ versus time will be a straight line with negative slope, a plot of $(Cr_2O_7^{2-})$ versus time will provide a curve in which the $Cr_2O_7^{2-}$ concentration will decrease rapidly at first and then the rate of disappearance of $Cr_2O_7^{2-}$ will slow down. Answer(d)

16-13. Which of the following equations correctly describes the relationship between the rate at which NO_2 and Cl_2 are consumed in the following reaction?

$$2NO_2 (g) + Cl_2 (g) \rightarrow 2NO_2Cl (g)$$

(a) $- d(NO_2)/dt = 1/2 \{- d(Cl_2)/dt\}$
(b) $- d(NO_2)/dt = 1/2 \{d(Cl_2)/dt\}$
(c) $- d(NO_2)/dt = - d(Cl_2)/dt$
(d) $- d(NO_2)/dt = 2 \{- d(Cl_2)/dt\}$
(e) $- d(NO_2)/dt = 2 \{d(Cl_2)/dt\}$

According to the stoichiometry of the reaction 2 moles of NO_2 are consumed for every mole of Cl_2 consumed. This means that the rate of disappearance of NO_2 will be twice as fast as the rate of disappearance of Cl_2. $- d(NO_2)/dt = 2 \{- d(Cl_2)/dt\}$ Answer(d)

16-14. Which statement is true?

(a) The rate of appearance of products of a chemical reaction is always equal to the rate of disappearance of reactants
(b) If a reaction follows a second-order rate law, it must have two steps in its reaction echanism
(c) The half-life for a first-order reaction is always larger than the half-life for a second-order reaction
(d) The half-life for a first-order reaction is independent of the initial concentration of the reactant
(e) The half-life for s second-order reaction is independent of the initial concentration of the reactant

For a first-order reaction $Ln\{(X)/(X)_0\} = - kt$

$$t_{1/2} = - ln(0.5)/k = 0.693/k$$

So the half-life for a first-order reaction is independent of the initial concentration of the reactant. Answer(d)

16-15. The instantaneous rate of appearance of water from the reaction,

$$4NH_3 (g) + 5O_2 (g) \rightarrow 4NO (g) + 6H_2O (g)$$

at some moment in time is 21.3 mm Hg per minute. What is the instantaneous rate of disappearance of NH_3 at the same moment in time?

According to the stoichiometry of the reaction 6 moles of H_2O are produced for every 4 moles of NH_3 consumed. This means that the rate of appearance of H_2O is 1.5 times as fast as the rate of disappearance of NH_3.

$$d(H_2O)/dt = 1.5 \ \{- d(NH_3)/dt\}$$

$$- d(NH_3)/dt = (1/1.5) \ d(H_2O)/dt = (21.3/1.5) \text{ mm Hg/minute} = 14.2 \text{ mm Hg/minute}$$

16-16. When NH_3 is treated with O_2 at elevated temperatures, the rate of disappearance of NH_3 is found to be 3.5×10^{-2}. The equation for the reaction is:

$$4NH_3 \ (g) + 5O_2 \ (g) \rightarrow 4NO \ (g) + 6H_2O \ (g)$$

Calculate the rate of appearance of NO and H_2O.

According to the stoichiometry of the reaction, 6 moles of H_2O are produced for every 4 moles of NH_3 consumed. This means that that the rate of appearance H_2O is 1.5 times as fast as the rate of disappearance of NH_3.

$$d(H_2O)/dt = 1.5 \ \{- d(NH_3)/dt\} = 1.5 \times (3.5 \times 10^{-2}) = 5.25 \times 10^{-2} \ M^{-1}s^{-1}$$

According to the stoichiometry of the reaction, 4 moles of NO are produced for every 4 moles of NH_3 consumed. This means that the rate of appearance of NO is equal to the rate of disappearance of NH_3.
$$d(NO)/dt = - \ d(NH_3)/dt = 3.5 \times 10^{-2} \ M^{-1}s^{-1}$$

16-17. For which reactant or product in the following reaction will the rate of change of concentration will be the largest?

$$6Fe^{2+} + Cr_2O_7^{2-} + 14H^+ \rightarrow 2Cr^{2+} + 6Fe^{3+} + 7H_2O$$

(a) Cr^{3+} (b) $Cr_2O_7^{2-}$ (c) Fe^{2+} (d) H^+ (e) H_2O

According to the stoichiometry of the reaction for every mole of $Cr_2O_7^{2-}$ consumed, 6 moles of Fe^{2+} will be consumed, 14 moles of H^+ will be consumed, 2 moles of Cr^{3+} will be produced, 6 moles of Fe^{3+} will be produced and 7 moles of H_2O will be produced. So the rate of change of concentration with time will be the largest for H^+.   Answer(d)

16-18. The rate of disappearance of MnO_4^- in the following reaction,

$$2MnO_4^- + 10I^- + 16H^+ \rightarrow 2Mn^{2+} + 5I_2 + 8H_2O$$

is 8.9×10^{-3} M/s. What is the rate of appearance of I_2 in M/s?

According to the stoichiometry of the reaction 5 moles of I_2 are produced for every 2 moles of MnO_4^- consumed. This means that the rate of appearance of I_2 is 2.5 times as fast as the rate of disappearance of MnO_4^-.

$$d(I_2)/dt = (5/2) \{- d(MnO_4^-)/dt\} = 2.5 \times (8.9 \times 10^{-3}) \text{ M/s} = 22.25 \times 10^{-3} \text{ M/s}$$

16-19. Which of the following statements correctly describes a reaction for which a plot of

log (X) versus time produces a straight line?

    (a) The rate constant for the reaction can be obtained from the value of the intercept o the vertical axis
    (b) The rate constant is proportional to the slope of the line
    (c) The rate of reaction does not depend on the concentration of X
    (d) The initial concentration of X can be calculated from the intercept on the horizontal axis
    (e) All of the above statements are true.

For a first-order reaction $\text{Ln} (X) = - kt + \ln (X)_0$

$\text{Log} (X) = \{- (1/2.303)k\}t + \log (X)_0$

A plot of log (X) versus time will be a straight line and its slope will be $= (- k/2.303)$.

The rate constant will be proportional to the slope of the line.   Answer(b)

16-20. If a plot of $\{1/(A)\}$ versus time produces a straight line, which of the following is true?

    (a) The reaction is first order in A
    (b) The reaction is second-order in A
    (c) The reaction is first-order in two reactants
    (d) The rate of reaction does not depend on the concentration of A
    (e) None of the above

For a second-order reaction $\{1/(A)\} = kt + \{1/(A)_0\}$

A plot of 1/(A) versus time will be a straight line whose slope will be equal to the rate constant.

                                              Answer (b)

16-21. The reaction, $N_2O_5 \rightarrow 2NO_2 + (1/2)O_2$ is first-order in N_2O_5 with a half-life of 19.25

min. How long would it take for the N_2O_5 concentration to decrease from 0.050 M to 0.030 M?

For a first-order reaction $\text{Ln} \{(X)/(X)_0\} = - kt$ and $t_{1/2} = 0.693/k$

$t_{1/2} = 19.25$ min.

$k = 0.693/19.25 = 0.036$

$t = - \ln\{(X)/(X)_0\} \times (1/k) = - \{\ln (0.03/0.05)\} \times (1/0.036) = 14.2$ min.

16-22. The reaction, $2N_2O_5 \rightarrow 4NO_2 + O_2$ obeys the rate law: Rate $= k (N_2O_5)$.

What is the relationship between the rate at which N_2O_5 disappears and O_2 appears? If half of

the initial N_2O_5 reacts in 1403 sec, what is the value of rate constant? If the initial concentration

of N_2O_5 is 1.00 M, what will be concentration of O_2 when t = 701 sec.?

According to the stoichiometry 1 mole of O_2 is produced for every 2 moles of N_2O_5 consumed. This means that the rate of disappearance of N_2O_5 will be twice as fast as the rate of appearance of O_2. $\qquad -d(N_2O_5)/dt = 2\ d(O_2)/dt$

Rate $= k\ (N_2O_5)$

The rate law for this reaction is first-order in N_2O_5.

For a first-order reaction $t_{1/2} = 0.693/k = 1403$ sec

$k = 0.693/t_{1/2} = 0.693/1403 = 4.94 \times 10^{-4}\,s^{-1}$

For a first-order reaction $\quad Ln\ \{(X)/(X)_0\} = -kt$

$(X)_0 = 1.00\ M$ and $t = 701$ sec.

$Ln\ (X) = -(4.94 \times 10^{-4}) \times (701) = -0.3463 \qquad\qquad (X) = 0.71\ M$

16-23. For the reaction, $OCl^-\ (aq) + I^-\ (aq) \rightarrow OI^-\ (aq) + Cl^-\ (aq)$

The following data were obtained in the presence of a large excess of I^- ion.

Time	(OCl^-)
0 s	0.0025 M
3 s	0.00137 M
6 s	0.00075 M
9 s	0.00041 M

(i)   What is the rate constant for this reaction?

Consider the first two data. If we assume that the rate law is pseudo first-order in OCl^-,

$Ln\ \{(X)/(X)_0\} = -k't$,

$k' = -\{ln\ (0.00137/0.0025)\}/(3) = 0.200$

Consider the first and the last data. If our assumption is correct then the k' should be equal to 0.200.

$k' = -\{ln\ (0.00041/0.0025)\}/(9) = 0.200$

So our assumption is correct and the reaction is a pseudo first-order reaction with k' = 0.200.

This value of k' is the actual rate constant multiplied by the concentration of the I^- ion that is present in excess. In other words the actual rate constant for the reaction will be less than 0.200.

(ii)   What is the half-life in seconds?

This is a pseudo first-order reaction.

$t_{1/2} = 0.693/k' = 0.693/0.200 = 3.465$ sec.

    (iii)   What is the concentration of OCl^- after 12 seconds.

This is a pseudo first-order reaction.

$Ln\{(X)/(X)_0\} = -k't = -(0.2) \times (12) = -2.4$

$Ln(X) = -2.4 + \ln(0.0025) = -2.4 - 6.0 = -8.4$

$(X) = 2.25 \times 10^{-4}$ M

16-24. The decomposition of H_2O_2 is first-order in H_2O_2 with a rate constant, in the presence of a platinum catalyst, of 2×10^{-2} s^{-1}. If 40 liters of H_2O_2 are used to fuel a rocket, how long is it before only 20 liters remain?

For a first-order reaction, $t_{1/2} = 0.693/k = 0.693/(2 \times 10^{-2}) = 34.65$ sec.

16-25. For a reaction A $\rightarrow$ B, the following data were obtained.

Time	(A)
0 s	0.380 M
10 s	0.319 M
20 s	0.268 M
30 s	0.225 M
40 s	0.189 M
50 s	0.158 M
60 s	0.133 M

    (i)   What is the rate constant, k, for the reaction?

Consider the first and the third data. If we assume that the rate law is first-order in A,

$Ln\{(A)/(A)_0\} = -kt$,

$k = -\{\ln(0.268/0.380)\}/(20) = 0.0175$

Consider the first and the last data. If our assumption is correct, k should be equal to 0.0175.

$k = -\{\ln(0.133/0.380)\}/60 = 0.0175$

So our assumption is correct and the reaction is first-order in A and the rate constant

$= 0.0175$ s^{-1}

(ii)   What is the half-life of this reaction in seconds?

This is a first-order reaction .

$t_{1/2}$ = 0.693/k = 0.693/0.0175 = 39.6 sec.

(iii)   What will be the concentration of A when t = 120 sec.?

This is a first-order reaction.

Ln $\{(A)/(A)_0\}$ = $-$ kt

Ln (A) = $-$ $\{(k)\times(120)\}$ + ln $(A)_0$

Ln (A) = $-$ (0.0175×120) + ln (0.380) = $-$ 2.1 $-$ 0.9676 = $-$ 3.0676

A = 0.047 M

16-26. For a second-order reaction, it takes 15 seconds for the initial concentration of a reactant to decrease from 0.60 M to 0.53 M. What is the initial half-life of this reaction?

The reaction is a second-order reaction.

$\{1/(X)\}$ $-$ $\{1/(X)_0\}$ = kt

15k = (1/0.53) $-$ (1/0.60) = 1.887 $-$ 1.667 = 0.22

K = 0.22/15 = 0.01467

Initial $t_{1/2}$ = $1/\{(k)(X)_0\}$ = $1/\{(0.01467)\times(0.60)\}$ = 113.6 sec.

16-27. A second-order reaction is found to have a rate constant of 0.135 $M^{-1}s^{-1}$. What is the half-life of this reaction if the initial concentration of the reactant is 0.75 M?

$t_{1/2}$ for a second order reaction = $1/\{(k)\times(X)_0\}$ = $1/\{(0.135)\times(0.75)\}$ = 9.88 sec.

16-28. Which of the following statements about the half-life of a reaction is true?

(a) The half-life does not depend on the order of the reaction
(b) The half-life of a first-order reaction increases with time
(c) The half-life of a second-order reaction is independent of concentration
(d) A zero-order reaction does not have a half-life
(e) None of the above are true

The half-life depends on the order of the reaction.

For a zero-order reaction $t_{1/2}$ = $(X)_0$/ (2k)

For a first-order reaction $t_{1/2}$ = 0.693/(k)

For a second-order reaction $t_{1/2}$ = $1/k(X)_0$

So the first four statements are wrong.   Answer(e)

16-29. The overall rate of a chemical reaction determined by:

(a) The fastest step in the reaction mechanism
(b) The first step in the reaction mechanism
(c) The slowest step in the reaction mechanism
(d) The last step in the reaction mechanism
(e) ΔG for the overall reaction

The slowest step in the reaction mechanism is called the rate limiting step in the reaction

because it literally limits the rate of appearance of products in the reaction. So the overall rate

of reaction is more or less equal to the rate of the slowest step in the mechanism.   Answer(c)

16-30. Which of the following rate laws suggests that the reaction probably occurs in a single

step?

(a) NO (g) + O_2 (g) → NO_2 (g) + O(g)   Rate = k (NO) (O_2)
(b) H_2 (g) + Br_2 (g) → $2HBr$ (g)   Rate = k (H_2) (Br_2)$^{1/2}$
(c) $(CH_3)_3CBr$ + OH^- → $(CH_3)_3COH$ + Br^- Rate = k [$(CH_3)_3CBr$]

For a single-step reaction, the rate law should agree with the stoichiometry of the reaction.

Only in (a) the rate law agrees with the stoichiometry of the reaction.   Answer(a)

16-31. Use the rate laws given below to determine which of the following reactions most likely

occurs in a single step?

(a) $2NO_2$ (g) + F_2 (g) → $2NO_2F$ (g)   Rate = k (NO_2) (F_2)
(b) $2NOCl$ (g) → $2NO$ (g) + Cl_2 (g)   Rate = k ($NOCl$)^2
(c) $2O_3$ (g) → $3O_2$ (g)   Rate = k (O_3)
(d) H_2 (g) + Br_2 (g) → $2HBr$ (g)   Rate = k (H_2) (Br)$^{1/2}$

For a single step reaction, the rate law should agree with the stoichiometry of the reaction.

Only in (b) the rate law agrees with the stoichiometry of the reaction.   Answer(b)

16-32. Nitrogen oxide reacts with hydrogen to form nitrogen and water vapor.

$$2NO \text{ (g)} + 2H_2 \text{ (g)} → N_2 \text{ (g)} + 2H_2O \text{ (g)}$$

The following mechanism has been proposed for this reaction.

Step 1: $2NO + H_2 → N_2 + H_2O_2$ (slow step)

Step 2: $H_2O_2 + H_2 → 2H_2O$ (fast step)

Use this mechanism to predict the rate law for this reaction.

The overall rate of a reaction is more or less equal to the rate of the slowest step, which is the

rate limiting step. The rate law for a single step should agree with the stoichiometry of that step. So   Rate = k $(NO)^2 (H_2)$

16-33. The reaction $2NO_2$ (g) + F_2 (g) $\rightarrow$ $2NO_2F$ (g) has the following rate law:

Rate = k (NO_2) (F_2)

Which of the following mechanisms provides the best explanation of

the observed rate law?

   (a) $2NO_2 + F_2 \rightarrow 2NO_2F$ one step
   (b) $NO_2 + F_2 \rightarrow NO_2F + F$ (fast)
       $NO_2 + F \rightarrow NO_2F$ (slow)
   (c) $NO_2 + F_2 \rightarrow NO_2F + F$ (slow)
       $NO_2 + F \rightarrow NO_2F$ (fast)
   (d) $F_2 \rightarrow 2F$ (slow)
       $2NO_2 + 2F \rightarrow 2NO_2F$ (fast)
   (e) None of these explains the observed rate law

The overall rate of a reaction is more or less equal to the rate of the slowest step, which is the

rate limiting step. The rate law for a single step should agree with the stoichiometry of that

step.

Rate for (a) = k $(NO_2)^2$ (F_2) Rate for (b) = k (NO_2) (F)

Rate for (c) = k (NO_2) (F_2) Rate for (d) = k (F_2) Answer(c)

16-34. The oxidation of iodide ion by hypochlorite ion in aqueous solution

$ClO^- + I^- \rightarrow Cl^- + IO^-$ has been postulated to occur by the following three-step mechanism:

   1. $ClO^- + H_2O \rightarrow HClO + OH^-$ (fast, equilibrium)
   2. $I^- + HClO \rightarrow HIO + Cl^-$ (slow, equilibrium)
   3. $OH^- + HIO \rightarrow H_2O + IO^-$ (fast, equilibrium)

Determine the rate law required by this mechanism.

The overall rate of a reaction is more or less equal to the rate of the slowest step, which is the

rate limiting step. The rate law for a single step should agree with the stoichiometry of that

step.

Rate = k' (I^-) $(HClO)$

Because the first step in this reaction is faster than the second, the first step should come to

equilibrium. When that happens,

k_f (ClO^-) = k_r $(HClO)$ (OH^-)

$(HClO)$ = $\{(k_f)/(k_r)\}$ (ClO^-) $(OH^-)^{-1}$

Rate $= k'$ (I^-) $\{(k_f)/(k_r)\}$ (ClO^-) $(OH^-)^{-1}$

Rate $= k$ (I^-) (ClO^-) $(OH^-)^{-1}$ where $k = k'\{(k_f)/(k_r)\}$

16-35. The reaction $3NO\,(g) \rightarrow N_2O\,(g) + NO_2\,(g)$ has the following experimental rate law:

Rate $= k\,(NO)^3$ Which of the following mechanisms is consistent with this rate law?

    (a) $2NO \rightarrow N_2O_2$ (slow, equilibrium step)
        $N_2O_2 + NO \rightarrow N_2O + NO_2$ (fast step)
    (b) $2NO \rightarrow N_2O_2$ (fast , equilibrium)
        $N_2O_2 + NO \rightarrow N_2O + NO_2$ (slow, equilibrium step)
    (c) Both of these mechanisms are consistent with this rate law.
    (d) Neither of these mechanisms are consistent with this rate law.

The overall rate of a reaction is more or less equal to the rate of the slowest step, which is the rate limiting step. The rate law for a single step should agree with the stoichiometry of that step.

Rate for (a) $= k\,(NO)^2$

Rate for (b) $= k'\,(N_2O_2)\,(NO)$

Because the first step in (b) is faster than the second, the first step should come to equilibrium. When that happens,

$k_f\,(NO)^2 = k_r\,(N_2O_2)$ $(N_2O_2) = \{(k_f)/(k_r)\}\,(NO)^2$

Rate for (b) $= k'\{(k_f)/(k_r)\}\,(NO)^2\,(NO) = k\,(NO)^3$ where $k = k'\{(k_f)/(k_r)\}$ Answer(b)

16-36. If the rate constant increases from $0.40\ M^{-1}s^{-1}$ at 25 °C to $0.80\ M^{-1}s^{-1}$ at 35 °C what is the activation energy of this reaction?

$Ln\,(k_1/k_2) = (E_a/R)\{(1/T_2) - (1/T_1)\}$

$Ln\,(0.4/0.8) = E_a\,(1/8.314)\,[\{1/(273+35)\} - \{1/(273+25)\}] = E_a\,(1/8.314)\,(-1.09 \times 10^{-4})$

$E_a = \{ln\,(0.5)\} \times 8.314/(-1.09 \times 10^{-4}) = 52859\ J/mol$ $E_a = 52.859\ kJ/mol$

16-37. Which of the following correctly describes the variation of rate constant, k_f, with temperature?

    (a) $k_f = \ln A - (E_a/RT)$ (b) $Ln\,k_f = \ln A + \{e^{-(Ea/RT)}\}$ (c) $Ln\,k_f = \ln A - (E_a/RT)$

    (d) $Ln\,k_f = \ln A + (E_a/RT)$ (e) $k_f = \ln A - (E_a/RT)$

Arrhenius equation: $k = A\,e^{-(Ea/RT)}$

        $Ln\,k = \ln A + \ln e^{-(Ea/RT)}$

$$Ln\ k = \ln A - (E_a/RT)$$ Answer(c)

16-38. The rate constant for a reaction is 0.40 $M^{-1}s^{-1}$. What is the order of the reaction?

For a first-order reaction the unit for k will be $(M/s)/M = s^{-1}$

For a second order reaction the units for k will be $(M/s)/M^2 = M^{-1}s^{-1}$

So the reaction is a second-order reaction.

16-39. The rate constant for the acid hydrolysis of sucrose (cane sugar) is $2.12 \times 10^{-4}\ M^{-1}s^{-1}$

at 27 ^{0}C and $8.46 \times 10^{-4}\ M^{-1}s^{-1}$ at 37 ^{0}C. What is the activation energy for this reaction?

$$Ln\ (k_1/k_2) = (E_a/R)\{(1/310) - (1/300)\}$$

$$Ln\ \{(2.12 \times 10^{-4})/(8.46 \times 10^{-4})\} = (E_a/8.314) \times (-1.07 \times 10^{-4})$$

$$E_a = (-1.384 \times 8.314)/(-1.07 \times 10^{-4}) = 1.08 \times 10^5\ J/mol = 108\ kJ/mol$$

16-40. The activation energy for an endothermic reaction is 134.4 kJ/mol. and $\Delta H = 10.5$

kJ/mol. What is the activation energy for the reverse reaction?

For an endothermic reaction the activation energy of the reverse reaction

= The activation energy of the forward reaction $- \Delta H = 134.4 - 10.5 = 123.9$ kJ/mol

16-41. The activation energy for an exothermic reaction is 115kJ/mol and $\Delta H = -3.1$ kJ/mol.

What is the activation energy for the reverse reaction?

For an exothermic reaction the activation energy of the reverse reaction

= The activation energy of the forward reaction $+ |\Delta H| = 115 + 3.1 = 118.1$ kJ/mol

16-42. What happens to the activation energy when temperature increased?

Increasing the temperature will not change the activation energy.

16-43. What happens to the activation energy when the concentration of the reactants are

increased?

Increasing the concentration of the reactants will not change the activation energy.

16-44. What happens to the activation energy when a catalyst is added?

Addition of a catalyst will decrease the activation energy.

16-45. If the temperature of a reaction increases which of the following statements is correct?

   (a) The activation energy will increase
   (b) The rate of reaction will increase
   (c) The rate constant will not increase

Increasing the temperature will not change the activation energy. Increasing the temperature will increase the rate of the reaction and the rate constant. **Answer(b)**

16-46. Calculate the fraction of HI molecules which will have energy equal to or greater than the E_a (activation energy) for the following reaction at 563 K, 573 K, 673 K and 773K.

$$2HI\ (g) \rightleftharpoons H_2\ (g) + I_2\ (g) \qquad E_a = 183\ kJ/mol$$

(a) At 563 K

Fraction of molecules which will have energy equal to or greater than $E_a = e^{-E_a/RT}$

$= e^{-183000/(8.314 \times 563)}$

$= 1.049 \times 10^{-17}$

(b) At 573 K

Fraction of molecules which will have energy equal to or greater than $E_a = e^{-E_a/RT}$

$= e^{-183000/(8.314 \times 573)}$

$= 2.076 \times 10^{-17}$

(c) At 673 K

Fraction of molecules which will have energy equal to or greater than $E_a = e^{-E_a/RT}$

$= e^{-183000/(8.314 \times 673)}$

$= 6.252 \times 10^{-15}$

(d) At 773 K

Fraction of molecules which will have energy equal to or greater than $E_a = e^{-E_a/RT}$

$= e^{-183000/(8.314 \times 773)}$

$= 4.301 \times 10^{-13}$

CHAPTER 17

COORDINATION COMPOUNDS

17-1. Predict the electron configuration of the Fe^{3+} ion.

Electron configuration of Fe : [Ar] $3d^6 4s^2$

The 4s electrons are lost first in forming the ion.

Electron configuration of Fe^{2+} : [Ar] $3d^6$

Electron configuration of Fe^{3+} : [Ar] $3d^5$

17-2. Calculate the charge on the transition metal in the following complexes:

   (a) $Na_2[Co(SCN)_4]$ (b) [Ni $(NH_3)_6](NO_3)_2$ (c) K_2PtCl_6

   (a) $Na_2[Co(SCN)_4]$: The charge on Co is +2
       +1   +2   −1
   (b) $[Ni(NH_3)_6](NO_3)_2$: The charge on Ni is +2
       +2   0   −1
   (c) K_2 $PtCl_6$: The charge on Pt is +4
       +1 +4 −1

17-3. Name the following compounds:

   (a) $K_4[Fe(CN)_6]$ (b) $Fe(acac)_3$ (c) $[Cr(en)_3]Cl_3$

   (d) $[Cr(NH_3)_5 H_2O] (NO_3)_3$ (e) $[Cr(NH_3)_4Cl_2]Cl$

   (a) Potassium hexacyanoferrate (II)

   (b) Tris(acetylacetonato)iron(III)

   (c) Tris(ethylenediamine)chromium(III)chloride

   (d) Pentaammineaquachromium(III)nitrate

   (e) Tetraamminedichlorochromium(III)chloride

17-4. Use valence bond theory to explain why Fe^{2+} forms $[Fe(CN)_6]^{4-}$ complex ion.

Electron configuration of Fe : [Ar] $3d^6 4s^2$

The 4s electrons are lost first in forming the ion.

Electron configuration of Fe^{2+} : [Ar] $3d^6$

In Fe^{2+}, 4s and 4p orbitals are empty. Fe^{2+} : [Ar] $3d^6$ $4s^0$ $4p^0$

Concentrating the 3d electrons in d_{xy}, d_{xz}, d_{yz} orbitals in this subshell provides the following electron configuration:

Fe^{2+} :

$\uparrow\downarrow$ $\uparrow\downarrow$ $\uparrow\downarrow$ __ __ __   __   __ __ __

          3d   4s   4p

The $3d_{x^2-y^2}, 3d_{z^2}, 4s, 4p_x, 4p_y, 4p_z$ orbitals are mixed to form a set of empty d^2sp^3 hybrid orbitals that point towards the corners of an octahedron. Each of these orbitals can accept a pair of non-bonding electrons from a [:C≡N:]⁻ ion to form a complex ion in which the iron atom has a filled shell of valence electrons.

$[Fe(CN)_6]^{4-}$:

$\uparrow\downarrow$ $\uparrow\downarrow$ $\uparrow\downarrow$ $\uparrow\downarrow$ $\uparrow\downarrow$ $\uparrow\downarrow$ $\uparrow\downarrow$ $\uparrow\downarrow$

          3d d^2sp^3

17-5. Explain why $[Co(NH_3)_6]^{3+}$ ion is a diamagnetic complex ion, whereas the $[CoF_6]^{3-}$ ion is a paramagnetic complex ion.

Electron configuration of Co : [Ar] $3d^7$ $4s^2$ In both complex ions the charge on Co is +3.

The 4s electrons will be lost first in forming the ion.

Electron configuration for Co^{3+} : [Ar] $3d^6$

The coordination number is 6. So the geometry will be octahedral. The ligand will split the degeneracy of the five 3d orbitals in Co^{3+} ion. Two orbitals (e_g orbitals: $d_{x^2-y^2}$, d_{z^2}) will have higher energy and three orbitals (t_{2g} orbitals: d_{xy}, d_{xz}, d_{yz}) will have lower energy.

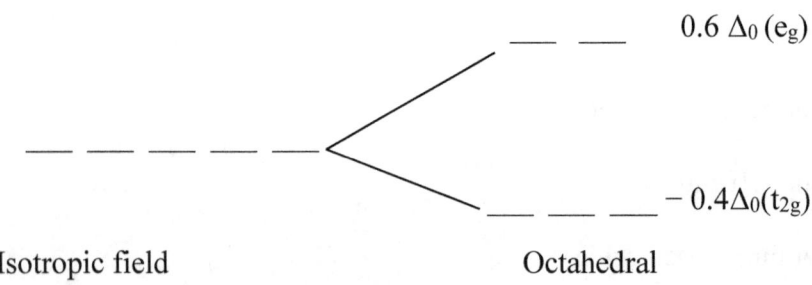

NH₃ is a strong field ligand. So the magnitude of splitting (Δ_o) will be high and the electron configuration will be (d^6- low spin):　　　___ ___ e_g

$$\underset{t_{2g}}{\uparrow\downarrow \quad \uparrow\downarrow \quad \uparrow\downarrow}$$

There will not be any unpaired electron and the $[Co(NH_3)_6]^{3+}$ ion will be a diamagnetic complex ion. F^- is a weak filed ligand. So the magnitude of splitting (Δ_o) will be lower and the electron configuration will be (d^6- high spin) :　　$\underline{\uparrow}\ \underline{\uparrow}\ e_g$

$$\underset{t_{2g}}{\underline{\uparrow\downarrow}\ \underline{\uparrow}\ \underline{\uparrow}}$$

There will be four unpaired electrons and the $[CoF_6]^{3-}$ ion will be a paramagnetic complex ion.

17-6. Explain why $[NiCl_4]^{2-}$ is a paramagnetic complex ion and $[PdCl_4]^{4-}$ is a diamagnetic complex ion.

$[NiCl_4]^{2-}$: charge on Ni is +2.

Electron configuration of Ni : $[Ar]\ 3d^8\ 4s^2$　　The 4s electrons are lost first in forming the ion.

Electron configuration of Ni^{2+} : $[Ar]\ 3d^8$

Ni^{2+} is a small ion and Cl^- is a large negative ion. So the ligand-ligand repulsions favor a tetrahedral geometry. In tetrahedral ligand field the $d_{x^2-y^2}$, d_{z^2} orbitals will have lower energy and d_{xy}, d_{xz}, d_{yz} will have higher energy. Chloride is a weak field ligand and Ni^{2+} is a weak field ion. So the electron configuration will be (d^8):

$$\underline{\uparrow\downarrow}\ \underline{\uparrow}\ \underline{\uparrow}\quad 0.4\ \Delta_t$$
$$\underline{\uparrow\downarrow}\ \underline{\uparrow\downarrow}\quad -0.6\ \Delta_t$$

Tetrahedral

There will be two unpaired electrons and the $[NiCl_4]^{2-}$ complex ion will be a paramagnetic complex ion.

$[PdCl_4]^{2-}$: the charge on Pd is +2

Electron Configuration of Pd: $[Kr]\ 4d^{10}$　　Electron configuration of Pd^{2+}: $[Kr]\ 4d^8$

CFSE calculations favor a square planar geometry. Ligand-ligand repulsions will not disfavor a square planar geometry because Pd^{2+} is a big ion. Pd^{2+} a strong field ion. So the magnitude of splitting (Δ) will be high even with a weak field ligand like Cl^-. So the electron configuration will be (d^8- low spin):

$d_{x^2-y^2}$ ___

1.0 Δ

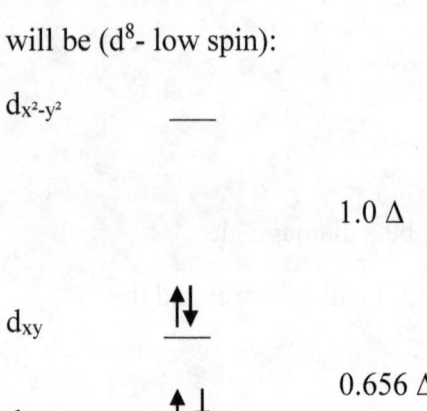

d_{xy}

0.656 Δ

d_{z^2}

0.086 Δ

d_{yz}, d_{xz}

Square planar

$d_{x^2-y^2}$ orbital will be left empty and there will be no unpaired electron and the $[PdCl_4]^{2-}$ complex ion will be a diamagnetic complex ion.

17-7. Will $[TiCl_6]^{3-}$ ion be paramagnetic or diamagnetic?

$[TiCl_6]^{3-}$: The charge on Ti : +3

Electron configuration of Ti: [Ar] $3d^2 4s^2$ The 4s electrons will be lost first in forming the ion.

Electron configuration of Ti^{3+} : [Ar] $3d^1$

Coordination number is 6. So the geometry will be octahedral. Chloride is a weak field ligand.

The electron configuration will be (d^1):   ___ ___ e_g

___ ___ ___ t_{2g}

There will be one unpaired electron and the complex ion will be paramagnetic.

17-8. Which of the following high-spin complexes would you expect to exhibit a Jahn-Teller distortion?

(a) $[Cr(H_2O)_6]^{2+}$ (b) $[MnCl_6]^{3-}$ (c) $[Fe(H_2O)_6]^{3+}$

$[Cr(H_2O)_6]^{2+}$: The charge on Cr : +2

Electron configuration of Cr : [Ar] $3d^5 4s^1$

The 4s electron will be lost first in forming the ion.

Electron configuration of Cr^{2+}: [Ar] $3d^4$

Coordination number is 6. So the geometry will be octahedral. H_2O is a weak field ligand and the electron configuration will be (d^4-high spin) :

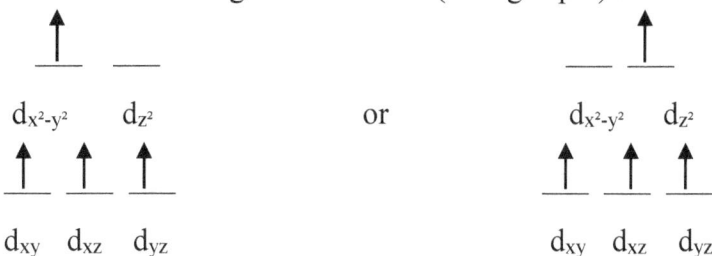

Jahn-Teller distortion can be expected if orbitals of the same energy level are occupied by different number of electrons. In this case only one of the e_g orbitals is occupied by a single electron . So Jahn-Teller distortion can be expected.

$[MnCl_6]^{3-}$: The charge on Mn : +3

Electron configuration of Mn : [Ar] $3d^5 4s^2$ The 4s electrons are lost in forming in the ion.

Electron configuration of Mn^{3+}: [Ar] $3d^4$

Coordination number is 6. So the geometry will be octahedral. Chloride is a weak field ligand. So the electron configuration will be (d^4- high spin) :

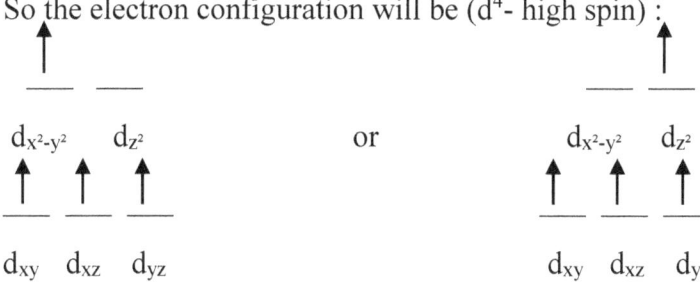

Jahn-Teller distortion can be expected if orbitals of the same energy level are occupied by different number of electrons. In this case only one of the e_g orbitals is occupied by a single electron. So Jahn-Teller distortion can be expected.

$[Fe(H_2O)_6]^{3+}$: The charge on Fe : +3

Electron configuration of Fe : [Ar] $3d^6 4s^2$

The 4s electrons will be lost first in forming the ion.

Electron configuration of Fe^{3+}: [Ar] $3d^5$

Coordination number is 6. So the geometry will be octahedral. H_2O is a weak field ligand and the electron configuration will be (d^5- high spin) :

$\underline{\uparrow}\ \ \underline{\uparrow}$ e_g

$d_{x^2-y^2}$ $\quad d_{z^2}$

$\underline{\uparrow}\ \ \underline{\uparrow}\ \ \underline{\uparrow}$ t_{2g}

d_{xy} $\ d_{xz}$ $\ d_{yz}$

Jahn-Teller distortion can be expected if orbitals of the same energy level are occupied by different number of electrons. In this case each e_g orbital and each t_{2g} orbital is occupied by a single electron. So Jahn-Teller distortion cannot be expected.

17-9. Calculate the magnetic moment (spin only) for the following complex ion : $[RuCl_6]^{2-}$.

$[RuCl_6]^{2-}$: The charge on Ru : +4

Electron configuration of Ru : $[Kr]\ 4d^7\ 5s^1$

The 5s electron is lost first in forming the ion.

Electron configuration of Ru^{4+} : $[Kr]\ 4d^4$

The coordination number is 6. So the geometry will be octahedral. Ru^{4+} is a strong field ion. So the magnitude of splitting (Δ_o) will be high even with a weak field ligand like chloride. So the electron configuration will be (d^4- low spin):

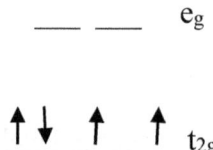

$\underline{}\ \underline{}$ e_g

$\underline{\uparrow\downarrow}\ \underline{\uparrow}\ \underline{\uparrow}$ t_{2g}

There will be two unpaired electrons and the magnetic moment (spin only)

$$= \sqrt{2(2+2)}\ \mu_B\ = \sqrt{8}\ \mu_B\ = 2.83\ \mu_B$$

17-10. How many geometric isomers and optical isomers are possible for M_{a2b2c2} (a,b, and c are monodentate ligands) ?

4 Geometric and 2 optical isomers are possible. 4 Geometric isomers:

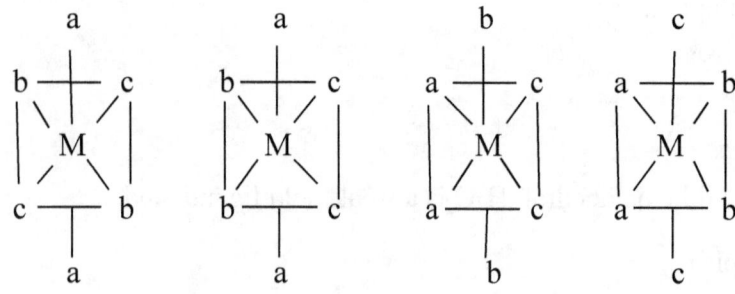

2 Optical isomers (1 pair of enantiomers – Non-superimposable mirror images of each other)

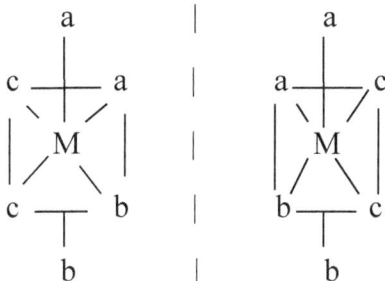

17-11. How many geometric isomers and optical isomers are possible for M_{a_2bcde} (a,b,c,d, and e are monodentate ligands)?

3 Geometric isomers and 12 optical isomers are possible.   3 Geometric isomers:

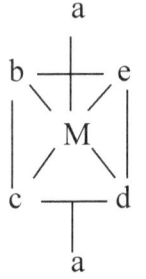 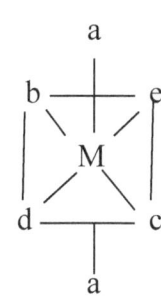 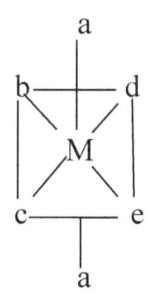

12 Optical isomers (6 pairs of enantiomers)

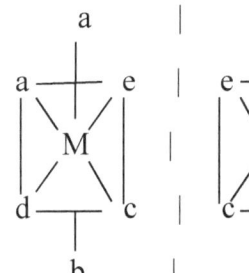 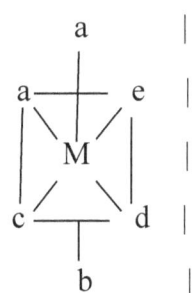 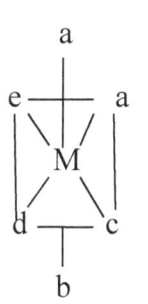 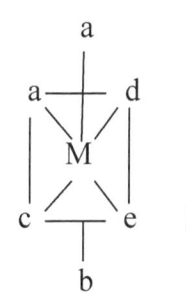 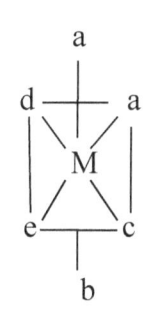

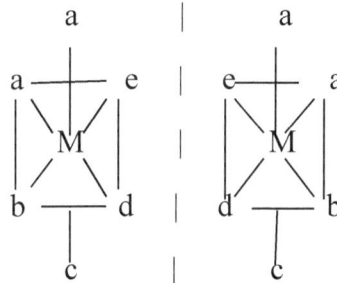 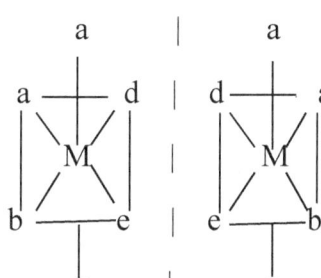 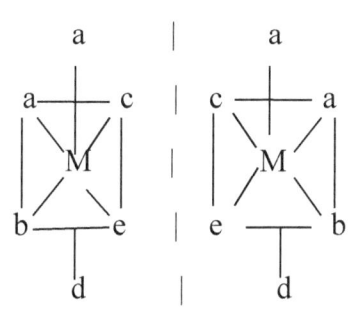

17-12. How many geometric and optical isomers are possible for square planar M_{abcd}(a,b,c, and d are monodentate ligands)?

3 Geometric isomers are possible .

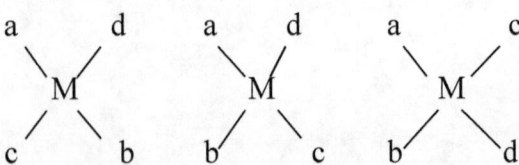

Optical isomers are not possible for square planar complexes.

17-13. How many geometric isomers are possible for hypothetical tetrahedral M_{abcd} (a,b,c, and d are monodentate ligands)?

Geometric isomers are not possible for tetrahedral complexes.

17-14. How many stereoisomers are possible for M_{abcdef} (a,b,c,d,e and f are monodentate ligands)?

30 Stereoisomers are possible (15 pairs of enantiomers). All are optical isomers.

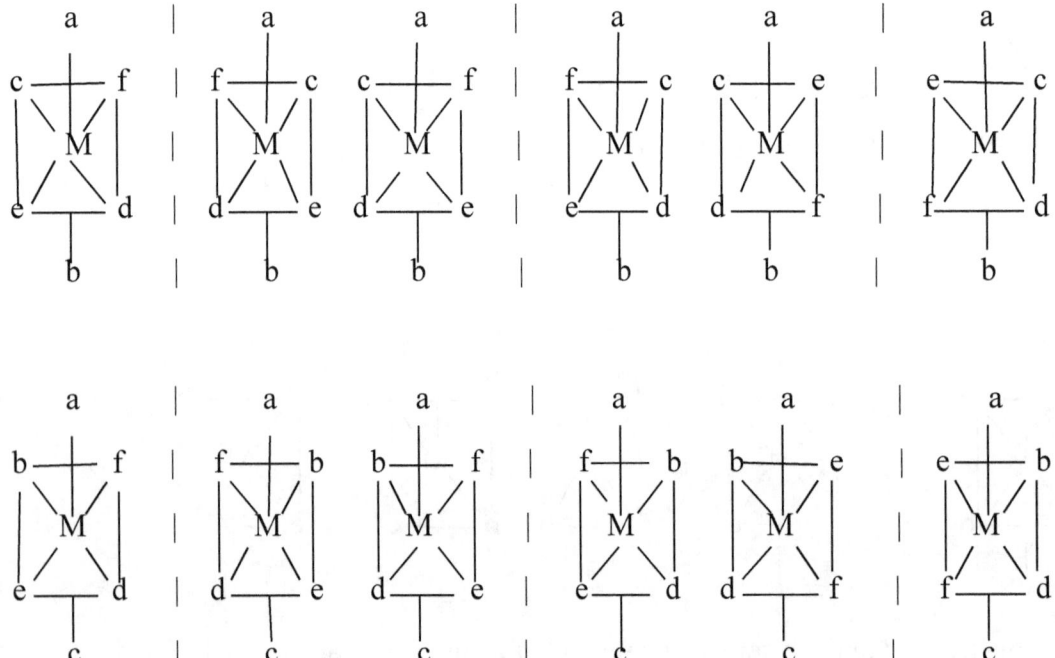

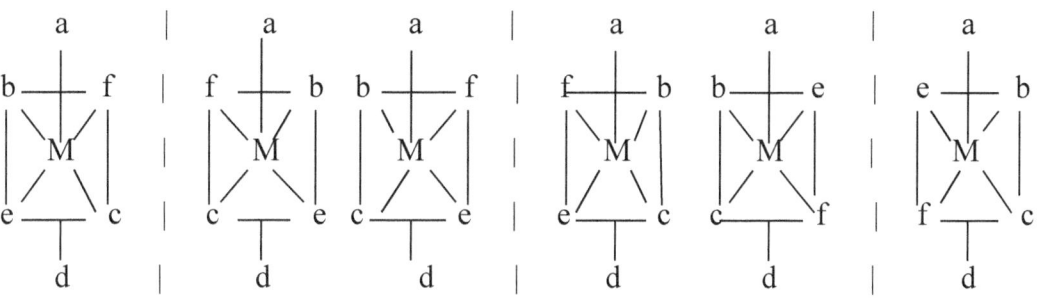

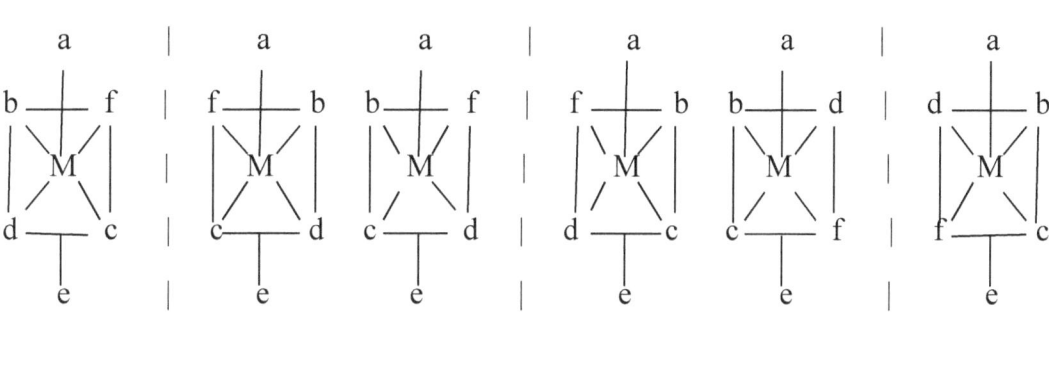

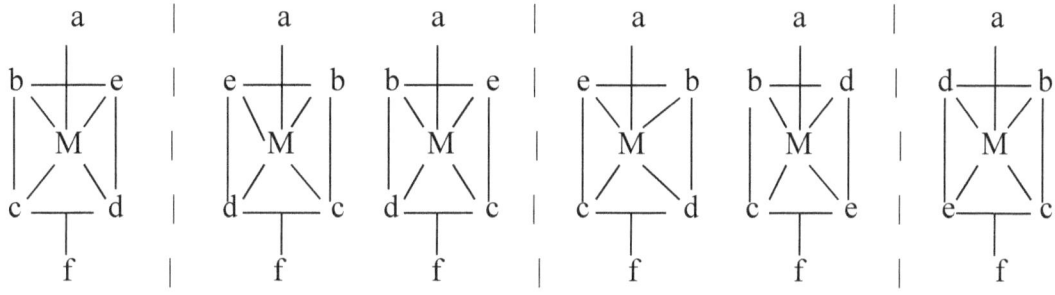

17-15. How many isomers are possible for square planar M(en)b₂ (en is a bidentate ligand and b is a monodentate ligand)?

Only one square planar structure is possible.

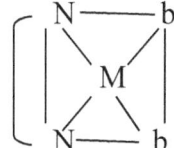

'en' is a bidentate ligand. So trans N␣␣N is not possible. So trans b—b also is not possible.

Δ_o Values for Some Complex Ions

Complex Ion	Δ_o(kJ/mol)	Complex Ion	Δ_o(kJ/mol)
$[Ti(H_2O)_6]^{3+}$	242.8	$[MnCl_6]^{4-}$	89.7
$[V(H_2O)_6]^{2+}$	150.7	$[Mn(H_2O)_6]^{2+}$	101.7
$[V(H_2O)_6]^{3+}$	226.0	$[MnCl_6]^{3-}$	239.2
$[CrCl_6]^{3-}$	155.5	$[Mn(H_2O)_6]^{3+}$	251.2
$[Cr(H_2O)_6]^{2+}$	166.2	$[Fe(H_2O)_6]^{2+}$	124.4
$[Cr(H_2O)_6]^{3+}$	208.1	$[Fe(H_2O)_6]^{3+}$	171.0
$[Cr(NH_3)_6]^{3+}$	257.1	$[Fe(CN)_6]^{4-}$	392.3
$[Cr(CN)_6]^{3-}$	318.1	$[Fe(CN)_6]^{3-}$	418.6
$[CoF_6]^{3-}$	155.5	$[Ni(H_2O)_6]^{2+}$	101.7
$[Co(H_2O)_6]^{2+}$	111.2	$[Ni(NH_3)_6]^{2+}$	129.2
$[Co(CN)_6]^{3-}$	416.2	$[IrCl_6]^{3-}$	299.0
$[RhCl_6]^{3-}$	244.0	$[Ir(NH_3)_6]^{3+}$	490.4
$[Rh(H_2O)_6]^{3+}$	322.9	$[Rh(NH_3)_6]^{3+}$	406.6
$[Rh(CN)_6]^{3-}$	544.2		

Δ_t Values for Some Complex Ions

Complex Ion	Δ_t (kJ/mol)
$[VCl_4]^{2-}$	108.0
$[CoCl_4]^{2-}$	39.5
$[CoBr_4]^{2-}$	34.7
$[CoI_4]^{2-}$	32.3

CHAPTER 18

NUCLEAR CHEMISTRY

INVALUABLE INFORMATION

Mass of Electron:	0.0005485799 amu
Mass of Neutron:	1.0086649 amu
Mass of Proton:	1.0072765 amu
Avogadro's Number:	6.022142×10^{23} mol^{-1}
Speed of light in vacuum:	2.99792458×10^{8} m/s
1 eV:	$1.6021892 \times 10^{-19}$ J
1 amu:	$1.6605389 \times 10^{-24}$ g
1 amu:	$1.6605389 \times 10^{-27}$ kg

$E = mc^2$

E for 1 amu:	$(1.6605389 \times 10^{-27}) \times (2.998 \times 10^{8})^{2}$ J
	1.492493×10^{-10} J
E for 1 amu:	$(1.492493 \times 10^{-10})/(1.6021892 \times 10^{-19}) \times (10^{-6})$ MeV
	931.5 MeV

$E = mc^2$

E per mole for 1 amu:	$(1.6605389 \times 10^{-27}) \times (2.998 \times 10^{8})^{2} \times (6.022142 \times 10^{23})$ J/mol
	8.988×10^{10} kJ/mol

Experimentally Measured Relative Masses of Selected Isotopes

^{8}B   : 8.024607 ^{10}B : 10.012937 ^{11}B : 11.009305 ^{12}C : 12.000000

^{60}Co : 59.933817 ^{90}Sr : 89.907738 ^{90}Rb : 89.914802 ^{92}Kr : 91.926156

^{141}Ba : 140.914411 ^{143}Xe : 142.935110 ^{144}Cs : 143.932077 ^{167}Os : 166.971550

^{171}Pt : 170.981240 ^{194}Hg : 193.965439 ^{194}Tl : 193.971200 ^{199}Pb : 198.972917

^{199}Bi : 198.977672 ^{206}Pb : 205.974465 ^{208}Bi : 207.976652 ^{208}Po : 207.981246

^{209}Fr : 208.99592 ^{212}At : 211.990745 ^{214}Bi : 213.998712 ^{214}Pb : 213.999797

^{216}Fr : 216.003198 ^{227}Ra : 227.029178 ^{228}Ac : 228.031021 ^{234}Pa : 234.043308

^{234}Th : 234.043601 ^{238}Pa : 238.054500

18-1. Provide examples for isotopes, isobars and isotones

Isotopes: $^{142}Nd, ^{143}Nd, ^{144}Nd$: All contain same number of protons (60 protons) but different mass numbers.

Isobars: $^{124}Sn, ^{124}Te, ^{124}Xe$: All have same mass numbers but different number of protons.

Isotones: $^{54}Cr, ^{55}Mn, ^{56}Fe$: All contain same number of neutrons (20 neutrons) but different number of protons.

18-2. Complete the following reaction and calculate the amount of energy released from this reaction in kJ/mol:

$$^{194}Te \rightarrow ? + {}_{+1}^{0}e^{+}$$

^{194}Te decays by positron emission. Proton is converted to a neutron with the emission of a positron and a neutrino (mass almost zero).

$$_{81}^{194}Te \rightarrow {}_{80}^{194}Hg + {}_{+1}^{0}e^{+}$$

Add 81 electrons to both sides. Positrons and electrons have same mass.

Δm = m[daughter atom] +2m[electron] – m[parent atom]

$\quad$ = 193.965439 + (2×0.0005486) – 193.971200

$\quad$ = – $4.6638×10^{-3}$ amu

ΔE = (– $4.6638×10^{-3}$)×($8.988×10^{10}$) kJ/mol = – $4.192×10^{8}$ kJ/mol

Energy Released = $4.192×10^{8}$ kJ/mol

18-3. Complete the following reaction and calculate the amount of energy released in kJ/mol.

$$^{214}Pb \rightarrow ? + {}_{-1}^{0}e^{-}$$

^{214}Pb decays by electron (β^{-}) emission. Neutron is converted to a proton with the emission of an electron and an antineutrino (mass almost zero).

$$_{82}^{214}Pb \rightarrow {}_{83}^{214}Bi + {}_{-1}^{0}e^{-}$$

Add 83 electrons to both sides.

Δm = m[daughter atom] – m[parent atom]

$\quad$ = 213.998712 – 213.999797

$\quad$ = – $1.085×10^{-3}$ amu

ΔE = (– $1.085×10^{-3}$)×($8.988×10^{10}$) kJ/mol= – $9.752×10^{7}$ kJ/mol

Energy Released = 9.752×10^7 kJ/mol

18-4. Complete the following reaction and calculate the energy released in kJ/mol.

$$^{171}Pt \rightarrow ? + _2^4He^{2+}$$

^{171}Pt decays by α emission.

$$_{78}^{171}Pt \rightarrow _{76}^{167}Os + _2^4He^{2+}$$

Add 78 electrons to both sides.

Δm = m[daughter atom] + m[^{4}He atom] − m[parent atom]

$\quad$ = 166.971550 + 4.002603 − 170.981240

$\quad$ = − 7.087×10^{-3} amu

ΔE = (− 7.087×10^{-3})×(8.988×10^{10}) kJ/mol= − 6.370×10^8 kJ/mol

Energy Released = 6.370×10^8 kJ/mol

18-5. Calculate the amount of energy released by the neutron-induced fission of ^{235}U to give ^{141}Ba, ^{92}Kr (mass = 91.926156) and three neutrons. Report your answer in kJ/mol and MeV/atom.

$$_{92}^{235}U + _0^1n \rightarrow _{56}^{141}Ba + _{36}^{92}Kr + 3\,_0^1n$$

Add 92 electrons to both sides.

Δm = m[^{141}Ba atom] + m[^{92}Kr atom] + 2m[neutron] − m[^{235}U atom]

$\quad$ = 140.914411 + 91.926156 + (2×1.008665) − 235.043930

$\quad$ = − 0.186033 amu

ΔE = (− 0.186033)×(8.988×10^{10}) kJ/mol= − 1.67×10^{10} kJ/mol

Energy Released = 1.67×10^{10} kJ/mol

Δm = − 0.186033 amu

ΔE = (− 0.186033)×(931.5) MeV = − 173 MeV/atom

Energy Released = 173 MeV/atom

18-6. Complete the following reaction.

$$^{40}K + _{-1}^0e^- \rightarrow ?$$

In this nuclear reaction ^{40}K decays by electron capture. An electron that surrounds the nucleus is pulled inside the nucleus and combined with a proton to make a neutron with the emission of a neutrino.

$$_{19}^{40}\text{K} + _{-1}^{0}\text{e}^- \rightarrow _{18}^{40}\text{Ar}$$

18-7. Calculate the changes in mass (in amu) and energy (in kJ/mol and keV/atom) that accompany the radioactive decay of (tritium) ^3H to ^3He and a β^- particle.

$$_1^3\text{H} \rightarrow _2^3\text{He} + _{-1}^{0}\text{e}^-$$

Add 2 electrons to both sides.

Δm = m[daughter atom] – m[parent atom]

   = 3.016029 – 3.016049

   = – 0.000020 amu

   = – 2.0×10^{-5} amu

ΔE = $(- 2.0 \times 10^{-5}) \times (8.988 \times 10^{10})$ kJ/mol = $- 1.8 \times 10^6$ kJ/mol

$\Delta E = (- 2.0 \times 10^{-5}) \times (931.5)$ MeV = $(- 2.0 \times 10^{-5}) \times (931.5) \times 10^3$ keV = $- 19$ keV/atom

18-8. Calculate the binding energy (in MeV) and the binding energy per nucleon for ^{238}U if the mass of this nuclide is 238.050788.

^{238}U contains 92 protons, 92 electrons and 146 neutrons.

Predicted Mass = $(92 \times 1.0072765) + (92 \times 0.0005485799) + (146 \times 1.0086649)$

$$= 239.9849828 \text{ amu}$$

Experimental Mass   = 238.050788 amu

Mass Defect   = 239.9849828 – 238.050788 amu = 1.9341948 amu

Total Binding Energy   = $(1.9341948) \times (931.5)$ MeV = 1801.7 MeV/^{238}U

Binding Energy per Nucleon   = $(1801.7)/(238)$ MeV = 7.57 MeV/nucleon

18-9. Determine the number of atoms in 10 mgs sample of ^{14}C (^{14}C mass = 14.003242).

Number of atoms = $\{(6.022 \times 10^{23})/(14.003242)\} \times (10 \times 10^{-3}) = 4.300 \times 10^{20}$ atoms

18-10. What mass of ^{14}C must be in a sample to have an activity of 20 mCi?

($t_{1/2}$ for ^{14}C is 5730 years)

$t_{1/2} = 5730 \times 365 \times 24 \times 60 \times 60$ seconds

k   = $(\ln 2)/(t_{1/2})$ = $(0.693)/(5730 \times 365 \times 24 \times 60 \times 60)$ = 3.835×10^{-12} s^{-1}

A   = kN

$1 Ci = 3.7 \times 10^{10}$ atoms/s

20 mCi $= 20 \times 10^{-3} \times 3.7 \times 10^{10}$ atoms/s

$\quad N = (20 \times 10^{-3} \times 3.7 \times 10^{10})/(3.835 \times 10^{-12})$ atoms $= 1.930 \times 10^{20}$ atoms

Mass of ^{14}C in the sample $= \{(14.003242)/(6.022 \times 10^{23})\} \times (1.930 \times 10^{20})$ g $= 4.49 \times 10^{-3}$ g

$$= 4.49 \text{ mgs}$$

18-11. If Radium 223 has a half-life of 10.33 days, what time duration would it require for the activity associated with this sample to decrease to 1.5% of its present value?

$\quad t_{1/2} = (0.693)/(k) \quad$ and $\quad \ln[(N)/(N)_0] = -kt$

$\quad k = (0.693)/(10.33) = 0.0671$ d^{-1}

$\quad \text{Ln}\{(1.5)/(100)\} \qquad = -kt$

$\qquad\qquad\qquad t = -\{\ln(0.015)\}/(0.0671) = -(-4.1997)/(0.0671) = 62.59$ days

18-12. Calculate the amount of energy (in kJ/mol and MeV/atom) released when deuterium and tritium fuse to give helium-4 and a neutron.

$$_1^2H + {}_1^3H \rightarrow {}_2^4He + {}_0^1n$$

$\quad \Delta m$ = mass of products − mass of reactants

$\qquad = 4.002603 + 1.008665 - 2.014102 - 3.016049$

$\qquad = -0.018883$ amu

$\quad \Delta E = (-0.018883) \times (8.988 \times 10^{10})$ kJ/mol $= -1.697 \times 10^9$ kJ/mol

Energy Released $= 1.697 \times 10^9$ kJ/mol

$\quad \Delta E = (-0.018883) \times (931.5)$ MeV/atom $= -17.6$ MeV/atom

Energy Released $= 17.6$ MeV/atom

18-13. How old is a mummy if the sample retains 70.7% of the activity of the living tissue?

^{14}C decays by first order kinetics; half-life $= 5730$ years.

$k = (\ln 2)/t_{1/2} \quad$ and $\quad \ln[(N)/(N)_0] = -kt$

$k = (0.693)/5730 = 1.21 \times 10^{-4}$ y^{-1}

$\text{Ln}[(70.7)/(100)] \qquad = -kt$

$t = -\ln(0.707)/(1.21 \times 10^{-4}) = -(-0.3467)/(1.21 \times 10^{-4}) = 2865$ years

Time since death $= 2865$ years

18-14. Calculate the maximum kinetic energy of the emitted β^- particle in the decay of ^{14}C.

$$_6^{14}C \rightarrow {_7^{14}}N + {_{-1}^{0}}e^-$$

Add 7 electrons to both sides.

$\Delta m = m$ (daughter atom) – m(parent atom)

$\quad = 14.003074 - 14.003242$ amu

$\quad = - 1.68 \times 10^{-4}$ amu

$\Delta E = (- 1.68 \times 10^{-4}) \times (931.5)$ MeV $= - 0.156$ MeV

The maximum kinetic energy of the emitted β^- particle $= 0.156$ MeV

18-15. Calculate the maximum kinetic energy of the positron emitted in the decay:

$$_{19}^{40}K \rightarrow {_{18}^{40}}Ar + {_{1}^{0}}e^+$$

Add 19 electrons to both sides. Positrons and electrons have same mass.

$\Delta m = m$(daughter atom) + 2m(electron) – m(parent atom)

$\quad = 39.962383 + (2 \times 0.0005486) - 39.963998$

$\quad = - 5.178 \times 10^{-4}$ amu

$\Delta E = (- 5.178 \times 10^{-4}) \times (931.5)$ MeV $= - 0.482$ MeV

The maximum kinetic energy of the emitted positron is 0.482 MeV

18-16. Isotope A requires 6 days to fall to (1/20) of its initial value. Isotope B has a half life 1.5 times that of A. How long will it take for isotope B to decrease to (1/16) of its initial value?

Isotope A: Ln $[(N)/(N)_0] = - kt$

If $(N)_0 = 1$, then $N = (1/20) = 0.05$

$t = 6$ days   Ln $(0.05) = - k \times (6)$

$k = - \{ln (0.05)\}/(6) = - (- 2.996)/6 = 0.499$ d^{-1}

$t_{1/2} = (0.693)/(0.499) = 1.389$ days

Isotope B: $t_{1/2} = (1.389) \times (1.5) = 2.0835$ days

Ln $[(N)/(N)_0]$ $= - kt$

If $(N)_0 = 1$, then $(N) = (1/16) = 0.0625$

Ln (0.0625) = − 2.773 and k = (0.693)/ $t_{1/2}$ = (0.693)/(2.0835) = 0.333 d^{-1}

t = (2.773)/(0.333) = 8.327 days

www.ingramcontent.com/pod-product-compliance
Lightning Source LLC
Chambersburg PA
CBHW080649190526
45169CB00006B/2039